# TRACEABILITY IN AGRICULTURE PRODUCTS

# TRACEABILITY IN AGRICULTURE PRODUCTS

**Miljan Grkinić**

Delve
PUBLISHING
www.delvepublishing.com

**Traceability in Agriculture Products**

*Miljan Grkinić*

**Delve Publishing**
4164 Lakeshore Road
Burlington, ON L7L 1A4
Canada
www.delvepublishing.com
Email: orders@arclereducation.com



ISBN: 978-1-77956-472-6 (Hardcover)

# ABOUT THE AUTHOR

**Miljan Grkinić** is a PhD student specializing in phytopathology and mycology at the University of Agriculture in Belgrade, where he also earned his bachelor's and master's degrees in Phytomedicine. With a strong academic and professional background in AI training, data annotation, and multilingual transcription, Miljan has contributed significantly to the development of AI models, particularly in natural language processing and speech recognition. His professional experience includes research in data-driven methods and statistical analysis, improving AI training pipelines for multilingual projects, and working on high-quality data annotation. Fluent in Serbian and English, with conversational proficiency in Spanish and Russian, he brings a diverse linguistic skill set to his work. Additionally, Miljan's volunteer translation work has enhanced his cultural and technical communication abilities. His hobbies include reading, sports, horticulture, and traveling, reflecting his passion for continuous learning and exploration.

# TABLE OF CONTENTS

# LIST OF FIGURES

# LIST OF TABLES

# PREFACE

In the modern world, where consumers are increasingly concerned about the safety, sustainability, and origins of their food, traceability in agriculture has emerged as a critical aspect of ensuring transparency and accountability in food systems. The agricultural sector, which forms the backbone of many economies, is undergoing a transformation, driven by the need for greater transparency, regulatory compliance, and technological innovation. As food supply chains become more complex and globalized, the ability to trace agricultural products from farm to table has become essential for maintaining food safety, ensuring product quality, and building consumer trust.

This book, *Traceability in Agriculture Products*, provides a comprehensive guide to the evolving landscape of traceability systems, their importance in the agricultural sector, and the technological and regulatory frameworks that support them.

The importance of traceability cannot be overstated. Chapter 1, *Introduction to Traceability in Agriculture* highlights the role traceability plays in ensuring food safety, reducing food fraud, and supporting sustainable farming practices. It also deals with the various types of traceability systems and the core components that make up an effective traceability framework. As agriculture moves toward more sustainable and eco-friendly practices, traceability is proving to be a vital tool in promoting responsible production and consumption patterns. The chapter further outlines the economic benefits that traceability systems bring to both producers and consumers, including improved market access and consumer confidence.

Chapter 2, *Regulatory Frameworks and Standards*, covers the international standards, such as ISO and HACCP, which play a pivotal role in harmonizing traceability practices across borders. Local and regional regulations, as well as the challenges associated with compliance, are also addressed, highlighting the need for a unified approach to traceability that transcends geographic boundaries. The chapter emphasizes that adherence to these standards is not just a matter of regulatory compliance but also a key factor in gaining consumer trust and expanding market reach.

Technology is at the heart of modern traceability systems. Chapter 3, *Technology in Traceability Systems*, explores the various technological innovations that have revolutionized how traceability is implemented in agriculture. From barcodes and QR codes to RFID technology and blockchain, this chapter discusses the tools and systems that are being used to track agricultural products throughout the supply chain. Blockchain, in particular, has emerged as a game-changer, offering unparalleled transparency and security in traceability. This chapter showcases how digital technologies are reshaping the way agricultural products are traced, improving efficiency, reducing errors, and enhancing the reliability of supply chains.

Beyond technology, the scope of traceability extends to all aspects of crop and livestock production. Chapters 4 and 5, *Traceability in Crop Production* and *Traceability in Livestock Production*, provide in-depth analyses of how traceability is applied in different segments of agriculture. From pesticide use and fertilizer tracking to livestock identification and health monitoring, these chapters illustrate the practical applications of traceability in ensuring the quality and safety of agricultural products.

Chapter 6 focuses on *Supply Chain Management and Traceability*, examining its necessity, tracking methods, technologies, and the integration of traceability systems into agricultural supply chains. It explores how traceability enhances sustainability in supply chains, especially through blockchain technology, fostering social and environmental benefits.

Chapter 7 sheds light on D*esigning Safety Evaluation and Risk Traceability Systems for Agricultural Products*, discussing the development of evaluation systems and models, and methods for designing traceability systems to ensure food safety and quality, backed by system implementation and testing.

Chapter 8 addresses *Traceability and Consumer Trust*, investigating how traceability influences consumer perceptions, purchasing behavior, and transparency, emphasizing its role in the growing demand for food safety and sustainability. It discusses how consumers influence traceability practices through demand for information and transparency in the food journey.

Chapter 9 looks at the *Economic Implications of Traceability*, focusing on cost-benefit analysis, challenges, and opportunities for agribusinesses. It explores how traceability contributes to sustainable economic growth and addresses challenges within agricultural value chains, with an emphasis on digital transformation.

Chapter 10 introduces *Innovations in Traceability for Sustainability*, examining how traceability supports sustainable agricultural practices, with a specific focus on carbon footprint tracking and blockchain technology. It discusses future trends and the integration of emerging technologies to monitor and improve sustainability.

Finally, Chapter 11 projects the *Future of Traceability in Agriculture*, analyzing emerging technologies like AI, IoT, and blockchain, along with regulatory changes, market demands, and evolving consumer expectations. It envisions the future of traceability systems and their increasing role in ensuring sustainability, safety, and transparency in the agricultural sector.

*Traceability in Agriculture Products* provides a holistic view of the current state and future potential of traceability in agriculture. Providing insights from regulatory frameworks, technological advancements, and consumer trends, this book aims to serve as a valuable resource for industry professionals, policymakers, and anyone interested in understanding the critical role that traceability plays in shaping the future of agriculture.

–Author

# CHAPTER 1

# INTRODUCTION TO TRACEABILITY IN AGRICULTURE

## Contents

# INTRODUCTION

The capacity to communicate details about a product throughout its manufacturing, processing, and delivery phases is known as traceability. A product's journey from the farm to the final customer can be found with effective traceability. It serves as a means of recording data at each stop along a product's journey and facilitating the sharing of that data as needed. Traceability is an important issue for all parties involved in the supply chain. This covers the following: food producers, suppliers, logistics, freight, food processing, and storage. To comprehend how commodities affect forests, develop, implement, and oversee successful solutions to mitigate the loss of forests, as well as contribute to more sustainable agricultural and food systems, traceability and transparency in supply chains are essential. In addition to discussing success factors and enabling conditions to enhance resource utilization and lessen forest loss, this section offers an updated evidence base to support cooperative discussions and initiatives on traceability and transparency.

## 1.1. DEFINITION AND IMPORTANCE OF TRACEABILITY

In the farm business these days, the term "traceability" in agriculture is frequently used. Its meaning is wide but generally speaking, it refers to the methods used to track produce throughout the entire production process. More information about production processes is being demanded by consumers more and more. A recent Trace One survey revealed that only 12% of consumers fully trust the safety of the food they eat, despite 91% of consumers believing it is important to know where their food comes from. Better product traceability programs and higher traceability standards are now required of farmers as a result of this. We have categorized the numerous justifications for investing in traceability into three primary areas such as:

1. **Create and Maintain Consumer Trust:** Eat Local and other consumer movements demonstrate the need for greater information about the origins of food. This is known as "farm-to-fork traceability" and places responsibility on farmers to keep accurate records of the whole history of the goods they send to market. The complexity of product labeling increases with the amount of information needed. By making unique labels for every distributor that contain lot numbers to track the produce all the way back to the field, a traceability program can lessen this burden. Utilizing farm management software to implement a traceability program can also increase consumer trust by organizing all of your data into reports. Reports facilitate the process of obtaining the certifications that

consumers seek by allowing auditors to view your entire production history in one location.

2. **Decrease Waste/Loss:** Utilizing a product traceability program is highly recommended as it can effectively reduce loss by improving inventory management. It is possible to better address issue areas when products can be tracked through the entire production process. In the event of a recall, you can track particular products by lot numbers back to contamination sites rather than having to remove all of your inventory. Ensuring that products are tracked accurately from the field to shipment reduces the possibility of sorting errors, allowing for consistent product packaging and labeling. Maintaining precise inventory records for packaging materials enables you to place precise reorders and enhances your ability to determine the optimal packaging types for both storage and retail. Additionally, traceability can increase the productivity of your farm's workforce. You can identify problem areas and address them as they emerge by connecting workers and work crews to production procedures.
3. **Improve Productivity and Profits:** Not only can an enhanced traceability program track workers to identify production bottlenecks, but it can also boost revenue by enabling more precise wage calculations, such as piece rate pay schemes. Piece rate computations can increase productivity incentives and help your farm save money. Additionally, traceability over time can aid in raising yields. Farmers can create the best production practices and make data-driven decisions with the help of traceability and an accurate farm record-keeping system. The agricultural sector will see a rise in the importance of traceability. To remain competitive in today's markets, farmers, and farm managers must establish and maintain accurate traceability programs (Aung and Chang, 2014; Zhang et al., 2018; Lv et al., 2023).

### 1.1.1. The Importance of Food Traceability in Agriculture

The Food and Agricultural Organization (FAO) states that India ranks second globally in the production of rice, wheat, sugarcane, groundnuts, vegetables, fruit, and cotton, and first globally in the production of milk, pulses, and jute. In addition to having a high level of agro-ecological diversity, the nation is a major producer of fish, poultry, livestock, spices, and plantation crops. With a GDP of $2.11 trillion, it ranks third in the world, behind the US and China. The government wants to raise the value of agricultural and processed food exports to $60 billion by 2022 from $38.49 billion in 2019. The term "food traceability" describes a group of technology procedures that include product and food identification and verification. It provides insight into the transit system across the different phases of production, processing, distribution, and consumption for farmers, companies, and consumers. In the agricultural value chain, food traceability guarantees moral, sustainable, and environmentally friendly production, making it a vital link in the global food trade. This technology is driving market growth and boosting global agricultural trade due to growing concerns about the safety and hygiene of agricultural products, food, and beverages; increases in the number of recalled products; strict export regulations; and demands for 100% transparency in food-supply chain systems.

At a compound annual growth rate (CAGR) of 9.1%, the global food traceability market is expected to reach a valuation of USD 26.11 billion by 2025, from an estimated $16.8 billion in 2020. Traceability technology could be adopted by India due to a rise in agricultural exports. The beverage, dairy, fisheries, fresh produce and seeds, meat, and livestock, seafood, spices, and high-value cash crops like cotton, cocoa, coffee, and condiments for flavors and fragrances are among the leading market sectors. In India, the majority of food stores are family-run businesses called Kiranas. The industry's capacity to satisfy the steadily rising demand for seasonal produce is being hampered by mismanaged inventories, unstable supply chains, and a lack of visibility in the cycles of fresh fruit and vegetable production and sales. Food traceability in the nation is characterized by arbitrary and inconsistent practices. Many fraudulent incidents, including adulteration, counterfeiting, diversion, theft, refilling, and tampering, often happen at every step of intricate supply chains where there is no oversight. The development of food traceability systems is being hampered by inadequate agricultural practices, shaky food safety regulations, stringent government marketing standardization, poorly developed and disorganized supply chains and market infrastructure, and a lack of cold chain facilities for small and medium-sized farmers and local businesses.

To reduce these errors and strengthen the supply chain management system's resilience, traceability is essential. Integrated systems that guarantee efficient operational control help prevent contamination of any kind. To help businesses combat the problems of overstocking and understocking, technology can revolutionize inventory management by providing real-time information about produce stocked in warehouses. The most popular traceability technology, despite developments being made in India, is the 2D barcode, which is still used more often than RFID, holograms, and nuclear techniques for production process monitoring. The export development authority for agricultural and processed food products (APEDA) is in charge of spearheading vegetable traceability programs. Utilizing artificial intelligence (AI), block chain technology, and data analytics in digital farm management guarantees food safety and traceability via a digital supply chain, tracks the entire agricultural produce cycle in a matter of seconds, presents a plethora of business development opportunities in addition to procurement strategies and sales forecasting, aids in obtaining premium prices for producers' products, and lowers losses related to spoilage. Robust technological solutions are imperative to ensure quality in the face of intense global competition in the food market.

### 1.1.2. Traceability in Agri-Food

Due to globalization, food now travels a greater distance from the producer to the final consumer. But it's getting harder to keep food safe and of high quality throughout the supply chain. The meat industry in particular has seen many crises in recent decades that have called into question its credibility. These crises are the reason for the previous point's mention. Many laws, regulations, and standards pertaining to food safety and quality have been created and enhanced in response to the mounting concerns. Traceability aims to meet standards for food quality and safety. It is defined by many standards as the capacity to track a particular food through every step of its preparation, processing, and delivery. Nonetheless, customers keep asking for more details regarding the provenance of goods and their constituents. Thus, traceability systems must be created and developed to receive food that is safe and of high quality. Nonetheless, a lot of businesses use their traceability systems to meet legal requirements, which are outlined in EU Regulation (EC) No 178/2002 in the case of Portugal. Businesses must monitor their products both internally and externally, but only for that purpose—not to dominate or appease customers.

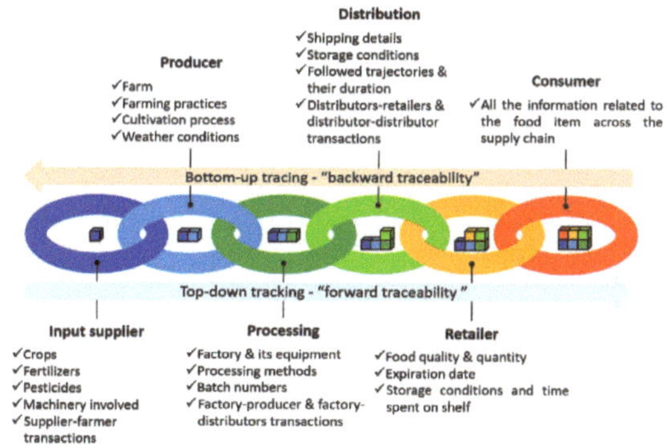


**Figure 1.1.** *Theoretical representation of food traceability.*

*Source: Tagarakis AC, Benos L, Kateris D, Tsotsolas N, Bochtis D. Bridging the Gaps in Traceability Systems for Fresh Produce Supply Chains: Overview and Development of an Integrated IoT-Based System. Applied Sciences. 2021; 11(16):7596. https://doi.org/10.3390/app11167596.*

Nonetheless, the development of traceability technology has been impacted by the products' origins' transparency. Because they have different meanings even though they are related, traceability and transparency are two terms that have been used interchangeably in the context of supply chains. As previously stated, traceability is the capacity to obtain data regarding any link in the supply chain. That is, as illustrated in Figure 1.1, what, how, where, why, and who. Conversely, supply chain transparency refers to the availability of data about the product and its operations across the supply chain. To put it briefly, supply chain transparency is made possible by traceability.

### 1.1.3. Types of Traceability

Traceability is divided into six groups:

- Product traceability;
- Process traceability;
- Genetic traceability;
- Traceability of entries;
- Traceability of diseases and pests; and
- Traceability of measurements.

Internal traceability refers to the documentation of all information and procedures about the product's processing inside an organization, including the link between a new product and its initial raw materials. This link is created using the traceable item's lot number. Even when the traceable item—like pallets or containers—is a part of a hierarchy, this principle still holds true. For external traceability to work, every product that a business produces must have a unique identification. Through product labels or documentation, all parties involved in the distribution chain must exchange the identification numbers of traceable items to maintain external traceability. Upstream (to the supplier) and downstream (to the customer) traceability are made possible by external traceability.

However, a novel strategy called end-to-end traceability is suggested by organizations like the Codex Alimentarius Commission and the GS1 Global Traceability Standard. After taking a closer look at the supply chain, we find that every business and organization is in charge of its traceability data. Accessing and combining the data from the various entities involved in the product life cycle will be required to achieve end-to-end traceability. Through all parties involved in its production, both upstream and downstream, this new approach will make it possible to track and locate an object over its whole life cycle. All businesses and chain operators must adhere to a set of standards so that their disparate systems and technologies can be integrated and communicated with one another to achieve the goal of end-to-end traceability in the supply chain.

### 1.1.4. Identification and Traceability Tools

The supply chain field has experienced rapid technological advancement as a result of the requirement to deliver safe food and instruments that lessen the likelihood of events that compromise food security. To this aim, traceability systems have been the primary focus of these applications. Since most traceability systems in use today are centralized, trust concerns like information falsification and tampering may arise. Moreover, centralized systems are susceptible to failure because a single point of failure has the potential to bring down the entire system. But thanks to developments in technology, traceability systems can now be decentralized and are less vulnerable to fraud and data manipulation thanks to technologies like blockchain. When data collection is done manually in traceability systems, it takes a lot of time because the employee has to log information at every stage of the activity or in every step of the product's application process before entering it into the computer system. Inaccuracies in data entry or information collection may result from this manual process. Identification technologies, such as barcodes, radio-frequency identification (RFID), near-field communication (NFC), and wireless sensor networks (WSN), are used for products and information capture to reduce human error and, consequently, improve the efficiency and accuracy of traceability processes along the supply chain. Consequently, a technological ecosystem that can track along the agri-food value chain is proposed in the work that is being presented. Block chain technology is presented as a solution to ensure the integrity of this traceability. By doing this, it will be feasible to guarantee that the final user can verify a specific product using the block chain's data. This will contain information about the people engaged in each process, the date each was completed, and other crucial details (Aung and Chang, 2014; Zhang et al., 2018; Lv et al., 2023).

### 1.1.5. Core Components of Traceability in Agriculture

In agriculture, traceability is the capacity to monitor the flow of agricultural products from the farm to the customer at each stage of production, processing, and distribution. It is an essential part of contemporary food supply chains and is necessary to meet regulatory standards, improve consumer confidence, and guarantee food safety. Transparency, accountability, identification, data collection, and record-keeping are the main components that make up traceability in agriculture. A comprehensive system for monitoring the movement of goods and guaranteeing the integrity of agricultural products during their journey is provided by these components working together. Starting at the point of production, data collection is one of the fundamental components of traceability. When it comes to raising crops, raising animals, or any other type of agricultural production, farmers gather comprehensive data. This comprises information on water use, livestock veterinary care, seeds, fertilizer, pesticides, and other inputs. Practices that impact the safety and quality of agricultural products can be monitored with the help of accurate data collection. To make sure that products are handled in a way that reduces the risk of contamination or spoiling, information is also documented regarding harvesting practices, storage conditions, and transportation.

Agricultural traceability also requires identification. A unique identifier, frequently in the form of a barcode, QR code, or RFID tag, is assigned to each product or batch of products. These identifiers are necessary to follow the flow of goods as they move through the various supply chain stages. To enable stakeholders to track a product back to the precise farm, field, or even production lot, for example, every batch of harvested crops or livestock is labeled with an identifier that links it to its origin. This procedure facilitates the maintenance of transparency and enables prompt resolution of any quality problems or safety issues that might develop later in the supply chain. Another essential component of traceability is record-keeping. At every stage of the agricultural process, from production to processing and distribution, meticulous records must be kept. These documents usually contain details about the product's origin, manufacturing date, and numerous processes it has gone through. For instance, thorough records are maintained on all ingredients used, equipment used, and processing methods in food processing facilities. This guarantees the ability to pinpoint the exact point in the supply chain where a contamination problem occurred, in addition to identifying the source. To guarantee adherence to food safety regulations, regulatory agencies frequently demand record-keeping, which facilitates a methodical and structured approach to traceability.

Systems for agricultural traceability that work must be transparent. It describes how all parties involved in the supply chain—farmers, processors, distributors, retailers, and consumers—openly exchange information. Each party will have access to the information they need to confirm the products' safety and quality thanks to transparency. For example, retailers use traceability systems to make sure the goods they sell adhere to legal and safety requirements. However, consumers are calling for more openness about where their food comes from, including how it was grown, processed, and shipped. Customers are better able to make decisions and have more faith in the agricultural supply chain when there is transparency in the traceability system. Accountability is yet another crucial component. Each link in the supply chain is accountable for keeping correct records and making sure the data they submit is trustworthy in a traceability system. Because of accountability, it is feasible to track a product back through the supply chain and determine its origin in the event of an issue, such as an outbreak of foodborne illness. Taking prompt corrective action, such as implementing stronger controls in processing facilities, improving production methods, or recalling products, is another aspect of accountability. Traceability thus improves the general safety and quality of the agricultural supply chain in addition to monitoring the movement of goods.

Agriculture's traceability has improved significantly as a result of technological developments. Real-time traceability data recording, storing, and sharing has been made simpler by digital tools like blockchain technology, GPS tracking, and cloud-based data storage systems. For example, blockchain provides an immutable, decentralized ledger where all agricultural product-related transactions can be tracked. This technology ensures that data cannot be changed or tampered with once it is entered, adding an extra layer of security and transparency. GPS tracking is frequently used to track the movement of goods, guaranteeing that they are delivered on schedule and in the right circumstances (e.g., keeping perishable goods cold-chained). All stakeholders can always get the information they require thanks to cloud-based systems that integrate traceability data from multiple sources. Regarding regulations, many nations have put in place standards and laws about food safety that require agriculture to use traceability systems. By guaranteeing that agricultural products are safe for consumption, these regulations aim to protect the general public's health. To identify and prevent potential hazards, companies in the food industry are required by the Food Safety Modernization Act (FSMA) in the United States to implement traceability systems. Comparably, the General Food Law of the European Union requires all food enterprises to be able to track their products through the supply chain both forward and backward, guaranteeing that any problems with food safety can be promptly identified and fixed.

### 1.1.6. Economic Benefits of Traceability for Producers and Consumers

Because traceability increases efficiency, opens doors to premium markets, lowers risks, and builds consumer trust, it has a significant impact on the financial results of farmers, supply chain participants, and consumers. Traceability systems facilitate improved decision-making, cost savings, and value creation throughout the supply chain by offering transparency and thorough tracking of agricultural products from farm to table.

Here's how traceability impacts each of these key groups economically:

1. **Economic Benefits for Farmers:**

a. **Access to Premium Markets:** Through traceability, farmers can certify their products following specific sustainability, safety, or quality standards, giving them access to high-end markets. Goods that adhere to higher standards, like those that are organic, Fair Trade, or sourced sustainably, can fetch a premium price from both retailers and customers. Farmers can increase their revenue by selling their products at a higher price by following these guidelines and supplying traceability information.

b. **Risk Reduction:** Farmers can reduce their risk of food safety problems and product recalls by using traceability. In the event of an issue, farmers possessing a well-established traceability system can promptly pinpoint the origin of the problem and restrict the extent of recalls. This lessens the possibility of facing legal repercussions, minimizes monetary losses, and safeguards their reputation. Money and resources can be saved by being able to swiftly isolate impacted batches as opposed to removing entire product lines from the market.

c. **Improved Farm Management:** Farmers can gather and evaluate information about crop performance, resource consumption, and production procedures with traceability systems in place. Better farm management decisions are made possible by this data, including how best to use inputs (pesticides, water, and fertilizers), which can reduce expenses and boost output. Long-term gains in efficiency result in increased yields, decreased waste, and more successful farming operations.

2. **Economic Impact on Supply Chain Actors:**

a. **Supply Chain Efficiency:** Efficiency is increased through traceability, which streamlines supply chain quality assurance, inventory control, and logistics. Real-time product tracking helps supply chain participants—processors, distributors, and retailers—manage their inventories more effectively, cutting down on waste, delays, and spoiling. Lower operating costs and increased profitability are the outcomes of this. Furthermore, the capacity to track goods all the way through the supply chain minimizes duplication and improves cooperation amongst various parties.

b. **Enhanced Product Differentiation:** Traceability gives processors and retailers a chance to set their products apart from rivals. They can justify premium pricing by promoting traceable products as being safer, better-quality, or more sustainable. This is particularly crucial in marketplaces where customers want more openness regarding the provenance and moral implications of the products they buy. For instance, businesses can charge more and foster greater brand loyalty if they can show that the products they sell come from ethically run businesses or environmentally conscious farms.

c. **Regulatory Compliance:** Companies can more easily adhere to environmental and food safety regulations when they have traceability. To stop the spread of foodborne illnesses and guarantee product authenticity, governments everywhere are enacting stricter regulations pertaining to food traceability. Traceability systems automate data collection and reporting, which lessens the administrative burden associated with complying with these regulations, which can be expensive. In addition to saving businesses money on fines and recalls, adhering to these regulations creates new export prospects in markets with high standards.

3. **Economic Advantages for Consumers:**

a. **Informed Purchasing Decisions:** Consumers who can trace products are better equipped to make educated decisions about what to buy. Concerns over food safety, quality, environmental sustainability, and ethical sourcing are growing among consumers today. Customers can select products that are consistent with their values by using traceability, which offers comprehensive information on these factors. Customers are frequently prepared to pay more for products that can be traced back to their source and are approved as being produced ethically or sustainably. Demand for traceable products is growing, which is advantageous for supply chain participants and farmers alike.

b. **Increased Trust and Loyalty:** Trust between food brands and consumers is increased through traceability. Customers feel more assured about the safety and caliber of the goods they buy when they are aware of the origins, production processes, and handling personnel of the food they eat. Businesses and merchants who invest in traceability systems stand to gain financially from this trust's ability to foster enduring brand loyalty. For instance, a business that uses transparent traceability to ensure the safety and provenance of its products is more likely to keep customers, which lowers acquisition costs and encourages repeat business.

c. **Safer and Higher Quality Products:** Traceability guarantees that products adhere to stringent safety standards from a health and safety standpoint. When problems with food safety occur, consumers gain from quicker and more precise recalls, which lowers their exposure to health hazards. Furthermore, because traceability permits stricter quality control throughout the production process, it can result in higher-quality products. For example, traceable products can be kept an eye on to make sure that the right storage conditions are met, preserving the freshness and nutritional value of perishable goods until they are consumed.

4. **Overall Market Implications:**

a. **Price Premiums for Verified Products:** Prices for goods that come with guarantees of sustainability, ethical sourcing, or superior quality are often higher due to the demand for traceable products. Due to their ability to differentiate their goods and increase prices for the added value of transparency, producers as well as retailers profit from this. The whole supply chain benefits when products bearing certifications or guarantees of traceability fetch higher prices in local and foreign marketplaces.

b. **Market Competitiveness:** Businesses that do not implement traceability systems run the risk of losing market share to those who do, as traceability becomes a consumer expectation. Consumer demand for transparency is driving the global agriculture market more and more, particularly in high-income nations. This change puts pressure on farmers, processors, and retailers to adopt traceability or else they may find themselves shut out of exclusive market segments or subject to more stringent regulatory oversight. On the other hand, companies that adopt traceability can improve their competitiveness and market position.

c. **Global Trade Opportunities:** Traceability creates opportunities for farmers and supply chain participants, especially in developing nations, in international markets. For agricultural imports, many importing nations, particularly those in Europe and North America, have stringent traceability regulations. Farmers and exporters can meet these international standards and expand their market reach and revenue potential by putting traceability systems in place.

### 1.1.7. Role of Traceability in Sustainable Agriculture

The ability to follow the production, processing, and delivery of agricultural goods from their point of origin to the final consumer is referred to as traceability in the agricultural industry. It is a system that guarantees transparency from farm to table throughout the whole supply chain. Environmental stewardship and sustainable agricultural practices both greatly benefit from traceability. Traceability enables improved resource management, waste reduction, ethical sourcing, and the encouragement of ecologically friendly practices by increasing the transparency and accountability of the production process. Traceability plays a major role in promoting sustainable agriculture by making better resource management possible. To minimize its negative effects on the environment while maintaining productivity, sustainable agriculture depends on the efficient use of resources like water, soil, and energy. Traceability systems enable farmers, regulators, and consumers to keep an eye on inputs like pesticides, fertilizers, and water usage by offering comprehensive information on how resources are used at every stage of production. Farmers can implement more effective practices, cut back on overuse, and prevent environmental degradation by keeping track of the quantity and kind of resources they use. For example, traceability systems and precision farming technologies enable the application of water or fertilizers only where and when needed, minimizing waste of resources and pollution.

Traceability also aids in identifying and advancing sustainable farming methods. It provides a framework for certifying goods made using ecologically friendly practices, like agroforestry, integrated pest management (IPM), and organic farming. Farmers are incentivized to adopt more environmentally friendly practices by the market when they can prove to consumers and retailers that their products meet sustainability standards by tracking agricultural products back to the farm. Traceability-based certification programs, like Rainforest Alliance and Fair Trade, frequently emphasize boosting soil health, decreasing deforestation, and advancing biodiversity. Farmers who adhere to these guidelines can thus access higher-value markets and support environmental preservation at the same time. Sustainability also depends on traceability since it is essential to cutting waste in the agricultural supply chain. Food waste is a major global problem that has a significant impact on both the depletion of natural resources and greenhouse gas emissions. Implementing traceability systems makes it simple to identify and address supply chain inefficiencies like overproduction, spoiling, and improper handling. For instance, more accurate demand forecasting is made possible by real-time tracking of agricultural products, which helps to avoid overproduction and excess waste. Comparably, tracking perishable items during their whole journey enables faster problem-solving, reduces food loss, and guarantees that goods arrive at the customer in ideal condition.

Furthermore, traceability promotes ethical sourcing by bringing transparency to the social and environmental aspects of agricultural production. Customers want more information about the products' origins as they become more environmentally conscious. Information regarding farming practices, working conditions, and the effects of production processes on the environment can be found via traceability systems. Because their operations are scrutinized, producers are encouraged to adopt more ethical and sustainable practices as a result of this level of transparency. To ensure that their supply chains adhere to sustainability standards, large corporations and retailers are utilizing traceability. This approach helps them minimize their environmental impact and improve their corporate responsibility profiles. Traceability plays a significant role in promoting environmental stewardship, especially in preventing deforestation and maintaining biodiversity, which is one of its most significant contributions to sustainability. Numerous agricultural products, including soy, cattle, and palm oil, are associated with deforestation, which has catastrophic effects on ecosystems and the climate worldwide. By tracking these goods back to their source, traceability systems can make sure that they don't come from places with a high level of biodiversity or illegally cleared forests. Traceability data is frequently used by NGOs and governments to uphold environmental laws and safeguard delicate ecosystems. For instance, traceability is used by the EU's Forest Law Enforcement, Governance, and Trade (FLEGT) initiative to confirm the legitimacy of imported timber, assisting in the fight against illicit logging and advancing sustainable forestry methods.

Traceability also helps in calculating and controlling agriculture's carbon footprint. Greenhouse gas emissions are mostly caused by agriculture, particularly by practices like raising livestock, growing rice, and using synthetic fertilizers. Farmers and businesses can measure the carbon emissions linked to their products and pinpoint areas for improvement by putting traceability systems in place. For example, monitoring a product's whole lifecycle—from manufacturing to shipping—can reveal the areas with the highest emissions and suggest opportunities for the adoption of more environmentally friendly practices.

Reducing the overall carbon impact of food production helps to align agricultural practices with global climate goals, such as those outlined in the Paris Agreement. Traceability not only helps the environment but also advances social sustainability by strengthening smallholder farmers' livelihoods and enhancing food security.

Many developing nations rely heavily on small-scale farmers to produce the majority of their food. These farmers may have easier access to markets and instruments to enhance their agricultural methods thanks to traceability systems. Traceability helps smallholder farmers adopt environmentally friendly practices that improve soil health, lessen reliance on toxic chemicals, and increase crop diversity by linking them to sustainable supply chains. Consequently, this enhances regional food systems, fosters climate change resilience, and advances long-term sustainability (Aung and Chang, 2014; Zhang et al., 2018; Lv et al., 2023).

## 1.2. KEY ELEMENTS AND STANDARDS IN TRACEABILITY

All supply chain participants must work together to implement traceability systems, which can be difficult and complex. A standard makes implementation easier for all parties involved in the supply chain by giving them a clear framework to adhere to. Traceability becomes more affordable and accessible for all stakeholders when practices are shared, reducing duplication of data collection and enhancing data reliability. The basis for technical interoperability, or the capacity of disparate systems to share and utilize data, is also standardization. Technology solution suppliers can create systems that work together without the need for specialized system integration by using a common methodology for data exchange, formatting, and collection. As a result, seafood companies can select the system that best suits their needs without having to worry about it conflicting with those of their suppliers or customers (Figure 1.2).

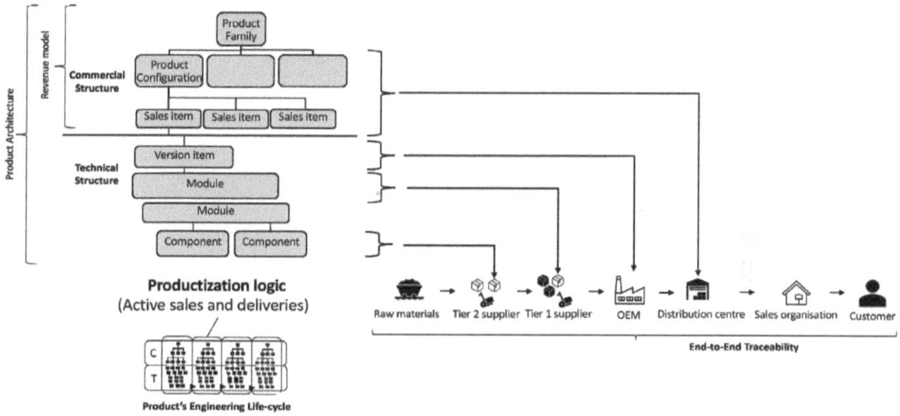


**Figure 1.2.** *Core elements of a traceability system.*

*Source: Harkonen J, Guerrero Rodriguez JM, Mustonen E. The Role of Productization in End-To-End Traceability. Eng. 2024; 5(4):2943-2965. https://doi.org/10.3390/eng5040153.*

Traceability is not a brand-new concept. It has been present in various forms in many supply chains, particularly in B2B industries with high transaction volumes. Still, until the pandemic, end-to-end traceability was not widely implemented. Thus far, it has been assumed that the end-user doesn't care where a product comes from as long as it delivers value. However, consumers can learn the product's origin and its journey through the supply chain thanks to traceability.

### 1.2.1. Identification and Labeling Systems

Systems for identification and labeling are essential to agricultural product traceability because they make it possible to monitor and record the movement of goods along the supply chain. From the farm to the end user, these systems guarantee that each product, batch, or lot is uniquely identifiable throughout its entire journey. Systems for product identification and labeling enable the tracking of a product's origin, history, processing techniques, and distribution route by giving each one a unique label or identifier. They are essential to guarantee the safety, quality, and transparency of agricultural products as well as the integrity of the agricultural supply chain. Assigning a distinct identifier to every product or batch is the fundamental component of identification systems. Barcodes, QR codes, RFID tags, and alphanumeric codes are just a few of the various ways this identifier can be created. These identifiers function as a physical or digital signature that connects a product to many important details, including the product's date of manufacture, processing information, and place of origin (farm or field). A batch of harvested apples, for example, could be labeled with a barcode that identifies the orchard in which they were grown, the pesticides that were applied, and the precise day that they were harvested. This identifier follows the product throughout the supply chain to make sure it can always be tracked back to the manufacturer.

An equally important component of these systems is labeling. Information is visible and easily accessible to consumers and supply chain participants when labels are attached to goods or packaging. Along with the unique identification, important information like batch numbers, expiration dates, country of origin, organic certification, and safety standard compliance may also be included. An organic label, for instance, guarantees to customers that the product satisfies particular health and environmental standards. Labeling becomes even more crucial when agricultural products are transported in large quantities or across international borders because it enables regulators and inspectors to confirm that the goods comply with legal regulations. Furthermore, labeling supports branding and marketing initiatives by enabling manufacturers to inform consumers about key product characteristics like non-GMO, fair trade, and sustainably sourced ingredients.

Large-scale, intricate supply chain management highlights the significance of identifying and labeling systems for traceability. Agribusiness goods frequently go through several phases, such as harvesting, processing, storing, and shipping. It would be impossible to track the flow of goods between various supply chain actors and across these stages without a trustworthy identifying system. For instance, cattle may be raised on a farm, sent to a processing plant, and then distributed to various retailers across different areas. Clear identification is necessary for each of these steps to preserve the integrity of the chain of custody. Identification and labeling systems allow stakeholders to promptly track impacted products back to their source in the event of a food safety issue, such as contamination or disease outbreak. Maintaining public health and lessening the effects of foodborne illnesses require the capacity to trace both forward—tracing the distribution of contaminated products—and backward—tracing the source of an issue.

Supporting regulatory compliance is one of the identification and labeling systems' other primary purposes. Strict traceability laws are in place in many nations and international trade organizations to safeguard consumers and guarantee the safety of food products. For example, companies are required by the General Food Law of the European Union to be able to track their products back and forth in the supply chain. In the same way, the U. S. Food manufacturers and processors are required by the FSMA to put in place traceability systems that can quickly detect and contain possible safety issues. Businesses abide by these regulations through systems of identification and labeling. Businesses can comply with regulations and stay out of trouble by making sure that each product has a unique identifier and is labeled with the relevant information. Effective labeling also makes it possible for government organizations to carry out audits and inspections more quickly, assisting in ensuring that goods that are put on the market are safe to eat. Identification and labeling systems are essential for improving transparency and customer trust in addition to compliance. Traceability is becoming an increasingly useful tool for satisfying consumer demands as they grow more concerned about the source of their food and the ethical and environmental standards that go into its production. Businesses can give customers comprehensive information about the goods they buy thanks to identification systems. For example, reading the QR code on a coffee package could provide details about the farm that produced the coffee beans, including the certifications the farm has received (e.g., fair trade or organic), as well as the methods employed in the environment. This degree of openness strengthens brand loyalty, fosters consumer trust, and enables manufacturers to set themselves apart in a crowded market.

Agriculture's identification and labeling systems have undergone a further revolution thanks to technological innovations like blockchain and RFID. For example, blockchain provides a safe, decentralized method of storing and exchanging traceability information. Every move or transaction that takes place throughout the supply chain is documented on the blockchain as soon as a product is given a unique identifier, which results in an unchangeable and impenetrable record. In addition to increasing traceability, this technology increases regulators' and consumers' trust in the accuracy and dependability of the information supplied. Similarly, tracking agricultural products is becoming easier with the use of RFID tags, especially in large-scale operations. Barcodes need to be manually scanned, but RFID tags can be automatically read by sensors, making goods tracking quicker and more effective (Aung and Chang, 2014; Zhang et al., 2018; Lv et al., 2023).

### 1.2.2. Integration of Digital Technologies in Traceability

Blockchain is a cutting-edge and novel technology that has taken the business world by storm and completely changed the way transactions and operations are carried out. Blockchain technology holds great potential to address current issues and reshape farm inventory management in the agriculture sector, where supply chain transparency, traceability, and efficiency are critical. The conventional methods of agriculture supply chains have been eclipsed by the integration of cutting-edge and customized blockchain solutions. It increases market access and gives smallholder farmers more power. All things considered, blockchain technology in agriculture can completely change how goods are produced, distributed, and consumed. According to a Statista report, the market for blockchain in agriculture and the food supply chain is anticipated to expand by $1.05 billion by 2026.

There's a lot to learn about blockchain's application in the agriculture sector, and it's not just a catchphrase; rather, it's a force for good in one of the most important sectors of the global economy: agriculture. Blockchain technology has shown promise in agriculture in recent years for solving a variety of issues, including data management, supply chain transparency, and food safety. Let's examine how it creates new possibilities like:

1. **Data Management and Analytics:** A lot of data is produced by agriculture, ranging from crop yields and market prices to weather patterns and soil conditions. You can quickly and easily streamline data collection, storage, and analysis with the aid of blockchain technology. It enables resource allocation to be optimized and informed decisions to be made by stakeholders. Researchers and farmers can work together more successfully to address agricultural challenges if data is securely shared on a decentralized network.
2. **Empowering Smallholder Farmers:** By giving smallholder farmers direct access to markets and facilitating equitable and transparent transactions, blockchain technology empowers them. Farmers can communicate directly with buyers using smart contracts and blockchain-based platforms, eschewing conventional middlemen and cutting transaction costs. Smallholders can access a larger consumer base and obtain more equitable prices for their goods thanks to this direct access to markets.

3.. **Food Safety and Quality Assurance:** By enabling prompt response to outbreaks of foodborne illness, blockchain technology in agriculture improves food safety. Additionally, it guarantees adherence to rules and guidelines. Higher authorities can swiftly identify the origins of tainted products thanks to the application of blockchain technology in agriculture. Furthermore, blockchain technology preserves quality assurance by logging a variety of data, including specifications, handling protocols, and temperature controls.

4. **Supply Chain Transparency and Traceability:** Stakeholders can follow agricultural products from farm to table thanks to blockchain technology. Blockchain is a great option for supply chain management because it offers transparency and traceability by logging every transaction and movement of goods on a decentralized ledger. Customers have access to comprehensive details regarding the food's origins, such as its growing location, processing method, and environmental impact.

### 1.2.3. Blockchain for Traceability

With the Internet of Things deployed, certain facilities can use technologies for data collection, processing, and filtering either locally or on the cloud; they can also store and analyze data to optimize processes; and they can create user-friendly interfaces and a variety of utilities tailored to installation and process requirements. Despite this, the interaction of numerous unreliable agents and devices—whose data needs to be unchangeable—means that security, transparency, and traceability are not sufficiently guaranteed. Blockchain-based systems have been created specifically to address this problem. Blockchain technology's fundamental benefits include decentralization, immutability, security, and transparency.

This is how the agricultural model's permissioned blockchain integration seeks to establish an information chain that is accessible to various parties, including associations, suppliers, clients, users, and government organizations. Each of these parties needs permission to access particular kinds of data within the chain. Every agricultural production facility takes into account many factors, such as the kinds of data that are available, the identities of authorized external actors, the data that are published, the procedures for receiving raw materials and sending completed goods, the way data transactions are displayed along the chain, and the methods used to achieve scalability. This model serves as the foundation for creating a suitable remedy. Nevertheless, depending on the crop or industry, there might be some adjustments and modifications; it is the expert's responsibility to offer a solution.

The architecture of our suggested model is layered (Figure 1.3), wherein the platform's functions and services are combined. Installed and linked to the processing and control layer are monitoring and actuator devices at an agricultural facility, whether it is a recently constructed or established one. A brief explanation of each of these layers is provided below. Human-machine interfaces (HMIs) are also installed in the physical layer, allowing technicians or farmers to act by entering requests and data. The processing layer interacts with the upper layer (IoT), performs control actions, and filters data. To oversee the various processes and carry out control and maintenance tasks, the edge layer additionally obtains data from the upper layer. Depending on the industry, type of cultivation, and environmental factors, each design may have a different IoT installation. In any event, it needs to offer the assistance required to transfer the data to the blockchain that is installed. Achieving data immutability through secure IoT communication protocols is the goal of integrating blockchain technology with IoT.

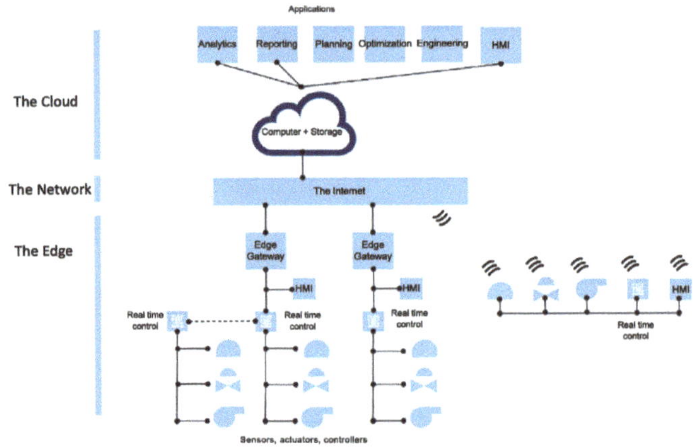


**Figure 1.3.** *An IoT-blockchain platform.*

*Source: By Paul McLaughlin, Rohan McAdam - The Undiscovered Country: The Future of Industrial Automation, CC BY-SA 4.0,* https://commons.wikimedia.org/w/index.php?curid=77601402.

A blockchain is integrated into the service layer and linked to the IoT system installed in every facility. In this layer, data from each facility (if multiple facilities are being managed) is communicated, and data is sent and received between the various applications of the top layer and the various local and cloud networks are managed. Applications are created and developed on various platforms (computers, mobile phones, business networks, etc.) at the application layer. ) by utilizing chaincodes to communicate with the blockchain. Each application's connection to the blockchain is made possible by these chaincodes. A relationship between the user and the kind of access permitted is established in each of them. Chaincodes can be programmed to execute automatically and to activate when certain conditions are met. For instance, if environmental conditions are met, a chaincode will be activated, an alert will be sent, or other defined control actions will be carried out. The architecture and functionality were previously explained. The blockchain is defined here, along with its integration with the Internet of Things, general implementation, and integration of crop singularities. Similar to choosing the right IoT technologies, there are several kinds and options for blockchain.

Researchers determined that the Hyperledger Fabric (HF) blockchain would be the most efficient technology to accomplish the specified goals for the implementation of the agricultural IoT traceability model. Distributed ledger technology (DLT) is implemented by HF in a permissioned blockchain, where entities must first undergo authentication to insert and monitor data using chaincodes. The kind of crop and agricultural facility determine the IoT solutions. Every scenario calls for a different set of embedded systems, sensors, and communication protocols. However, in all situations, the IoT must cover the following functionalities:

- Capture the data of the physical variables necessary for managing the processes (sensors).
- Install the actuators to allow manual or automatic control (actuators) using the access interfaces.
- Develop the necessary HMI interfaces so that data, actions, events, or tasks are accessible (mobile, web).
- Deploy data sensorization, control, and storage networks both locally and shared in the cloud and integrated and connected to the blockchain.
- Offer maintenance, storage, operation, upgrade, and testing services.
- Provide the necessary documentation and operating manuals.

Blockchain networks provide additional functionality to IoT facilities. They offer digital records that cannot be altered, veracity in the data, and the possibility to create algorithms that define business models—or relationship models—between various supply chain participants. These algorithms, sometimes known as "smart contracts," offer a very potent instrument for generating predetermined benefits. About the Internet of Things, the facility can modify the kind of blockchain design. However, it must provide the following functionalities in all installations:

The blockchain technology allows for verification without being dependent on third parties.

- The data structure in a blockchain is append-only. Since the transactions stored in each block are recorded in an unalterable way using hash functions and, in addition, each block contains the digest of the previous one, the data cannot be deleted or altered.
- It uses protected cryptography.
- All transactions and data are attached to a block after the process of verification.
- Transactions are recorded in chronological order. Thus, all blocks in the blockchain are time-stamped.
- It is distributed across every node participating in the blockchain.
- The transactions stored in the blocks are contained in different computers participating in the chain. Hence, it is decentralized. It also ensures that lost data, if any, can be recovered.
- It reduces the risk of duplicate entry or fraud.
- With smart contracts, businesses can pre-set conditions. Automatic transactions are triggered only when the conditions produce use software algorithms.

As the model develops, it recognizes the breadth of application and the ability of both blockchain and IoT to interact to produce more robust solutions than either technology could provide alone. Figure 1.4 illustrates the main blocks that the blockchain develops following the previously mentioned functionalities, as well as the workflow of the actions carried out between the various actors. It provides an understanding of each part of the suggested IoT blockchain platform and explains how the system functions. The application client offers a user interface through which one can submit transaction proposals to the blockchain network to use services like user and device registration. A transaction must be registered with a certificate that contains the private keys needed to sign it before it can be submitted.

The process of reading or writing data from the blockchain network is called a transaction. To register a new device or create a new task, various devices and authorized users (technicians and farmers) can submit a transaction. To carry out operations, the blockchain network receives the request from the IoT facility. Additionally, it can send the device's gathered sensing data or status updates back to the client after receiving the request from the device. The device owner's identity is verified, and transactions can be submitted directly from the physical device linked to that particular owner. The status or sensing data is then appended to the ledger and cross-referenced with the smart contract's defined conditions. A notification will be sent out or another task will be carried out if the values captured reach predetermined levels. The diagram illustrates how the blockchain network uses tamper-proof logic (smart contracts) to manage sensitive data, such as information on machine handling, environmental conditions, timestamps of cultivation processes, and access to protected facilities, while on-premises and cloud services host client applications for data input and visualization.

**Figure 1.4.** *Workflow between applications: blockchain and IoT systems.*

*Source: By Paul McLaughlin, Rohan McAdam - The Undiscovered Country: The Future of Industrial Automation, CC BY-SA 4.0, https://commons.wikimedia.org/w/index.php?curid=77601401.*

In general, a blockchain performs four actions on the transactions carried out with the client applications, the actors involved, and the IoT platform:

1. **Identity and Access Management, or IAM:** Blockchain technology in identity management allows all nodes in the network to have access to the same source of truth regarding which credentials are legitimate and who verified the authenticity of the information contained in the credentials without disclosing the actual data. Owners of identities, identity issuers, and identity verifiers are the three parties involved. Personal credentials can be issued for an identity owner (the user) by the identity issuer, a reliable entity like the local government. The entity that issues credentials certifies the accuracy of the personal information contained therein. If the identity owner needs to verify statements made about them to a third party (the verifier), they can keep those credentials in their personal identity wallet.
2. **Consensus Mechanism:** When distributed processes or multi-agent systems, like cryptocurrency, need to agree on a single data value or network state, this fault-tolerant mechanism is employed in computer and blockchain systems. Among other things, record-

keeping benefits from it. Different algorithms for consensus mechanisms operate on various policies.

3. **Distributed Ledger:** Many nodes or computing devices make up this database. Every node copies the ledger exactly and stores it. Every network node that participates updates itself on its own. Among distributed ledger technologies are blockchains.
4. **Smart Contracts:** Smart contracts are self-executing programs primarily hosted on blockchain platforms. They are defined by pre-coded features and attributes, which ensure transparency, traceability, and immutability during execution. The features and attributes of a smart contract are established in advance by the code's author. Transactions facilitated through smart contracts become irreversible, transparent, and traceable. A smart contract can function independently, regardless of how many other smart contracts it interacts with. While smart contracts are a relatively new technology with significant potential, they are not without flaws. For instance, the underlying code must be flawless and error-free to prevent vulnerabilities. Blockchains that integrate with Internet of Things (IoT) platforms naturally inherit management attributes such as distributed networks, consensus assurance, and identification capabilities. The design of customized smart contracts enhances their applicability and usefulness. For example, Hyperledger Fabric is a blockchain platform that aligns well with IoT requirements. The model's effectiveness is influenced by the properties of the blockchain, the types of smart contracts employed, and their connection to IoT services. Such designs can provide farms and other industries with enhanced traceability (Aung and Chang, 2014; Zhang et al., 2018; Lv et al., 2023).

## 1.3. TRACEABILITY IN GLOBAL AGRICULTURE SUPPLY CHAINS

The entire set of information and processes that can preserve the required details about a product and its constituent parts throughout all or a portion of its manufacturing and use chain is known as a traceability system.

When goods and materials are received from suppliers, processed, and then distributed as final goods, a traceability system keeps track of and documents the process. Thus, the identification of items that travel along the supply chain forms the foundation of all traceability systems.

The basic characteristics of traceability systems are as follows:

- Identification of units/batches of all ingredients and products;
- Registration of information on when and where units/batches are moved or transformed; and
- A system linking these data and transferring all relevant traceability information with the product to the next stage or processing step.

These qualities, i.e. regardless of the process or product involved, identification, information, and the connections between supply chain participants are common. The amount of information recorded, the distance (forward or backward) the system tracks the information, and the accuracy with which the system can identify the movement of a specific product are some areas where the traceability systems may vary. In actuality, traceability systems are documentation tools that display the route taken by a specific product from suppliers to customers via intermediary stages (Figure 1.5). Traceability systems may identify additional information in addition to the product (e.g., species, country of origin, and best-by date) that is connected to the product. Paper-based systems, bar codes, and RFIDs are examples of traceability systems. Data can be collected at low operating costs thanks to technologies for automatic identification and data capture.

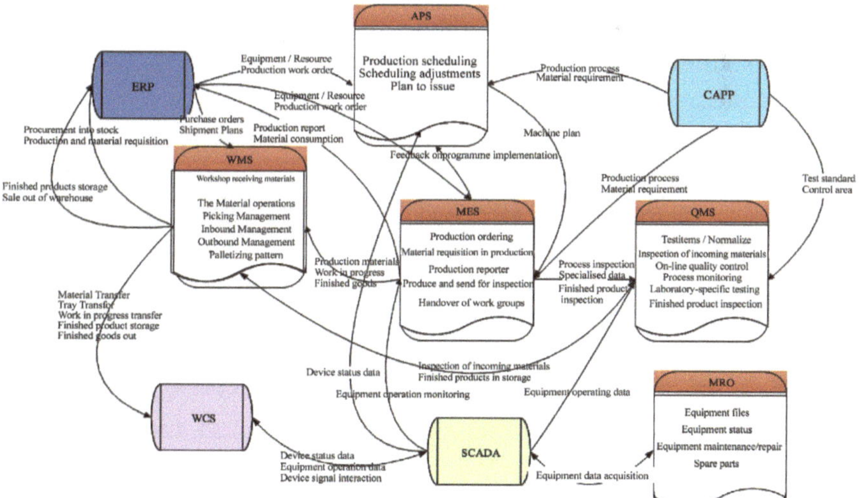


**Figure 1.5.** *Conceptual framework of a traceability system.*

*Source: Xiao L, Sun W, Chang S, Lu C, Jiang R. Research on the Construction of a Blockchain-Based Industrial Product Full Life Cycle Information Traceability System. Applied Sciences. 2024; 14(11):4569. https://doi.org/10.3390/app14114569.*

### 1.3.1. Traceability as a Tool for Compliance with International Trade Regulations

Traceability guarantees that products meet the safety, quality, and sustainability requirements needed for export, which aids agricultural producers and exporters in adhering to international trade regulations. From production and processing to distribution, it offers a clear and comprehensive record of the whole supply chain, which is crucial for fulfilling the strict specifications set by importing nations. There are serious financial repercussions for breaking these rules, such as rejected shipments, lost market access, fines, and reputational harm to a business. Traceability aids in compliance and mitigates the possible consequences of noncompliance with international standards.

1. **Ensuring Compliance with Safety and Quality Standards:**

a. **Food Safety:** Food safety is a major concern in the global trade of agricultural products. To stop the spread of contaminants, illnesses, and pests, many nations have strict sanitary and phytosanitary (SPS) regulations that must be followed when importing agricultural goods. Traceability systems keep track of every step of the production process, recording the use of inputs like water, fertilizer, and pesticides as well as handling procedures during processing and delivery. Agricultural producers can demonstrate that they are in compliance with safety standards and that their products are safe to consume by keeping thorough records that show no harmful contaminants are present.

b. **Quality Assurance:** In addition to food safety, many importing nations have quality requirements for agricultural products, including freshness, size, and appearance. Exporters can ensure that their products meet the target market's specific quality requirements by using traceability to document the growth, harvesting, and storage of their products. For instance, traceability aids in cold chain monitoring for perishable goods, guaranteeing that goods are kept at the proper temperature during transit to preserve quality.

2. **Supporting Compliance with Environmental and Sustainability Standards**

a. **Sustainability Certifications:** Environmental sustainability standards are mandatory for agricultural imports into many markets, especially in North America and Europe. Systems for tracing product origins are essential for verifying that goods are sourced responsibly. Products like soy, beef, and palm oil, for instance, are frequently linked to greenhouse gas emissions, deforestation, and biodiversity loss. Proof that these products are not connected to unlawful deforestation or environmental harm may be required by importing nations or regions. By tracking these products' origins, traceability systems make sure they originate from sustainable sources and adhere to legal requirements like the European Union's deforestation-free certification.

b. **Environmental Compliance:** Environmental standards set by importing nations must be met by a large number of agricultural products. The use of specific pesticides, water use, or emissions may be restricted by these rules. By documenting input usage and tracking the effects of production on the environment, traceability guarantees that all environmental laws are followed. Environmental standards violations can result in shipment rejections or bans, which can cost exporters a lot of money.

3. **Facilitating Compliance with Ethical and Labor Standards**

a. **Social and Labor Regulations:** A lot of international trade agreements now contain provisions pertaining to labor and social standards, especially to stop child labor, forced labor, and hazardous working conditions. Exporters and farmers must show that these unethical practices are not present in their supply chains. By revealing information about working conditions and recording labor practices, traceability systems help exporters demonstrate compliance. In markets like the US and the EU with stringent labor laws, where noncompliance may lead to import prohibitions or boycotts by customers, this helps guarantee that goods can be sold.

b. **Fair Trade and Ethical Sourcing:** Ethically sourced goods are becoming more and more in demand from consumers and governments, particularly for commodities like cotton, cocoa, and coffee. Traceability systems are used by Fair Trade certifications to confirm that farmers are paid fairly for their goods and that products are made in fair working conditions. Exporters can gain access to high-value markets where customers value ethical production by demonstrating their adherence to these ethical sourcing standards through traceability.

4. **Consequences of Non-Compliance**

a. **Rejected Shipments:** Shipments are subject to rejection as one of the direct results of breaking international trade laws. A product may be refused entry into the importing nation if it does not satisfy the necessary safety, quality, or sustainability standards. This leads to monetary losses because of product waste, transportation expenses, and potential reputational damage. For example, a shipment of fruits or vegetables may be rejected or destroyed if an importing nation discovers pesticide residues above permissible limits.

b. **Loss of Market Access:** Market access may be lost as a result of persistent infractions or noncompliance with trade regulations. Importing nations may put up trade restrictions on goods from noncompliant suppliers, such as increased tariffs, quotas, or complete prohibitions. For instance, exporters may be blacklisted and unable to access important markets if they violate laws about labor practices or environmental sustainability. A producer's earnings and prospects for expansion may be severely impacted by losing access to markets with strong demand.

c. **Financial Penalties and Fines:** For exporters who violate trade laws, governments, and regulatory agencies may levy fines and penalties. For safety infractions, environmental violations, or labor abuses, these sanctions may include expensive fines. Legal disputes over non-compliance, shipment returns, and product recalls can also come with high costs. The financial penalties imposed on major agricultural exporters may have a long-term effect on their profitability.

d. **Reputation Damage:** The reputation of an exporter or agricultural producer can be seriously harmed by noncompliance with international trade regulations. Customers and retailers demand high standards in today's global marketplace, where accountability and transparency are valued more and more. Consumer boycotts, bad press, and strained business ties may result from an exporter's alleged violations of safety, environmental, or ethical standards. It may take years to repair a damaged reputation, and businesses risk losing important alliances with distributors and retailers who value sustainability and compliance.

**5. The Role of Traceability in Mitigating Non-Compliance Risks**

a. **Real-Time Monitoring:** Real-time product monitoring across the supply chain is made possible by traceability systems, which enable the prompt identification of possible problems before they reach crucial phases, like entering international markets. This aids manufacturers and exporters in making sure they adhere to all relevant laws during the entire production, processing, and distribution process.

b. **Quick Response to Issues:** Traceability enables a prompt response in the event of a food safety issue or regulatory violation by pinpointing the precise cause of the issue. By exhibiting responsibility and acting quickly to address issues, exporters are able to minimize the extent of product recalls, minimize losses, and preserve market trust.

c. **Proof of Compliance:** The data and documentation needed to demonstrate adherence to international trade laws are produced by traceability systems. Importing nations frequently ask for these documents as part of their regulatory inspections. Exporters can easily supply this information by having an effective traceability system, preventing delays and possible disruptions to trade.

### 1.3.2. Role of Traceability in Mitigating Supply Chain Disruptions

To reduce risks and disturbances in global agricultural supply chains, traceability is essential, especially in the face of major obstacles like pandemics, climate change, and geopolitical conflicts. The complexity of supply chains, involving numerous actors in different regions, has made transparency and accountability crucial. Traceability systems offer real-time data that aids in the prevention, management, and recovery from disruptions. They also provide the infrastructure required to track agricultural products from farm to consumer. Traceability increases the robustness of international agricultural supply chains by promoting effective problem identification, maximizing resource utilization, and guaranteeing regulatory compliance.

Increased transparency and visibility are two of the main ways traceability reduces supply chain risks. Agricultural products frequently travel through multiple middlemen, including distributors, retailers, and processors, in intricate supply chains before they are purchased by the final customer. Supply chain participants have access to comprehensive data regarding the product's origin, processing, and journey through each phase of the supply chain thanks to traceability systems. When supply chains are disrupted, this visibility is particularly crucial because it enables stakeholders to promptly pinpoint and isolate problem areas. For example, traceability systems can identify the source of contamination or quality issues, allowing for prompt corrective actions and targeted recalls. This lessens the potential harm to consumer confidence as well as the financial losses brought on by widespread recalls.

Traceability systems are essential for keeping agricultural supply chains operational during major world emergencies like pandemics. The vulnerabilities of global supply chains, particularly in the food and agriculture sectors, were made evident by the COVID-19 pandemic. Delays, shortages, and price swings were the results of major disruptions brought on by lockdowns, travel restrictions, and labor shortages. Strong traceability systems in supply chains, however, made them more adaptable. Companies were able to reroute shipments, improve inventory management, and guarantee that necessary goods arrived at their destination on time thanks to traceability, which provided real-time data on the location and status of goods. Furthermore, traceability systems assisted in controlling the safety issues arising from the virus's spread by enabling businesses to keep an eye on worker health, hygienic procedures, and the handling of goods throughout the supply chain.

In a similar vein, changes in precipitation patterns, growing seasons, and extreme weather events pose an increasing threat to global agricultural supply chains due to climate change. Traceability facilitates improved supply chain planning and resource management, which helps to reduce these risks. Traceability data, for example, enables businesses to promptly find alternate suppliers or regions that can meet demand when a climate-related event, such as a drought or flood, impacts crop yields in one area. Traceability systems also make it possible to track agricultural inputs like pesticides, fertilizers, and water more precisely, which enables farmers to modify their methods in response to shifting environmental conditions. Food security in the face of climate-related disruptions depends on this flexibility. Traceability also helps reduce risks by enhancing adherence to global trade laws, which are impacted by geopolitical unrest. Changing trade policies, import/export restrictions, and tariffs frequently affect agricultural supply chains, which can cause disruptions in the flow of goods. The ability to trace agricultural products guarantees that they fulfill the sustainability, safety, and quality requirements set forth by various nations. For instance, strict regulations on agricultural product traceability are in place in many importing nations, especially in North America and Europe, to stop the importation of goods associated with illicit activities like deforestation or unethical labor practices. With traceability systems, businesses can quickly adjust to new regulations or trade barriers brought about by geopolitical issues by supplying the required certifications and documentation, guaranteeing ongoing market access and reducing disruptions.

To manage supply chain disruptions associated with food safety and public health emergencies, traceability is also essential. Traceability enables prompt identification and product recall in the event of a foodborne illness outbreak or contamination, halting the disease's spread and safeguarding the public's health. This is especially crucial for global supply chains, as the possibility of contamination increases due to the frequent long-distance and cross-border transportation of goods. Systems for tracing back contamination to its source, identifying contaminated batches, and removing them from distribution before they reach consumers are made possible by traceability data. This maintains consumer confidence in the safety of the food supply while also lowering the financial losses linked to widespread recalls. Traceability systems support sustainability and ethical sourcing in addition to mitigating short-term risks, strengthening the long-term resilience of global agricultural supply chains. Unsustainable practices are the root cause of many disruptions in agricultural supply chains, including labor disputes, water shortages, and deforestation. By monitoring and confirming the social and environmental effects of their supply chains, traceability enables businesses to make sure that their products are sourced ethically and in accordance with global sustainability standards. For instance, traceability enables businesses to make sure that their goods don't fuel illegal deforestation, a major cause of biodiversity loss and climate change. Traceability lessens the possibility of disruptions caused by resource depletion or environmental degradation by encouraging sustainable practices.

Traceability also improves cooperation between supply chain participants, which promotes a more responsive and flexible system. Coordinated responses during disruptive times depend heavily on the supply chain's ability to share data. To improve decision-making and problem-solving efficiency, traceability systems enable communication between farmers, processors, distributors, and retailers. For example, supply chain actors can swiftly find alternate sources or reroute shipments to ensure continuity if a specific region experiences a sudden shortage as a result of a climate event or political unrest thanks to traceability.

## 1.3.3. Key Issues in the Traceability of Agricultural Products

Traceability is the ability to track every step of a process, from the acquisition of raw materials to production, use, and disposal, to determine "where, when, and by whom the product was made." The traceability system can react rapidly to issues, gather information about the entire process, including the manufacturing and distribution of the problematic product, investigate the issue and its location, identify the underlying cause, and then address the issue in a targeted way. Traceability has three primary objectives: (1) improved supply chain management; (2) improved product differentiation and quality assurance; and (3) improved ability to detect non-compliant product offerings. The food supply chains' traceability procedure is depicted in Figure 1.6.

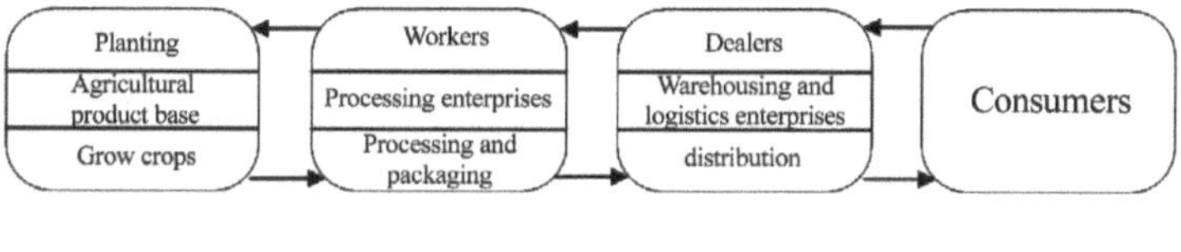


**Figure 1.6.** *Traceability process in food supply chains.*

*Source: Lv G, Song C, Xu P, Qi Z, Song H, Liu Y. Blockchain-Based Traceability for Agricultural Products: A Systematic Literature Review. Agriculture. 2023; 13(9):1757. https://doi.org/10.3390/agriculture13091757.*

The traceability industry has grown quickly due to the increasing demand for commodity traceability. However, because there isn't a quick and reliable way to get information about the provenance of the product, the current systems lack transparency and the trust of consumers. Furthermore, there is no guarantee of a consistent information flow along the agricultural product chain with the current traceability systems. Emerging technology like blockchain offers a fresh approach to traceability. In 2008, blockchain technology surfaced as a fundamental element of the Bitcoin cryptocurrency. Blockchains allow distributed ledgers to function as transactional ledgers without requiring a centralized, reliable authority. Serial recording is made possible by cryptographic time stamping, and ledger-recorded updates are unchangeable.

Although agricultural product traceability has significantly improved with the development of blockchain and traceability technologies, numerous issues have emerged at this historic juncture as a result of technological advancements and interactions with various social environments. For instance, there are many questions regarding the implementation of a decentralized system, and people are particularly worried about whether information has been altered, particularly in light of the 2020 outbreak of the novel coronavirus and concerns about food safety. The use of blockchain technology in agriculture offers answers to contemporary issues and is also strongly tied to the state of the environment and society today. The established decentralized system can effectively address issues when the central institution's credibility is called into question. Each block in the blockchain system is organized chronologically in a linear fashion, and each block's data is open and transparent, preventing information manipulation and fostering mutual trust. Food safety can be ensured and the unlawful use of chemicals in production can be avoided by using blockchain technology to store all food information in a transparent and shared system for all supply chain participants.

Let's demonstrate how blockchain can successfully stop industrial processes from using too many chemicals such as: (i) Traceability and transparency: Blockchain technology offers a decentralized, unchangeable ledger that can capture and preserve data about the complete supply chain. It is feasible to monitor the source, flow, and use of chemicals from the stage of raw materials to the finished product by incorporating this technology into the production process. Accountability is improved and stakeholders can confirm the legitimacy and conformance of the chemicals used thanks to this transparency; (ii) Smart contracts and automation: Predetermined guidelines and requirements for the use of chemicals can be established through the implementation of smart contracts on the blockchain. These agreements can automate compliance checks, setting off alarms or even stopping production if chemicals found to be uncertified or in excess are found. This simplifies the monitoring and management of chemical usage and lessens the need for manual checks; (iii) Certification and verification: By using blockchain technology, a reliable system for confirming the legitimacy of chemicals can be established. The blockchain allows for the easy access and tamper-proof storage of audit reports and certificates. This guarantees that only approved chemicals that satisfy the necessary requirements are utilized during the manufacturing process; and (iv) Data sharing and collaboration: Blockchain makes it possible for supply chain participants, such as manufacturers, suppliers, regulators, and customers, to securely and effectively share data. Collaboration is made easier by this information sharing, which permits real-time monitoring, early detection of possible problems, and swift action to stop the excessive or illegal use of chemicals.

The use of blockchain technology for food safety aligns with the Chinese central government's initiative to "promote the application of blockchain technology in food safety supervision." The new coronavirus outbreak that started in early 2020 has had a significant impact on agriculture worldwide. Agricultural products with excessively long transportation intervals risk losing their transportation records, and the COVID-19 pandemic has slowed the movement of agricultural products. However, because the results of the detection of dangerous substances like bacteria and viruses attached to food are difficult to disclose, the public may be concerned about the safety of food that is transported from high-risk areas. The blockchain-based agricultural traceability system can effectively address the aforementioned issues. The transportation data will remain intact when stored in the cloud database, and the traceability system can precisely follow the movements of agricultural products from the start of production. In the event of an epidemic, the blockchain-based traceability system can precisely locate the disinfection site to guarantee the quality and safety of agricultural products if they need to be safely disinfected during transit. Table 1.1 lists the current blockchain-based fixes for issues with traceability.

**Table 1.1.** *Blockchain Solutions to Traceability Problems*

| Problem | Solution |
|---|---|
| How can we set up | Chronological distributed |
| a system of honesty | databases are used by blockchain |
| without a central | where blocks are linked to each |
| authority? | other in a proper linear manner, |
| | and it is impossible to be |
| | deleted. |
| How to prevent | A newly added block cannot be |

| | |
|---|---|
| agricultural data | deleted, and it will become a |
| from being tampered | permanent record that can be |
| within this information | accessed and verified by |
| age? | everyone on the network. |
| How can we coordinate | To adapt more widely, |
| our system with policies? | our distributed system is based |
| | on important international |
| | traceability system standards, |
| | Regulation (EC) No 178/2002. |
| | In a specific country or |
| | region, our system rules can be |
| | adjusted to suit local policies. |
| Why it is important to | During critical emergency periods, |
| build the system under | it can make unseen products |
| the decline of agri- | traceable and produce products of |
| economy circumstances | high quality. It is an approach in |
| because of COVID-19? | disease warning. |

*Source: Lv G, Song C, Xu P, Qi Z, Song H, Liu Y. Blockchain-Based Traceability for Agricultural Products: A Systematic Literature Review. Agriculture. 2023; 13(9):1757. https://doi.org/10.3390/agriculture13091757.*

For instance, the use of blockchain technology in the supply chain, in conjunction with technologies like the IoT, enables the realization of the complete product record from the supply chain to the end user.

This includes the source of raw materials and production processes, processing links, warehousing information, inspection batches and logistics turnover, third-party quality inspection, customs entry and exit, anti-counterfeiting verification, and the ability to trace the entire process back to the consumer.

Regarding the trust issue, however, the blockchain supports the development of a cross-institution traceability system by using technical tools like consensus processes and time stamps to achieve the goals of data traceability and non-tampering. To break the information island, provide information support, and achieve some degree of transparency in the production process, third-party supervision agencies and consumers are simultaneously incorporated into the supervision system.

## 1.3.4. Applications of Blockchain in the Traceability of Agricultural Products

51% of consumers are willing to pay more for safe and healthy food, and this number is still rising, according to the Deloitte report. Customers expect verifiable traceability as a crucial component of quality and safety. A system that can offer details on the origin and life cycle of commodities is required to satisfy this demand. However, as precision agriculture technology has become more widely available, people's interest in agricultural data has grown in recent years. We are all aware that agricultural data particularly that obtained from joint yield monitors are often disorganized. The validity of the data worries analysts, particularly since other people may have an impact on the data quality at different points along the data path. Furthermore, handling large data sets presents many difficulties, chief among them being the presence of imprecise, unstructured, and incomplete data. Because it can confirm, audit, and transmit the confidentiality of data and information, blockchain technology—a decentralized ledger that enables block-encrypted data transactions—has garnered increasing attention from academic and business communities.

### *1.3.4.1. Fish*

One of the primary food-based industries with the highest value is seafood. But there hasn't been adequate regulation of the fishery. Fish stocks around the world are overfished to the tune of over 30%. In the meantime, it is challenging to assess the quality of seafood in practice because essential certificates of origin and temperature records are not fully visible to customers. Thus, it is challenging to precisely assess the quality of fish in practice because there is a dearth of trustworthy food and quality tracking data. A significant development for global fisheries is the use of blockchain technology to enhance tuna traceability to curb illicit and unsustainable fishing methods in the Pacific Island tuna industry. The authors collaborated with the Sydney Fish Market (SFM) to create the BeFAQT system, which tracks the provenance and quality of fish using blockchain technology. The application of blockchain encryption in the fish supply chain has raised privacy concerns. To address these issues and enable the sharing of trusted and private information among supply chain participants, a multi-layer blockchain architecture built on attribute-based encryption (ABE) is suggested. Field tests demonstrate that BeFAQT can deliver reliable and thorough real-time fish provenance and quality tracking data. Actors and blockchain specialists in the fish farming sector participated in semi-structured interviews as part of the exploratory study. The primary inference to be made from the findings is that blockchain can boost business value by utilizing the effectiveness of information to improve industry visibility and operational efficiency through improved planning and prediction through data analysis. A supply chain-related evaluation of blockchain technology is developed, and it is subsequently extended to a broader context. Thus, the evaluation of blockchain technology is helpful for businesses or individuals who wish to learn more about blockchain technology or who intend to use it in the future.

By detecting issues in real-time and putting solutions in place, digital technologies like blockchain can assist in achieving specific Sustainable Development Goals of livelihoods, food security, and the environment. The study examines the design of a blockchain-centric food supply chain to support the Sustainable Development Goals, using the Thai fishery as a backdrop. The primary conclusions suggest that the supply chain has data asymmetry, which can help achieve the SDGs. It is suggested to use a blockchain-based fish farm platform to guarantee the accuracy of agricultural data. The platform is made to give fish farmers safe storage for storing vast amounts of unchangeable agricultural data. To lower the possibility of error or manipulation, smart contracts are used to automatically carry out the different fish farm operations. The practical implementation of blockchain technology (BCT) in a multi-tiered supply chain for frozen fish products in Portugal is being investigated to examine the impact of BCT adoption in the context of multi-tiered supply chains (SC). Data was gathered using a combination of qualitative and quantitative methods. Suppliers, transporters, and retailers from a single frozen fish SC participated in three semi-structured interviews. Customers were more inclined to purchase fish with traceability information, according to the findings. A private Ethereum blockchain-based solution is suggested to efficiently track and manage products in the fishery supply chain. Fishery supply chain operations are efficiently managed by the solution in a decentralized, transparent, traceable, secure, private, and reliable manner. Additionally, five smart contracts and the solution architecture for automating the supply chain processes in the fishing industry are presented. To demonstrate that the suggested solution is reliable and secure, a security analysis is conducted.

### *1.3.4.2. Fruits*

Reducing food contamination is a significant challenge for the fruit retail industry.. Every year, nearly 10% of people worldwide get sick after eating tainted food, and this number is sharply rising. Therefore, to ensure the food supply for society both domestically and internationally, a blockchain-based retail fruit chain is required. Because of this, agricultural product traceability necessitates a technology that facilitates information exchange throughout the supply chain and may guarantee traceability for final consumers. The findings demonstrate that pineapple supply chain traceability can benefit from blockchain technology, offering valuable insights into its applications in the retail industry.

In addition to business procedures and competition, the study offers insights into the Malaysian fruit retail market from a variety of angles, including retailer-customer collaboration. To preserve its future global market share, the Malaysian retail sector urgently needs industrial innovation, technological transformation, and new growth avenues like implementing blockchain systems. The fresh fruit supply chain currently faces challenges like information, supply-demand relationships, tamper-proofing, and regulatory traceability due to its wide distribution, numerous subjects, and intricate transactions. Blockchain technology has the potential to improve the fresh fruit supply chain when it is implemented. From the time fruit is picked until it is consumed, the blockchain system gathers and uploads trustworthy data using IoT-related technologies. The blockchain does have certain drawbacks, though, like its dependence on the Internet of Things, its lack of advanced fruit preservation technology, its ambiguous legal oversight, and more. The "double chain and double storage" blockchain model was created for the fruit and vegetable supply chain to restrict storage growth and enhance the blockchain's query efficiency. The Hyperledger-based double chain double storage model can greatly increase the efficiency of the blockchain storage query by taking into consideration the openness, security, and privacy of enterprises in all links of the transaction information.

### *1.3.4.3. Vegetables*

In addition to being green, healthy, and high in nutrients, vegetable agricultural products are also very popular with consumers. However, there is a short shelf life for fruits, vegetables, and agricultural products, low storage temperature requirements, and a high risk of food safety incidents. A method for on-chain and off-chain data storage based on "database + blockchain" is proposed to address the issues of high data load pressure and inadequate privacy security during the process of data growth in the blockchain traceability system. The SHA256 algorithm is used to upload the hash value of the publicly visible data to the blockchain system from the local database in the supply chain. Related businesses can share the private data that has been encrypted using the CBC encryption algorithm thanks to its storage on the blockchain. Use hyperledger fabric to create a blockchain network that can handle the issues with the vegetable supplier while utilizing the key features of blockchain, such as transparency and privacy. Using Kafka, a blockchain network with nine channels was established. According to this performance, the vegetable suppliers may enhance their ledger capabilities within a day using a multi-channel blockchain network.. This section discusses field productivity and crop health. VegIoT Garden is a scalable, modular, low-cost IOT/smart agriculture (SA)-oriented management platform built on commercial off-the-shelf (COTS) devices. It is intended to help farmers improve the commodities they produce, identify important problems, and address irrigation-related issues. It also looks for potential correlations between climate characteristics.

### *1.3.4.4. Eggs*

Use cases for implementing blockchain in food delivery are presented, with a focus on egg supply chains in the Midwest of the United States, to track the products from farm to fork using IoT-enabled technologies and blockchain technology. A transparent and traceable food supply chain will enable consumers to get the information they require to make educated food decisions. Transparency and traceability help food supply chain participants build stronger client relationships, boost productivity, and lower costs associated with product losses, fraud, and food recalls.

### *1.3.4.5. Pork*

Walmart, a massive retailer, is leveraging blockchain technology to address supply chain food safety concerns in response to widespread food contamination incidents. Walmart tested a blockchain-based system in 2016 to keep an eye on pork products sold in China. Walmart tracked and verified pork products from a farm owned by Chinese meat producer Jinluo using blockchain technology in May 2017. Walmart's blockchain solution shortened the time needed for pork traceability by using a farm-to-table methodology. Moreover, there is a noticeable increase in customer loyalty. To better understand the potential impact on the sustainability of food supply chains, a thorough evaluation of blockchain traceability in food supply chains is carried out through a case study of the pork supply chains in Vietnam. This allows for an extensive study of the application of blockchain-enabled food supply chains in developing countries.

### *1.3.4.5. Tea*

All organizations involved in tea traceability must participate in the traceability process.. The groups involved are complicated, though, and the growing number of traceability links will lead to many issues. For example, adding links to every individual involved will undoubtedly increase the complexity of the traceability system and management's effectiveness. As a result, it's critical to figure out how to swiftly and efficiently locate the tea traceability data in each link. To achieve the reliable traceability of agricultural products, a blockchain-based smart agriculture system integrating Ethereum and wireless sensor networks has been developed. Information about farm product traceability that is authentic, trustworthy, and safe is given to customers via the OR code. To achieve effective food quality and safety control, a holographic database covering the whole tea industry chain—from farm to table—is also developed. A blockchain technology

(BCT) practice model powered by RFID technology is proposed to help manage the complexities of circular tea supply chain management (CTSCM) and expedite resource deployment, reduce waste and enhance supply chain sustainability. The suggested model can offer a more thorough comprehension of resource utilization, industry procedures, and inventory performance. In the field of agricultural product traceability, a machine learning (ML), blockchain, and IoT-based tea credible traceability system (MBITTS) is designed and put into use to achieve system automation and credibility. Important traceability-related data can be automatically gathered by IoT devices. The information is gathered and put into a blockchain system for querying, processing, and storing. Data entering the system is guaranteed to be secure due to the blockchain's distributed, decentralized, and immutable nature. When an ML data verification mechanism is added to the current blockchain-based agricultural product (tea) traceability systems, it can ensure up to 99% accuracy of the chain's information. Through the coordination of tea planting, product processing, transportation, and sales, the tea production and supply chain (PSC) can generate benefits for all participating businesses. A blockchain-enabled decentralized framework is suggested to accomplish full chain tea counterfeiting oversight and automated environmental management. Furthermore, an optimal irrigation strategy for automatic planting environment adjustment is suggested, and an enhanced adaptive weighted data fusion algorithm is created to achieve accurate IoT data and effective resource allocation. The outcomes of the experiments demonstrate that the suggested framework can accomplish resource-saving automation with high throughput and efficiency.

### *1.3.4.6. Crop*

A blockchain-based crop insurance solution is proposed to address the intricate and financially unviable issues with the current crop insurance system. One of the main barriers to using this facility at the moment is the expense of managing insurance. This fee can be significantly decreased by correctly utilizing the Ethereum-based blockchain. The Golden Seed Breeding Cloud Platform (GSBCP), an architecture to store high-throughput crop breeding data effectively and securely, is demonstrated by utilizing the enhanced blockchain technology in a breeding information management system. With a large amount of data, this storage architecture can greatly increase the efficiency of GSBCP, especially when breeders query breeding data. Connecting with and supplying farmers with information has never been easy. An affordable, blockchain-based security framework was presented as a solution to these issues, to foster a farmer community and crowdsource the data that farmers produce for the benefit of the farmer community. To give farmers incentives, a revenue model is also included.

It would make much more sense to incorporate a deep neural network-based model into the suggested framework, one that could identify potential abnormalities in the crops and provide potential remedies. An IoT-based crop monitoring and classification system is suggested for the automatic detection, logging, and tracking of real-time parameters that significantly affect crop quality and yield. A smart agriculture technique is proposed to boost crop yields and secure all the important data about seed type, fertilizer, and crop costs. Farm data is gathered via IoT sensor devices; machine learning algorithms are used to forecast crop yields; and critical information about crop costs, fertilizers, and seed varieties is stored in blocks for data and communication security. The suggested technique offers better accuracy and profit from seed selection to transaction when compared to the conventional methods. Transactions powered by edge computing and blockchain technology improve security and decrease latency. Agriculture's sustainability and traceability are guaranteed by the suggested system (Aung and Chang, 2014; Zhang et al., 2018; Lv et al., 2023).

## REFERENCES

1. Aung, M. M., & Chang, Y. S. (2014). Traceability in a food supply chain: Safety and quality perspectives. *Food Control, 39*, 172–184. https://doi.org/10.1016/j.foodcont.2013.11.007.
2. Badia-Melis, R., Mishra, P., & Ruiz-García, L. (2015). Food traceability: New trends and recent advances. A review. *Food Control, 57*, 393–401.
3. Curto, J. P., & Gaspar, P. D. (2021). Traceability in food supply chains: Review and SME-focused analysis—Part 1. *AIMS Agriculture and Food, 6*, 679–707.
4. Curto, J. P., & Gaspar, P. D. (2021). Traceability in food supply chains: SME-focused traceability framework for chain-wide quality and safety—Part 2. *AIMS Agriculture and Food, 6*, 708–736.
5. Kamble, S. S., Gunasekaran, A., & Sharma, R. (2020). Modeling the blockchain-enabled traceability in agriculture supply chain. *International Journal of Information Management, 52*, 101967. https://doi.org/10.1016/j.ijinfomgt.2019.05.023.
6. Lv, G., Song, C., Xu, P., Qi, Z., Song, H., & Liu, Y. (2023). Blockchain-based traceability for agricultural products: A systematic literature review. *Agriculture, 13*(9), 1757. https://doi.org/10.3390/agriculture13091757.
7. Mas, F. D., Massaro, M., Ndou, V., & Raguseo, E. (2023). Blockchain technologies for sustainability in the agrifood sector: A literature review of academic research and business perspectives. *Technological Forecasting and Social Change, 187*, 122155. https://doi.org/10.1016/J.TECHFORE.2022.122155.
8. Patelli, N., & Mandrioli, M. (2020). Blockchain technology and traceability in the agrifood industry. *Journal of Food Science, 85*, 3670–3678. https://doi.org/10.1111/1750-3841.15477.
9. Stranieri, S., Riccardi, F., Meuwissen, M. P., & Soregaroli, C. (2021). Exploring the impact of blockchain on the performance of agri-food supply chains. *Food Control, 119*, 107495.
10. Zhang, Y., Yang, Z., Yang, R., Jin, K., Lin, B., & Si, P. (2018). Traceability system of farm produce based on blockchain. *Technology and Intelligence Engineering, 3*, 4–13.

# CHAPTER 2

# REGULATORY FRAMEWORKS AND STANDARDS

## Contents

# INTRODUCTION

For food safety, quality, and sustainability to be guaranteed throughout the supply chain, agricultural product traceability regulations are crucial. These frameworks provide vital information on the origin, handling, processing, and distribution of food items by tracking and documenting the journey of agricultural products from farm to table. An increasing number of consumers are becoming conscious of environmental issues and food safety, which makes efficient traceability systems crucial. The fundamental tenet of traceability regulations is that accurate records of agricultural products must be kept by all parties involved, including producers, processors, distributors, and retailers. In the event of food safety incidents, such as contamination outbreaks, this accountability facilitates prompt responses and enables authorities to track impacted products back to their origin. In numerous areas, regulatory organizations like The European Food Safety Authority (EFSA) and the U.S. Food and Drug Administration (FDA) have developed regulations requiring traceability for a range of agricultural products. Standardized systems, such as barcodes, RFID tags, or digital platforms that enable real-time tracking and information sharing among supply chain participants, are frequently mandated by these regulations.

It is also addressed in the regulations on how reliable data management systems are. Producers must keep thorough records that contain data on cultivation techniques, inputs (such as seeds, fertilizer, and pesticides), and harvest specifics. This data supports sustainability practices and safety standard compliance in addition to facilitating traceability. In addition, processors and distributors must maintain documentation regarding handling and storage conditions to guarantee the safety and compliance of products during their consumer journey. Traceability frameworks are significantly shaped by international standards in addition to national regulations. Guidelines that promote safe and high-quality food products are provided by organizations like the Food and Agriculture Organization (FAO) and the World Health Organization (WHO) through the Codex Alimentarius. These global guidelines promote cross-border harmonization and facilitate trade while guaranteeing that food products fulfill safety and quality standards. Agriculture traceability systems are also changing as a result of emerging technologies. For instance, blockchain technology increases stakeholder trust and transparency by providing a decentralized, unchangeable record of transactions. These technologies empower consumers to make informed decisions about purchases and promote ethical consumption by giving them access to information about the origins and handling of their food. Although there are many advantages, there are difficulties in putting traceability systems into practice. Obstacles may arise for smallholder farmers in particular because of their restricted means, lack of access to technology, and ignorance of legal requirements. Legislators need to take these difficulties into account and encourage inclusive approaches that benefit all parties involved to prevent traceability frameworks from unfairly burdening smaller producers. Food safety, consumer confidence, and sustainable practices all depend on the regulatory frameworks for traceability in agricultural products. Through the implementation of innovative technologies and the establishment of well-defined guidelines, these frameworks facilitate the efficient tracking of agricultural products by various stakeholders along the supply chain. Strong traceability systems will become even more crucial as the demand for safe, high-quality food rises globally, necessitating the adoption of these laws by all parties involved in the agriculture industry.

## 2.1. INTERNATIONAL STANDARDS (ISO, HACCP, GS1)

International standards are essential for developing best practices in agricultural product traceability, which improves food efficiency, quality, and safety. Important frameworks are GS1, HACCP (Hazard Analysis Critical Control Point), and ISO standards.

### 2.1.1. ISO Standards

Globally recognized standards are created by the International Organization for Standardization (ISO) to assist organizations in ensuring quality and safety in a variety of industries, including agriculture. For instance, ISO 22000 is a standard for food safety management systems that specifies the needs for any chain of food establishment. It combines the concepts of HACCP, highlighting the necessity of efficient traceability to keep an eye on potential threats to food safety. Farmers can get a competitive edge in the market by implementing ISO standards and proving their dedication to food safety and quality.

ISO standards play a critical role in improving quality, safety, and efficiency in a variety of industries, but especially in food safety and agriculture. The International Organization for Standardization created these standards, which give businesses a framework for putting best practices into effect and promoting consistency and dependability in their operations. Food safety management systems are the subject of ISO 22000, one of the most important ISO standards for agriculture. The goal of this standard is to guarantee that risks to food safety are efficiently managed along the entire food supply chain by integrating the concepts of Hazard Analysis Critical Control Point (HACCP).

Organizations must follow the guidelines provided by ISO 22000 to create, implement, maintain, and continuously enhance a food safety management system. It highlights the necessity of precise documentation and communication, both of which are necessary for traceability. Organizations can improve consumer trust and safety by promptly identifying the root cause of any food safety issues by keeping accurate records of processes, inputs, and handling procedures. This is especially important now that foodborne illnesses can have serious health effects and cause large financial losses (Figure 2.1).

**Figure 2.1.** *The International Organization for Standardization (ISO) develops globally recognized standards.*

*Source: By International Organization for Standardization - Own work based on: File:Final ISO Red Square.png, Public Domain, https://commons.wikimedia.org/w/index.php?curid=67291827.*

Beyond food safety, ISO standards are also being adopted. Other ISO standards, like ISO 9001, concentrate on quality management systems and give businesses a more comprehensive framework to improve customer satisfaction and operational effectiveness. When ISO 9001 and ISO 22000 are used in tandem, there can be a synergistic effect that guarantees not only that food safety hazards are addressed but also that overall quality management practices are optimized. By assisting companies in adhering to international best practices and meeting regulatory requirements, this all-encompassing approach increases their competitiveness in the global market.

Furthermore, by standardizing procedures internationally, ISO standards promote trade between nations. Organizations can ease cross-border transactions by adhering to internationally recognized benchmarks and proving them to partners and customers by following these standards. This is particularly crucial in the agriculture industry, as goods frequently travel through intricate supply chains with numerous nations and parties involved. ISO standards improve quality and safety while also supporting agricultural sustainability initiatives. Environmental management systems are the focus of standards like ISO 14001, which encourage businesses to reduce their environmental impact by implementing sustainable practices. Agricultural producers can show their dedication to environmental stewardship and improve their traceability systems by incorporating these standards into their operations. There are difficulties in putting ISO standards into practice, especially for smallholder farmers and low-resource enterprises. Nonetheless, many programs are designed to assist these parties in implementing ISO frameworks. Smaller organizations can benefit from training programs, technical support, and streamlined certification procedures to help them manage the challenges of ISO compliance and make sure that everyone can take advantage of these standards.

The agricultural industry's adoption of safe, effective, and sustainable practices is greatly influenced by ISO standards. ISO standards help organizations meet the demands of both regulatory bodies and consumers by providing frameworks that support food safety, quality management, and environmental responsibility. The importance of ISO standards will only increase as the world's agricultural landscape changes, highlighting the necessity of continuing adherence to these crucial principles. The agriculture sector can improve traceability, leading to safer food systems and greater consumer confidence by implementing ISO standards consistently (Charlebois et al., 2014).

### 2.1.2. Definition of ISO

The ability to track the origin of materials and parts, the processing history, the distribution and location of the product after delivery, and the history, application, and location of the object under consideration are all included in the ISO definition of traceability. In addition to the basic need to be able to physically track goods from point of origin to point of destination and back again, traceability also entails being able to give details about the product's composition and history. These additional traceability factors are crucial for food safety, quality, and labeling.

This International Standard's scheme does not require perfect traceability, i.e. that a specific retail product should be able to be tracked from its point of origin to its final destination, or back to a single vessel or catch. Practically speaking, it is acknowledged that the mixing of units is likely to happen at several points in the distribution chains, such as when products are processed from raw materials and graded at auction markets before being sold. The food industry is changing the trade units where this kind of mixing takes place. The business must keep track of the identifier of each received trade unit that can be entered into a newly created trade unit, and vice versa, to ensure traceability. After that, it is possible to link the specific product to a limited number of ships or catches, and vice versa. The information specifications are unable to list every piece of information that might be needed in every scenario due to the wide range of fish products, the international distribution networks that handle these products, and the various legal requirements. A general foundation for traceability is provided by this International Standard. Businesses are given the flexibility to record additional data in their own non-standardized files that are keyed to the same unit IDs.

The data remains the property of the food company that produced it but can be accessed when required by law for traceability purposes (in the event of a food safety issue) or through a business-to-business agreement. Standardized names, structures, and content allow information to be easily shared across business-to-business distribution chains and guarantee that terms and meanings are understood by all. Although it is not the focus of this International Standard, commercial agreements for companies to exchange information through distribution chains are to be encouraged, especially when it comes to the information the trade wants to be visible at the various points of transaction in the chains. Although electronic data representation and communication were taken into consideration during the design of this International Standard, it is not mandatory to use this way. Paper systems can meet the requirements, but they will not provide the clear advantages of business efficiency, such as quick communication.

## 2.1.3. Why Do We Need ISO Standards for Agriculture?

The foundation of life, agriculture provides the world's population with food and the resources it needs to exist and flourish. Using farmland efficiently is the key to sustainable agriculture. Standards, which offer guidelines and best practices for equipment, tools, and farming techniques, can be useful instruments in achieving this.

Global positioning systems (GPS), irrigation, agricultural machinery, animal welfare, and sustainable farm management are all covered by ISO standards for agriculture. They guarantee that all components of the supply chain, from the farm to the fork, adhere to appropriate standards of quality and safety while also supporting efficient farming practices. The sustainability and good environmental management that lead to a better future are also promoted by ISO standards for agriculture, which set globally accepted solutions to global problems.

## 2.1.4. Who Benefits From ISO Standards for Agriculture?

ISO's agriculture standards primarily benefit farmers, manufacturers, and food producers. ISO standards guarantee smooth production and promote international trade by offering guidelines for globally accepted best practices and specifications for tools, machinery, and other products and processes. The production of food is made more efficient, sustainable, and traceable thanks to ISO standards.

Additionally, they offer a universal language across borders, increasing the potential for the exchange of agricultural products and food. Customers can thus choose from a wider variety of options in addition to receiving a high-quality, safe product. By establishing benchmarks for the functional safety, traceability, and quality of everything from agricultural machinery and materials to finished food products, ISO standards assist organizations in meeting regulatory and industry requirements. They offer a solid foundation for carrying out public policy because they were created through consensus and international expertise.

## 2.1.5. HACCP Standards

To create efficient traceability systems that improve food safety and quality, agriculture must adhere to HACCP (Hazard Analysis Critical Control Point) standards. HACCP, which was created as a methodical approach to detecting and managing risks to food safety, prioritizes prevention over response. HACCP principles enhance food safety and can support traceability systems when integrated with comprehensive data management tools.. Finding Critical Control Points (CCPs) in the production process is the foundation of HACCP. To guarantee food safety, these are crucial times when possible risks—whether biological, chemical, or physical—can materialize and need to be closely watched. For instance, farmers and processors must evaluate the risk of contamination when handling raw materials and put control measures in place to lessen that risk. Careful record-keeping is necessary for this process, covering every stage from sourcing input to processing, packaging, and distribution. This paperwork is essential for traceability because it allows manufacturers to monitor how their goods move through the supply chain.

Producers are required by HACCP standards to keep thorough records of every action about the CCPs. These documents contain details about raw material sources, processing temperatures, time logs, and quality assurance. In the event of a safety issue, such as an outbreak of foodborne illness, producers can promptly track products back to their origins by maintaining thorough records. To reduce the impact of such incidents, rapid traceability is essential because it enables prompt recalls and focused consumer and regulatory communication. Furthermore, the food supply chain's transparency is improved when HACCP and traceability are combined. Customers need to know more about the safety and quality of the food they eat, and traceability gives them the answers they need. Agricultural products gain the confidence and trust of consumers when they can be traced back to their source. Producers are better positioned to satisfy regulatory requirements and consumer expectations when they follow HACCP standards and can demonstrate effective traceability.

Another crucial component of putting HACCP standards into practice for traceability is training. Workers need to be informed about the significance of food safety as well as the particular procedures needed to preserve traceability. All staff members are guaranteed to understand their responsibilities for upholding HACCP principles and keeping accurate records through regular training. Within the company, a culture of food safety and accountability is fostered by this collective awareness.

It is impossible to exaggerate the importance of technology in supporting traceability and HACCP. Producers can find it easier to maintain accurate and current information by streamlining data collection with the help of digital record-keeping systems and software. The potential of technologies like blockchain to improve traceability by offering a transparent and safe record of product movements is being investigated more and more. Agricultural producers can increase the effectiveness and dependability of their traceability systems by utilizing these technologies. Even with the obvious advantages, there may be difficulties when integrating HACCP standards with traceability. Particularly smaller producers might find it difficult to acquire the resources needed for thorough documentation and technology adoption. Nonetheless, these producers can get assistance and resources from numerous industry associations and governmental organizations to help them deal with the challenges of HACCP compliance and traceability.

To create efficient traceability systems in agriculture, HACCP standards are essential. Producers can improve food safety and foster consumer trust by integrating technology, keeping thorough records, and identifying crucial control points. The integration of traceability and HACCP principles promotes regulatory compliance and risk management for food safety, which in turn makes the food supply chain safer and more transparent. The significance of HACCP standards in guaranteeing traceability will only increase as the agricultural landscape changes, highlighting the necessity of continuous dedication to food safety procedures (Khandke et al., 1998).

## 2.1.6. Product Traceability: HACCP

The internationally recognized HACCP method is used to manage food safety. It has been approved by the Codex Alimentarius Commission (Khandke et al., 1998). This instrument can be employed to systematically recognize hazards specific to specific products and procedures and present strategies for mitigating those risks to ensure the security of perishable commodities.

Good Manufacturing Practices (GMPs) procedures that successfully control general hazards to food safety must be followed for the HACCP Plan to be implemented within the industry. The chain needs to meet the structural and operational requirements outlined by GAP and GMP. HACCP's primary goals are to prevent, minimize, and get rid of food hazards and give consumers a safe product.

### *2.1.6.1. Principles of HACCP*

- Conduct a hazard analysis.
- Identify the critical control points.
- Establish critical limits.
- Establish monitoring procedures.
- Establish corrective actions.
- Establish verification procedures.
- Establish record-keeping and documentation procedures.

The various important supply chain participants are discussed about how HACCP operates in the perishable food chain. The Good Agricultural Practices (GAP) mentioned earlier are relatively new. Only a few farmers might be involved in it because of pressure from overseas buyers. However, it must be made abundantly evident that each and every participant in the perishable food supply chain bears responsibility for guaranteeing food safety from farm to fork (Figure 2.2; Khandke et al., 1998).

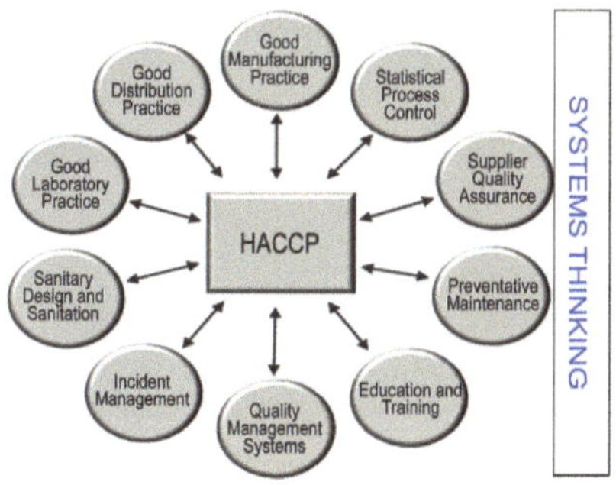


**Figure 2.2.** *Flowchart showing process HACCP implementation.*

*Source: Lee JC, Daraba A, Voidarou C, Rozos G, Enshasy HAE, Varzakas T. Implementation of Food Safety Management Systems along with Other Management Tools (HAZOP, FMEA, Ishikawa, Pareto). The Case Study of Listeria monocytogenes and Correlation with Microbiological Criteria. Foods. 2021; 10(9):2169. https://doi.org/10.3390/foods10092169 .*

## 2.1.7. Implementation Plan for Perishable Food Supply Chain Using HACCP in Developing Nations

The step-by-step procedure outlined above is essential for effective implementation of the HACCP plan within the establishment. The plan must be based on a solid foundation of GMPs and procedures that effectively control the general hazards to food safety.

### *2.1.7.1. Guidelines for Certified Growers*

A few recommendations based on the HACCP standard have been made to farmers in response to the demand for safe and high-quality produce. The guidelines facilitate the transparent flow of information and assist farmers in integrating into the channel. Additionally, it facilitates the farmers' horizontal integration into the food chain. To get a traceable chain in the channel, you need to take the next few steps.

1. **Step 1:** Follow the GAP principle: This task involves providing a general description of all fresh produce, ingredients, processes, and methods followed by the grower, starting from the pre-planting activity till harvesting. In this process, the grower's day-to-day activity must be noted or tabulated on an activity table.

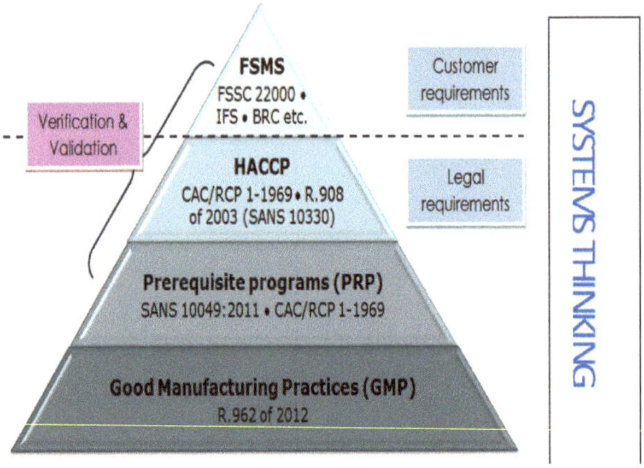


**Figure 2.3.** *Flow diagram for management.*

*Source: Lee JC, Daraba A, Voidarou C, Rozos G, Enshasy HAE, Varzakas T. Implementation of Food Safety Management Systems along with Other Management Tools (HAZOP, FMEA, Ishikawa, Pareto). The Case Study of Listeria monocytogenes and Correlation with Microbiological Criteria. Foods. 2021; 10(9):2169. https://doi.org/10.3390/foods10092169.*

2. **Step 2:** Develop a flow diagram: Creating a flow diagram is the following procedure. The flow diagram's objective is to give a concise, understandable summary of the processes involved in producing perishable produce. All of the growing process's steps must be included in the diagram. It is the sequential process that a grower follows following GAP guidelines, starting with the soil selection process and ending with harvesting (Figure 2.3; Charlebois et al., 2014).
3. **Step 3:** Application of hazard principle: Finding the list of growing processes and enumerating all possible risks not already mitigated by GAPs, including the list of processes, is the first step. Three criteria should be used to categorize the risks: physical, chemical, and biological. It is also necessary to tabulate a potential source of hazards. For instance, in the context of pest management, the possible risks that should be noted include improper dilution of pesticides, overuse, improper sourcing, and the use of prohibited chemicals.
4. **Step 4:** Determine the Critical Control Points and Establish Critical Limits: A CCP is a step or process that can be managed to prevent, eliminate, or reduce a hazard to acceptable levels. Using a CCP Decision Tree is one way to identify CCPs. A critical limit is the point at which a hazard must be managed to stop, manage, eradicate, or lessen its likelihood to a level that is acceptable. To make sure that boundaries are fulfilled, it must be watched. For instance, if excessive pesticide use is found, that procedure ought to be brought to light for critical examination. The scientific method must be used to determine the critical limit for pesticide use while taking the primary crop into account.
5. **Step 5:** Establish Corrective Actions, Documentation, and Record Keeping: Corrective Actions are necessary after the critical control point and critical limits have been determined. When operations depart from Critical Limits, which guarantee safety and appropriateness, they are necessary. Corrective Actions need to pinpoint the issue, address it, and look into ways to keep it from happening again. Ultimately, the foundation of any standardization is record keeping and documentation. When using a HACCP system for auditing purposes—whether for internal audits, verification processes, AQIS audits, or other external audits—accurate record-keeping is crucial.

### 2.1.7.2. Guidelines for Standardized Agribusiness

Agribusiness is made up of the operations that are conducted following harvest until the product is delivered to the end user. Packing, distribution, shipping, and retailing are the main steps in this lengthy chain. Vertical integration is the process of arranging these tasks. Effective application of standard procedures is necessary for a well-organized supply chain management of perishable produce. The rules for the standardized agribusiness are as follows:

1. **Step 1:** Follow GMP principles: Providing a general description of all fresh produce, ingredients, and chain players' procedures and methods from post-harvesting through the retail outlet is part of this task. An activity table needs to be used to record or tabulate the procedures that each important player follows in his daily activities. The series of product descriptions and processes include:

a. The source of the raw material;

b. The preservation method;

c. The packaging (e.g., vacuum packed, plastic liner in cardboard cartons, etc.);

d. Transportation of the product (including the method of transportation);

f. Storage conditions (e.g., frozen, refrigerated, at ambient temperature, etc.); and

g. The product standards applicable to the product (according to the Food Safety and Security Act, 2006).

2. **Step 2:** Develop a flow diagram: The creation of a flow diagram is the next step. From the farmer's farm gate to the retailer, the flow diagram's objective is to give a concise, understandable summary of the processes involved in the operation of the perishable produce. Every stage of the post-harvest, distribution, and retailing processes must be included in the flow diagram. According to GMP guidelines, it is a methodical process that is followed in the chain from the soil selection process to the grower's harvest.

3. **Step 3:** Application of hazard principle: Here, all potential risks in each process must be listed, along with a list of the activities involved in each functionality. The biological, chemical, and physical properties of the product should be taken into consideration when classifying the hazards. It is also necessary to tabulate potential sources of hazards, such as incorrect grading, non-standard material handling, and over-waxing (Figure 2.4).

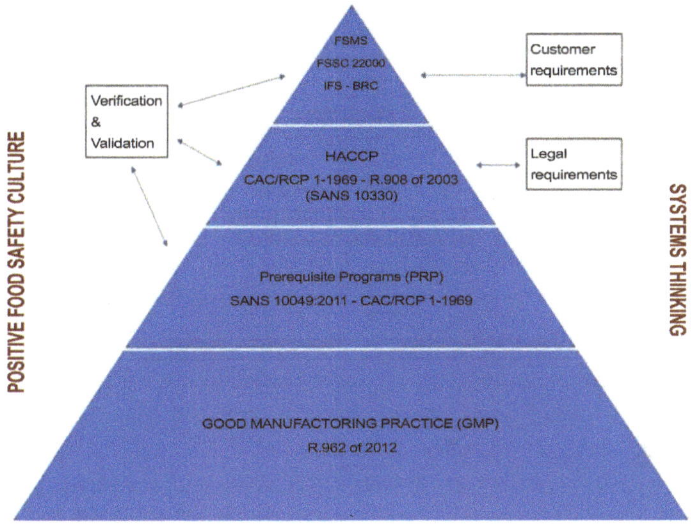


**Figure 2.4.** *The FSMS building blocks and the systems thinking approach.*

*Source: Lee JC, Neonaki M, Alexopoulos A, Varzakas T. Case Studies of Small-Medium Food Enterprises around the World: Major Constraints and Benefits from the Implementation of Food Safety Management Systems. Foods. 2023; 12(17):3218. https://doi.org/10.3390/foods12173218.*

4. **Step 4:** Determine the Critical Control Points and Establish Critical Limits: A CCP is arrived at based on the requirement of the customer. Process that are not in control should be identified and listed for further analysis.
5. **Step 5:** Establish Corrective Actions, Documentation, and Record Keeping: When operations surpass Critical Limits, corrective measures are necessary to guarantee safety and appropriateness. Corrective measures must pinpoint the issue and resolve it. The HACCP system and the distribution channel are both made more effective and efficient by accurate record-keeping (Charlebois et al., 2014).

## 2.1.8. GS1 Global Traceability Standards (GTS)

A non-profit organization called GS1 creates international standards for data synchronization and barcodes in business communication. A framework for locating, gathering, and exchanging product information throughout the supply chain is offered by the GS1 system. Traceability in agriculture is made easier by GS1 standards, which allow producers and retailers to exchange precise and consistent product data. GS1-standardized barcodes and QR codes make it simple to track agricultural products from farm to consumer, increasing supply chain efficiency and transparency.

The GTS was developed by GS1 to become the accepted benchmark for businesses implementing traceability systems. A community of over 800 businesses from Asia, Europe, and the Americas—representing retailers, suppliers, GS1 Member Organizations, and solution providers across numerous industries—developed GTS. Regardless of the number of businesses involved or the enabling technologies used, GTS enables traceability systems globally for both small and large organizations across the supply chain.

Other international standards like the ISO, Safe Quality Food (SQF), Global Food Standards, Global Food Consortium (BRC), Food Marketing Institute, and GLOBAL G are not in competition with the GS1 GTS. HACCP certifications, or other credentials. In actuality, by giving businesses and organizations the means to attain the necessary degrees of traceability, GS1 assists them in meeting the requirements of these standards and certifications.

### 2.1.9. Scope

The GS1 Global Traceability Standard is intended for use across end-to-end supply chains and is relevant to all events that span the lifecycle of a traceable object, including:

- The transformation and processing of raw materials, ingredients, intermediate products, components, and components into the product.
- Aggregation and disaggregation of products and linkage to assets (e.g., returnable assets).
- Transport and distribution, including cross-border trade.
- The maintenance, repair, and overhaul operations across multiple cycles of usage or service of the product.
- Consumption of products, including dispensing and administering.
- The disposal and destruction of the product and the recycling of materials.

This document is predicated on the idea that every organization will have unique goals in mind when deciding which traceability tools and systems to use to propel their business forward.

It is imperative for every organization to guarantee system interoperability throughout their supply chains in order for them to thrive. This chapter focuses on traceability's data management aspects. Using a straightforward model that functions across well-known and reliable chains of custody or ownership, it lists and makes reference to the prerequisites for gathering and exchanging data. To give applications enough business context to use the data effectively, traceability data is collected and shared across the "who, what, when, where, and why" dimensions. Additionally, it offers a framework for data sharing across increasingly intricate supply chains, where parties may need to build trust before sharing data and locate and obtain information from businesses that are not their direct trading partners.

This text is impartial to both products and sectors. The concepts are applicable to supply chains in a variety of industries, such as apparel, pharmaceuticals, food, and beverage, medical devices, logistics for humanitarian aid, and technical equipment and components. Additionally, this document aims to be technology neutral. It is founded on the fundamental ideas of "Identify-Capture-Share," which form the basis of the GS1 system of standards.

These fundamental ideas clarify how traceability solutions can be made possible by the GS1 system of standards.

Many data capture and sharing technologies as well as associated GS1 standards, such as GS1 barcodes, EPC/RFID, GDSN, EDI, and EPCIS, are mentioned in the document. The Global Food Safety Initiative (GFSI), other certification schemes, and other international standards that address traceability requirements are not intended to be competitors of the GS1 Global Traceability Standard; rather, it is meant to supplement and complete these other standards. The GTS assists businesses and organizations in understanding "how to" comply with these standards by using standardized traceability data, while these standards specify "what" should be done (Charlebois et al., 2014).

## 2.1.10. GS1 Traceability Implementation

Key traceability principles and implementation grid link the principles to enabling technologies and GS1 System tools (Figure 2.5).

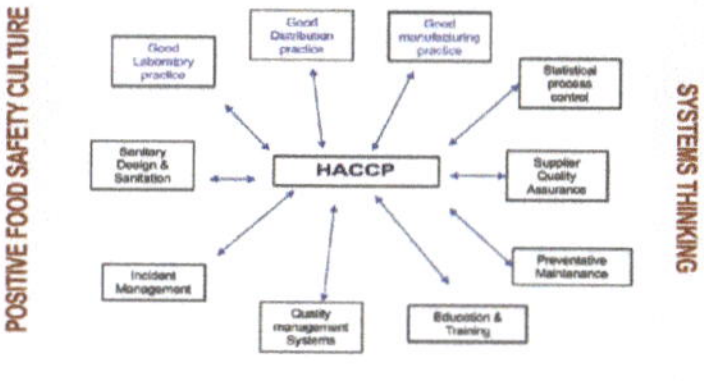


**Figure 2.5.** *Traceability principles and implementation grid.*

*Source: Lee JC, Neonaki M, Alexopoulos A, Varzakas T. Case Studies of Small-Medium Food Enterprises around the World: Major Constraints and Benefits from the Implementation of Food Safety Management Systems. Foods. 2023; 12(17):3218. https://doi.org/10.3390/foods12173218.*

### 2.1.10.1. Traceability Data Management in Production

Identification management in a production environment is characterized by:

- Several supplier locations (GLN 1–3), which send pallets of materials (SSCC 1–4).
- At reception, materials are stored and/or ordered for the production process.
- At the production site (GLN 4), consumer units (GTIN 1) are produced in separate batches (each identified with a distinct Batch Number).
- In the packaging step, consumer units (GTIN 1 and its Batch Number) are packed into standard grouping units (GTIN 2).
- In the next two steps - storage and preparation for shipping, pallets are created (SSCC 5–7) and dispatched to customer destinations (GLN 5–6) (Figure 2.6).

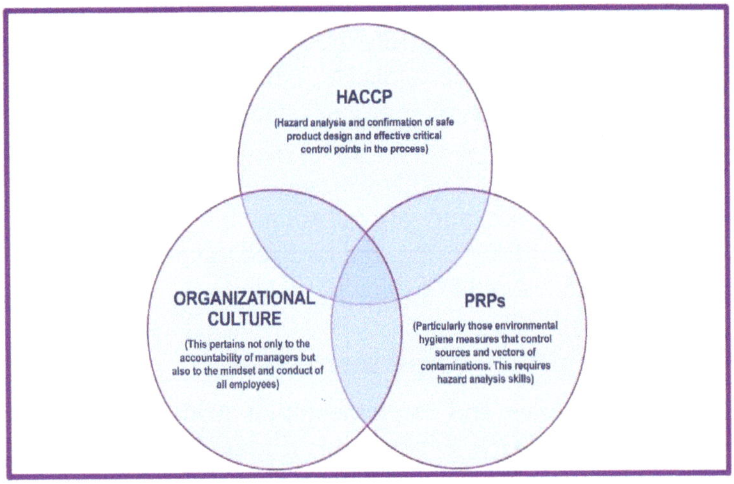


**Figure 2.6.** *The management in a production environment.*

*Source: Lee JC, Neonaki M, Alexopoulos A, Varzakas T. Case Studies of Small-Medium Food Enterprises around the World: Major Constraints and Benefits from the Implementation of Food Safety Management Systems. Foods. 2023; 12(17):3218. https://doi.org/10.3390/foods12173218.*

### 2.1.10.2. Key Business Rules – Production Environment:

- **Reception:** The SSCC of an incoming pallet is recorded and linked to the GLN of the supplier. Each time the pallet is moved, its SSCC is recorded and linked to the GLN of its new location (e.g., to storage or production).

- **Production:** Under ideal conditions, the SSCC of the pallet and/or GTIN + Batch Number of materials used in the production process are recorded and linked to the GTIN of the product made and its production batch. At the end of the production process, standard trade item groupings are made from individual products. A new GTIN is assigned and linked to the production Batch Number.
- **Packaging, Storage, and Expedition:** The GTIN of a standard trade item grouping is linked to the SSCC of the pallet onto which it is packed. The SSCC of an outbound pallet is linked via scanning to the GLN of its destination. While the destination GLN is always recorded for traceability purposes, it is not mandatory to display it on the physical label of the pallet.

### *2.1.10.3. Traceability Data Management in Distribution*

The steps below show the use of GS1 standards for identifying locations (GLN) and logistic units.

(SSCC) in a distribution environment, which is characterized by:

- Several supplier locations (identified with GLN 1–3), which send pallets of finished products (identified with SSCC 1–4).
- At the distribution centre (GLN 4) reception, pallets are stored and sent to the order picking process.
- In the order-picking step, orders are fulfilled either by shipping uniform pallets, cross docking or the creation of mixed pallets. They are either carried forward unchanged (uniform pallet identified with SSCC 1) or newly created (mixed pallets identified with SSCC 5–7) with products originating from different pallets (SSCC 2–4).
- In the last two steps - storage and preparation for shipping, both uniform (SSCC 1) and mixed pallets (SSCC 5–7) are dispatched to customer/point-of-sale destinations (identified with GLN 5–6) (Figure 2.7).

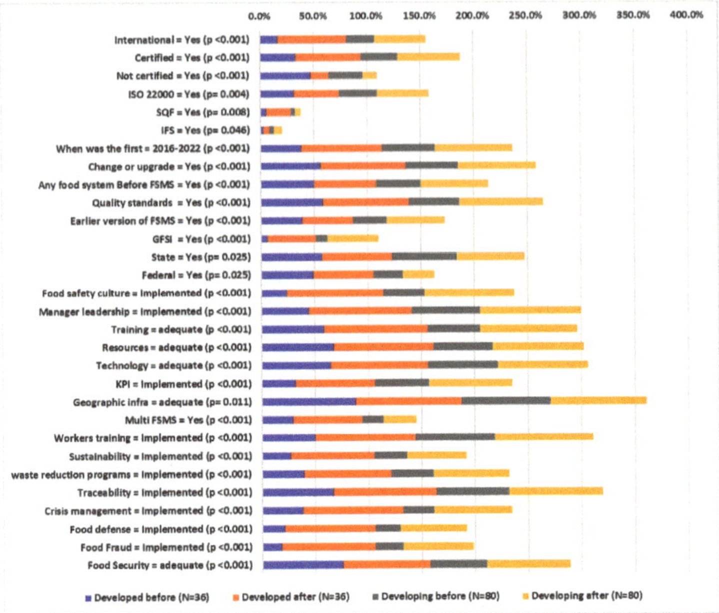


(a)

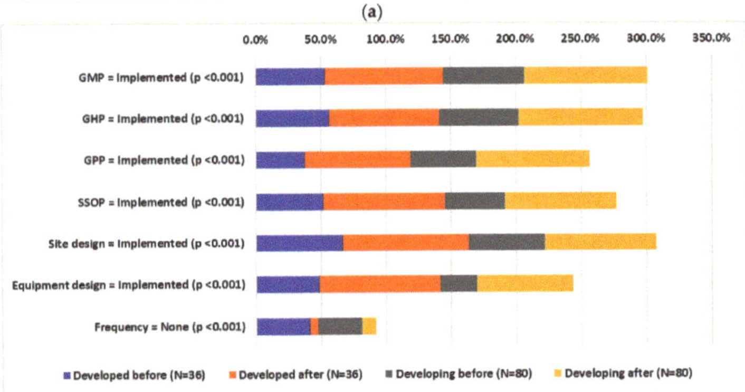


(b)

**Figure 2.7.** *a) Attitudes (FSMS element details) towards the constraints and benefits of implementing FSMS standards before and after implementation in developed and developing countries. Only attitudes with significant changes are presented. (b). Attitudes (Prerequisites) towards the constraints and benefits of implementing FSMS standards before and after implementation in developed and developing countries. Only attitudes with significant changes are presented.*

*Source: Lee JC, Neonaki M, Alexopoulos A, Varzakas T. Case Studies of Small-Medium Food Enterprises around the World: Major Constraints and Benefits from the Implementation of Food Safety Management Systems. Foods. 2023; 12(17):3218. https://doi.org/10.3390/foods12173218.*

### *2.1.10.4. Key Business Rules – Distribution Environment*

#### *2.1.10.1.1. Reception*

The SSCC of an incoming pallet is recorded and linked to the GLN of the supplier. Each time the pallet is moved, its SSCC is recorded and linked to the GLN of its new location (e.g., to storage, order-picking or distribution).

#### *2.1.10.1.2. Order-Picking and Distribution*

- The SSCC of an unmodified pallet picked for distribution from the storage area or cross-docked without any storage is recorded and linked to the GLN of its destination.
- Standard trade item groupings from various pallets are combined into a newly created pallet. In this instance, a new SSCC is given to it and connected to the SSCC numbers of every other pallet that was utilized in its construction, as well as, if relevant, the GTIN and Batch Number of every standard trade item grouping that was employed. This may require a significant amount of work, but it can be avoided by applying a "time window," which each company will define at the time a product is packed. Pallets that are created during this window of time and those that are used up within the same period of time can be linked. The destination's GLN is connected to and recorded by the SSCC.

### 2.1.11. International Regulatory Requirements for Food Traceability

According to the WTO Agreement on the Application of Sanitary and Phytosanitary Measures, any sanitary and phytosanitary measures (including traceability requirements) introduced by World Trade Organization (WTO) member states must meet the following requirements:

- Not to discriminate arbitrarily or unjustifiably between WTO member states;
- Have no more restrictions on trade than is necessary for WTO member states to reach an appropriate level of sanitary or phytosanitary protection, given their technical and economic capacities; and
- Not to apply rules in a manner that constitutes a disguised restriction on international trade.

The WTO encourages the adoption by the member-states of legal acts harmonized with relevant international standards, guidelines, and recommendations established by:

- The Codex Alimentarius Commission for Food Safety;
- The International Office of Epizootics (OIE) for animal health and zoonoses; and
- The Secretariat of the International Plant Protection Convention (IPPC) for plant health.

The Codex Alimentarius Commission defines traceability as the capacity to track the movement of a food article through designated stages of production, processing, and distribution. This definition is found in the CAC/GL 60–2006 standard. Furthermore, the principles of traceability are defined by the Codex Alimentarius Commission in the aforementioned standard as an essential tool for competent government authorities to use during inspections.

The ISO defines "traceability" as "the ability to track the movement of food or feed at certain stage(s) of production, processing, and distribution." This definition is included in the standard. Thus, it can be inferred that the capacity to track and identify specific products as they move through the supply chain is a crucial component of any traceability system.

The basic elements of the traceability system are as follows:

- Product identification;
- Information required to track the movement of the product/item through the supply chain; and
- Establishing a link between the product/item and the supply chain participants.

The type of product and production sector can affect the amount of information required to guarantee traceability. Furthermore, there are various ways to record this data (such as using a paper form, barcodes, or specific computer software). The idea that has been embraced by the majority of nations today calls for the "one step back—one step forward" traceability requirement to be included in national legislation. This is a fundamental requirement for guaranteeing efficient traceability for any kind of product and in any production area (Charlebois et al., 2014).

## 2.1.12. EU Regulatory Requirements for Food Traceability

### *2.1.12.1. General Food Traceability Requirements*

Regulation (EC) No 178/2002 establishes the legislative requirements for traceability in the European Union. The Preamble to this regulation specifically highlights the significance of implementing traceability protocols for food and feed in paragraphs 28 and 29 are:

- **(28)** Past experiences have demonstrated that in situations where it is impossible to trace food or feed, the internal market's ability to function may be compromised. Therefore, to avoid the possibility of needless wider disruption in the event of food safety issues, businesses involved in the production of food and feed must establish a comprehensive system of traceability. This will enable targeted and accurate withdrawals to be made or information to be provided to consumers or control officials.
- **(29)** To guarantee traceability at every stage of an investigation, it is imperative that a food or feed business, including an importer, be able to identify at least the business from which the food, feed, animal, or substance that may be incorporated into a food or feed has been supplied. "Traceability" is defined as "the ability to trace and follow a food, feed, food-producing animal, or substance intended to be, or expected to be incorporated into, a food or feed, through all stages of production, processing, and distribution" in Article 3 of Regulation (EC) No 178/2002.

Minimum regulatory requirements for traceability are determined in Article 18 of the Regulation (EC) No 178/2002:

## *2.1.12.2. Article 18. Traceability*

- The traceability of food, feed, food-producing animals, and any other substance intended to be, or expected to be, incorporated into a food or feed shall be established at all stages of production, processing, and distribution.
- Food and feed business operators shall be able to identify any person from whom they have been supplied with a food, a feed, a food-producing animal, or any substance intended to be, or expected to be, incorporated into a food or feed. To this end, such operators shall have in place systems and procedures that allow for this information to be made available to the competent authorities on demand.
- Food and feed business operators shall have in place systems and procedures to identify the other businesses to which their products have been supplied. This information shall be made available to the competent authorities on demand.
- Food or feed which is placed on the market or is likely to be placed on the market in the Community shall be adequately labelled or identified to facilitate its traceability, through relevant documentation or information in accordance with the relevant requirements of more specific provisions.

- Provisions to apply the requirements of this Article in respect of specific sectors may be adopted following the procedure laid down in Article 58(2).

A "one step back—one step forward" strategy, or the ability to recognize the business operator from whom they have received a product and the business operator to whom they have supplied a product, is mandated by Article 18 of Regulation (EC) No 178/2002 for operators of food businesses. Operators of food businesses must set up the systems and processes required to meet this requirement and give the relevant authorities the data required to guarantee traceability.

In accordance with the Guidance on the Implementation of Some Articles of Regulation (EC) No 178/2002 (hereinafter, "the Guidance"), food business operators must:

- Have a system in place enabling them to identify the immediate supplier(s) and immediate customer(s) of their products;
- Establish a link "supplier-product" to identify which products were supplied from a specific supplier; and
- Establish a "customer-product" link to identify which products were supplied to which customers.

Furthermore, to facilitate traceability, food products must be appropriately labeled or identified, according to part four of Article 18. Regulation (EC) No 178/2002 places more emphasis on introducing traceability systems than it does on outlining how these objectives are to be met. As a result, Article 18 does not specify what data must be kept and recorded by the operator of a food business to meet the traceability requirement.

Meanwhile, the Guidance provides an explanation of which data should be recorded and stored by the food business operator in pursuance to Article 18 of the Regulation (EC) No 178/2002:

- Name, address of supplier, and identification of products supplied;
- Name, address of customer, and identification of products delivered;
- Date and, where necessary, time of transaction/delivery; and
- Volume, where appropriate, or quantity.

The Regulation (EC) No 178/2002 lacks time constraints regarding the retention of records containing the data required to guarantee traceability. The guidelines explain the durations for which pertinent information must be kept: "A five-year period is adequate for maintaining traceability records.." Records, however, must be retained for a minimum of six months following the date of manufacture or delivery for food items with "use by" dates shorter than three months, as well as items like fruits, vegetables, and non-pre-packaged goods that are shipped straight to the end user. Records should be retained for the duration of the shelf life plus an additional six months for other food products that have a "best before" date. A general rule of five years applies to products (e.g., wine, vinegar, chewing gum) that do not have a specified durability date.

Regulation (EC) No 178/2002 has general requirements for traceability. Specific traceability requirements for certain types of food products in the EU are outlined in specific legislation. For instance:

- Regulation (EU) No 931/2011 laying down specific traceability requirements for food of animal origin;
- Regulation (EU) No 1337/2013 setting forth specific traceability requirements for certain types of meat;
- Regulation (EU) No 1224/2009 setting forth specific traceability requirements for fish products; and
- Regulation (EU) No 404/2011 with detailed traceability requirements for fish products (Joint FAO/WHO Food Standards Programme, 2011).

## 2.2. LOCAL AND REGIONAL TRACEABILITY REGULATIONS

The ability to follow the production, processing, and delivery of food products from farm to table is known as traceability in agricultural products. Due to increased consumer awareness of food safety, quality, and sustainability, this idea has become extremely important in recent years. Local and regional laws have a significant impact on how traceability systems are designed, how safety standards are followed, and how well crisis management is enabled.

### 2.2.1. Importance of Traceability

Traceability serves multiple purposes, including:

- **Food Safety:** Rapid identification and isolation of contaminated products during foodborne illness outbreaks.
- **Quality Assurance:** Verification of product quality and adherence to labeling claims, such as organic or non-GMO.
- **Consumer Confidence:** Enhancing transparency in the food supply chain, thus building trust between consumers and producers.
- **Regulatory Compliance:** Meeting legal requirements set by government bodies and ensuring adherence to international standards.

## 2.2.2. Local Traceability Regulations

Local laws differ greatly based on the nation, state, or local government. These laws are frequently customized to the unique farming methods, consumer needs, and local market circumstances.

### *2.2.2.1. United States*

In the United States the significance of traceability is emphasized by the Food Safety Modernization Act (FSMA). Although it mainly targets big food companies, local laws frequently expand on these federal directives. Additional traceability regulations, especially for fresh produce, have been implemented by states like California, which demand thorough documentation of every stage of the supply chain.

The Department of Food and Agriculture in California, for instance, requires farmers to keep track of pesticide applications, harvest dates, and shipment information. In a major agricultural state, this local focus aids in effectively managing food safety risks.

### *2.2.2.2. European Union*

The General Food Law (Regulation (EC) No 178/2002), which mandates that all food businesses establish traceability systems, contains the EU's traceability regulations.

Traceability requirements for organic products are further specified by local laws, such as the EU's Organic Regulation. To ensure that organic products fulfill organic certification requirements, these regulations require that they be traceable from farm to consumer. EU-wide regulations are also supplemented by the laws of EU member states. For example, the "Loi Egalim" in France, which requires food products to be traceable and places a strong emphasis on local sourcing, encourages transparency in food supply chains.

### *2.2.2.3. Asia*

Local traceability laws have become more popular in nations like South Korea and Japan as a result of worries about food safety. The Food Sanitation Act of Japan mandates thorough documentation of food items, including precise monitoring of agricultural inputs, processing, and distribution. For certain agricultural products, like fruits and vegetables, traceability systems are required by the Agricultural Products Quality Control Act in South Korea. Farmers are required to keep track of their product journey from the farm to the market and register them.

### 2.2.3. Regional Traceability Regulations

Recognizing that food systems extend beyond local borders, regional regulations frequently result from partnerships between several authorities. These laws can improve traceability by giving a unified framework over wider regions.

#### *2.2.3.1. North America*

The partnership in North America between the U.S. has resulted in programs such as the North American Free Trade Agreement (NAFTA) and its successor United States-Mexico-Canada Agreement (USMCA), the. These agreements prioritize harmonizing food safety standards, including traceability, even though their primary focus is trade. Guidelines set by the Canadian Food Inspection Agency (CFIA) are in line with U.S. rules, encouraging agricultural products' cross-border traceability. This alignment improves safety measures and makes compliance easier for producers.

#### *2.2.3.2. European Union*

The European Union is a singular example of regional regulation. The European Commission has put into effect many directives designed to improve traceability. Agricultural products are impacted by the European Regulation on the Registration, Evaluation, Authorization, and Restriction of Chemicals (REACH), especially with regard to the use of fertilizers and pesticides. In addition, traceability requirements are included in the EU's Common Agricultural Policy (CAP), which mandates that farmers document their operations to be eligible for subsidies. This rule guarantees that public funds are used to promote safe and sustainable agricultural practices.

#### *2.2.3.3. African Union*

The African Union (AU) has acknowledged that better traceability systems are necessary to improve food safety and security in Africa. To ease trade within the continent, the AU's Comprehensive Africa Agriculture Development Programme (CAADP) encourages investments in agricultural infrastructure, such as traceability systems. Traceability requirements are among the standardized regulations for agricultural products being developed by regional organizations like the East African Community (EAC). This strategy seeks to promote intraregional trade while enhancing food safety and quality.

### 2.2.4. Challenges in Implementing Traceability Regulations

While local and regional regulations are essential for effective traceability, several challenges impede their successful implementation:

1. **Cost and Complexity:** Implementing traceability systems can be costly, particularly for smallholder farmers who may lack the resources and technical knowledge. Compliance with local regulations can burden these producers, potentially discouraging participation.
2. **Technological Barriers:** The adoption of advanced technologies, such as blockchain and digital traceability platforms, requires infrastructure and training that may not be accessible in all regions. Inadequate technology can lead to gaps in traceability.
3. **Inconsistent Standards:** Variability in regulations across jurisdictions can create confusion for producers operating in multiple areas. The lack of harmonized standards can hinder effective traceability and complicate compliance efforts.
4. **Consumer Awareness:** While consumers demand transparency, there is often a lack of understanding of traceability systems. Without consumer education, the benefits of traceability may not be fully realized, limiting market incentives for producers to invest in these systems.

## 2.2.5. Innovations in Traceability

Despite the challenges, innovations are emerging that enhance traceability in agriculture:

1. **Digital Platforms:** Technologies such as blockchain and Internet of Things (IoT) devices allow for real-time tracking of agricultural products. These platforms enhance transparency and streamline compliance with local and regional regulations.
2. **Mobile Applications:** Mobile technology is being leveraged to enable farmers to record data easily. Applications designed for traceability help producers maintain records and share information with consumers and regulatory bodies.
3. **Collaborative Networks:** Partnerships among producers, regulators, and technology providers are fostering the development of cohesive traceability systems. Collaborative efforts can enhance data sharing and streamline compliance across regions.

## 2.2.6. Future Directions

As the global food system continues to evolve, the importance of robust traceability regulations will only increase. Future directions for local and regional traceability regulations may include:

1. **Strengthening Standards:** Ongoing collaboration between jurisdictions can lead to the development of harmonized traceability standards, simplifying compliance and enhancing food safety.
2. **Enhancing Consumer Engagement:** Educating consumers about traceability and its benefits can drive demand for transparent food systems, encouraging producers to invest in traceability initiatives.
3. **Embracing Sustainability:** Integrating sustainability metrics into traceability regulations can promote environmentally friendly practices while ensuring food safety and quality.
4. **Investing in Technology:** Governments and private sector stakeholders should invest in technology that facilitates traceability, making it accessible to producers of all sizes.

Regulations governing traceability at the local and regional levels are essential to guaranteeing the sustainability, safety, and quality of agricultural products. These rules improve the general effectiveness of food supply chains in addition to protecting consumers. Even though implementation still presents challenges, continued innovation and cooperative efforts can open the door to more efficient traceability systems. Agriculture's future will be greatly influenced by the development of traceability laws as the need for transparency in food systems grows globally (Joint FAO/WHO Food Standards Programme).

## 2.3. COMPLIANCE REQUIREMENTS FOR AGRICULTURAL TRACEABILITY

Agricultural traceability compliance standards are crucial for guaranteeing food safety, quality, and supply chain transparency. Adherence to regulatory frameworks set by national and regional authorities is usually one of these requirements. As an illustration, in the U. S. . Food products can be tracked from farm to table thanks to the FSMA, which requires record-keeping. Traceability along the entire food supply chain is also required by the General Food Law of the European Union. It is essential to maintain accurate records, which include thorough records of agricultural inputs, processing techniques, and distribution information. Producers are required to keep thorough logs that document the journey of every product, including unique identifiers like barcodes or QR codes to enable real-time tracking.

To ensure food safety, compliance also entails applying HACCP principles and fulfilling international standards such as ISO 22000. Employee education is crucial because it gives workers the skills they need to properly maintain documentation and use traceability technologies. Frequent third-party inspections and internal audits assist in confirming compliance and pinpointing areas in need of development. To promptly address food safety incidents, organizations must also establish crisis management protocols, which should include explicit recall procedures and communication plans. Traceability compliance can be improved by investing in technology, training, and harmonized regulations, despite obstacles like costs and inconsistent standards across regions. All things considered, a strong traceability system boosts consumer confidence in the agriculture industry while also safeguarding consumers.

The ability to follow and trace agricultural products from their point of origin through each stage of production, processing, and distribution to the final consumer is known as agricultural traceability. The significance of traceability has grown in recent years as worries about food sustainability, quality, and safety have grown. Ensuring that food products meet safety standards and that stakeholders can effectively respond to foodborne illnesses or quality issues requires compliance with traceability regulations. This in-depth analysis of agricultural traceability compliance requirements covers a wide range of regulatory frameworks, record-keeping procedures, identification, and labeling laws, standard compliance, technology integration, training requirements, auditing procedures, and crisis management procedures.

## 2.3.1. Regulatory Frameworks

### *2.3.1.1. National and Regional Regulations*

To ensure food safety and quality, national governments often set regulatory frameworks that serve as the starting point for compliance with agricultural traceability. The FSMA in the US places a strong emphasis on traceability for big food companies, especially those that deal with fresh produce. Guidelines for record-keeping and tracking have been developed by the FDA to guarantee quick response times in the event of an emergency involving food safety. Traceability across the food supply chain is required by the General Food Law (Regulation (EC) No 178/2002) in the European Union. This rule highlights how crucial it is to identify and monitor food items to guarantee consumer safety and create a system of accountability for manufacturers, processors, and distributors.

### *2.3.1.2. Local Regulations*

Although national frameworks offer a broad framework for guidance, local laws can differ greatly. States like California, for instance, have unique regulations that expand upon federal statutes and address regional agricultural practices, environmental issues, and consumer preferences. Such localized laws frequently call for thorough records of farming inputs, pest control procedures, and harvesting techniques.

#### *2.3.1.3. Record-Keeping*

One essential element of traceability compliance is record-keeping. Producers are able to follow the path of their products through the supply chain with the help of precise and thorough records. This entails recording all inputs utilized, procedures followed, and handling conditions for products.

1. **Types of Records Required:**

a. **Input Records:** Producers must maintain records of all agricultural inputs, such as seeds, fertilizers, pesticides, and veterinary drugs. This documentation ensures that all inputs are safe, compliant with regulations, and used appropriately.

b. **Process Documentation:** Farmers should keep detailed logs of cultivation practices, including planting dates, pest management strategies, and harvesting methods. This information is critical for demonstrating adherence to safety standards.

c. **Distribution Tracking:** Records must include details about the distribution process, including shipping dates, recipient information, and storage conditions. This traceability is vital for efficiently managing recalls in case of contamination or quality issues.

#### *2.3.1.4. Identification and Labeling*

For traceability to be effective, proper identification is necessary. Agricultural products ought to be given distinct identifiers that can be scanned at different stages of the supply chain, like barcodes or QR codes. All parties involved in the product's journey will always have access to the most recent information thanks to this technology, which also makes real-time tracking possible.

1. **Labeling Requirements:** Labels play a crucial role in traceability. They should contain:

a. **Origin Information:** Clear identification of where the product was grown or produced.

b. **Batch Numbers:** Information that allows for the tracking of specific batches of products back to their origin.

c. **Certification Labels:** Indications of compliance with standards such as organic certification, which require stringent traceability measures.

### 2.3.2. Compliance with Standards

The credibility of traceability systems is increased when they adhere to international standards. A successful traceability system must meet the requirements outlined in the ISO 22000 standard for food safety management systems. It is imperative for organizations to establish procedures that guarantee all phases of manufacturing and delivery follow safety guidelines.

### *2.3.2.1. Technology Adoption*

1. **Digital Tracking Systems:** Technology integration can greatly enhance traceability procedures and ensure compliance. Real-time tracking of agricultural products is made possible by digital platforms and software solutions, which enhance the precision and effectiveness of record-keeping. By automating data collection, these systems can lower the risk of errors that come with keeping manual records.
2. **Blockchain Technology:** Blockchain technology has come to light as a possible traceability solution. With its decentralized and unchangeable transaction records, it gives all parties involved access to the same data. Producers can increase transparency and give consumers more assurance about the safety and authenticity of their food by utilizing blockchain technology.

### *2.3.1.2. Training and Capacity Building*

Employee training is crucial to guaranteeing adherence to traceability regulations. Workers need to be aware of the significance of maintaining accurate records, the particular documentation procedures, and the technology being utilized.

### *2.3.1.3. Developing Skills*

It is recommended that organizations allocate resources towards capacity-building initiatives that provide employees with the requisite skills to effectively implement traceability systems. Staff members can stay informed about rules and best practices with the support of regular training sessions, workshops, and resources.

### *2.3.1.4. Auditing and Monitoring*

1. **Internal Audits:** Conducting regular internal audits is vital for assessing compliance with traceability protocols. These audits should evaluate the effectiveness of record-keeping processes, technology use, and adherence to safety standards. Internal audits can help identify areas for improvement and ensure ongoing compliance.
2. **Third-Party Inspections:** It is possible to further improve credibility by hiring outside auditors. An objective evaluation of adherence

to industry and regulatory standards is given by these inspections. Certifications that might be required for entry into particular markets or supply chains can also be facilitated by third-party audits.

#### *2.3.1.5. Crisis Management Protocols*

1. **Developing Recall Procedures:** Clearly defined recall procedures are crucial in the event of a food safety issue. To track impacted products back through the supply chain, organizations need to set up procedures. The particular batch numbers, manufacturing dates, and distribution locations of tainted goods must all be determined.
2. **Communication Plans:** In times of crisis, communication that works is essential. Companies should create plans for quickly alerting all relevant parties, such as customers, merchants, and government agencies. Being open and honest during a recall can preserve customer confidence and lessen possible harm to the brand.

### 2.3.3. Challenges in Compliance

#### *2.3.3.1. Cost Implications*

Implementing traceability systems can be costly, particularly for smallholder farmers who may lack the necessary resources. Compliance with regulations often involves investing in technology, staff training, and record-keeping systems.

#### *2.3.3.2. Technological Barriers*

Access to technology can be a barrier to effective compliance. Some regions may lack the infrastructure necessary for implementing digital tracking systems, leading to gaps in traceability.

#### *2.3.3.3. Inconsistent Standards*

Variability in regulations across jurisdictions can create confusion for producers operating in multiple areas. The lack of harmonized standards may hinder effective traceability and complicate compliance efforts.

#### *2.3.3.4. Future Directions*

1. **Strengthening Regulations:** As the demand for food safety and quality assurance continues to grow, there is a need for stronger regulatory frameworks. Governments and industry organizations must collaborate to establish harmonized traceability standards that simplify compliance for producers while ensuring consumer safety.

2. **Embracing Sustainability:** Future compliance requirements should also incorporate sustainability metrics. Encouraging practices that promote environmental stewardship can enhance the overall effectiveness of traceability systems while addressing consumer demand for sustainable products.
3. **Expanding Technology Use:** Investing in emerging technologies will be critical for improving traceability compliance. As digital solutions become more accessible, organizations should explore innovative tools that facilitate efficient tracking and record-keeping.
4. **Consumer Engagement:** Educating consumers about traceability and its benefits can drive demand for transparent food systems. Initiatives that promote awareness and understanding of traceability can encourage producers to invest in compliance efforts, ultimately benefiting the entire supply chain.

Agribusiness traceability compliance standards are crucial for guaranteeing food safety, quality, and transparency. Agricultural stakeholders can effectively navigate the complexities of traceability by embracing technology, implementing identification and labeling protocols, adhering to regulatory frameworks, and keeping accurate records.

Compliance efforts are further strengthened by making investments in crisis management procedures, auditing, and training. Notwithstanding obstacles, agricultural traceability has a bright future thanks to chances for stricter laws, environmentally friendly practices, and technological breakthroughs. The agriculture industry can maintain consumer trust and promote a safer, more transparent food system by putting compliance first (Joint FAO/WHO Food Standards Programme, 2011).

## REFERENCES

1. Bihn, E. A., Gravani, R. B., & Matthews, K. R. (2006). Role of good agricultural practices in fruit and vegetable safety. *Microbiology of Fresh Produce*, 21–53.
2. Bollen, A. F., Riden, C. P., & Cox, N. R. (2007). Agricultural supply system traceability, part I: Role of packing procedures and effects of fruit mixing. *Biosystems Engineering, 98*(4), 391–400.
3. Bollen, A. F., Riden, C. P., & Opara, L. U. (2006). Traceability in postharvest quality management. *International Journal of Postharvest Technology and Innovation*, 93–105.
4. Burn, A. J., Coaker, T. H., & Jepson, P. C. (1987). *Integrated Pest Management* (No. 04; SB950, B8.). London: Academic Press.
5. Charlebois, S., Sterling, B., Haratifar, S., & Naing, S. K. (2014). Comparison of global food traceability regulations and requirements. *Comprehensive Reviews in Food Science and Food Safety, 13.*
6. Codex Alimentarius Commission. (2011). *Codex Alimentarius: Joint FAO/WHO Food Standards Programme*. FAO/WHO.
7. European Parliament and Council. (2002). Regulation (EC) No 178/2002 of the European Parliament and of the Council of 28 January 2002 laying down the general principles and requirements of food law, establishing the European Food Safety Authority and laying down procedures in matters of food safety. *Official Journal of the European Union, L 139.*
8. Gupta, A. K. (1989). Scientists' views of farmers' practices in India: Barriers to effective interaction.
9. Hobbs, J. E., & Young, L. M. (2000). Closer vertical co-ordination in agri-food supply chains: A conceptual framework and some preliminary evidence. *Supply Chain Management: An International Journal, 5*(3), 131–143.
10. Hulebak, K. L., & Schlosser, W. (2002). Hazard analysis and critical control point (HACCP) history and conceptual overview. *Risk Analysis, 22*(3), 547–552.
11. Jevsnik, M., Hlebec, V., & Raspor, P. (2006). Meta-analysis as a tool for barriers identification during HACCP implementation to improve food safety. *Acta Alimentaria, 35*(3), 319–353.
12. Kader, A. A., & Rolle, R. S. (2004). The role of post-harvest management in assuring the quality and safety of horticultural produce (Vol. 152). FAO.
13. Khandke, S. S., & Mayes, T. (1998). HACCP implementation: A practical guide to the implementation of the HACCP plan. *Food Control, 9*(2), 103–109.

14. Mora, C., Menozzi, D., Giacomini, C., Cantoni, C., Massari, M., & Morelli, G. (2007). The attitude of Italian consumers towards GM foods. In *Food Culture: Tradition, Innovation and Trust—A Positive Force for Modern Agribusiness* (pp. 1–13). Wageningen University and Research.
15. Mortimore, S., & Wallace, C. (1998). *HACCP: A Practical Approach* (Vol. 1). Springer.
16. Opara, L. U. (2003). Traceability in agriculture and food supply chain: A review of basic concepts, technological implications, and future prospects. *Journal of Food Agriculture and Environment, 1*, 101–106.
17. Raspor, P. (2008). Total food chain safety: How good practices can contribute. *Trends in Food Science & Technology, 19*(8), 405–412.
18. Srinivasan, A. (2006). *Handbook of Precision Agriculture: Principles and Applications*. Binghamton, NY: Food Products Press.
19. Swanson, B. E. (2008). *Global Review of Good Agricultural Extension and Advisory Service Practices*. FAO.
20. van Rijswijk, W., & Frewer, L. J. (2008). Consumer perceptions of food quality and safety and their relation to traceability. *British Food Journal, 110*(10), 1034–1046.
21. van Wagenberg, C. P. A., Backus, G. B. C., van der Vorst, J. G. A. J., & Urlings, B. A. P. (2009). A framework for the design and analysis of incentive systems for food safety control in supply chains. In *113th EAAE Seminar, A Resilient European Food Industry and Food Chain in a Challenging World* (pp. 1–13). Wageningen University and Research.
22. Watada, A. E., Ko, N. P., & Minott, D. A. (1996). Factors affecting quality of fresh-cut horticultural products. *Postharvest Biology and Technology, 9*(2), 115–125.
23. Wilson, D. W., & Beers, P. T. (2001). Global trade requirements and compliance with World Trade Organization agreements: The role of tracing animals and animal products. *Scientific and Technical Review of the Office International des Epizooties, 20*(2), 379–384.
24. Zhang, N., Wang, M., & Wang, N. (2002). Precision agriculture: A worldwide overview. *Computers and Electronics in Agriculture, 36*(2), 113–132.
25. Zoroja, J., Omejec, D., & Pejić, M. B. (2017). Integrated model of traceability: Tracking information for food safety. *Poslovna Izvrsnost—Business Excellence, 11*, 139–153.

# CHAPTER 3

# TECHNOLOGY IN TRACEABILITY SYSTEMS

## Contents

# INTRODUCTION

The agriculture industry is undergoing a revolution thanks to technology in traceability systems, which improves efficiency, safety, and transparency across the supply chain. The ability to follow the path of agricultural products from point of origin to point of consumption—known as traceability—allows stakeholders to keep an eye on every stage of the production, processing, and distribution procedures. Technology—which includes tools like blockchain, data analytics, and Internet of Things devices—is essential to gathering, storing, and disseminating information about products, including their location, handling conditions, and environmental impact. Real-time data is accessible to farmers, processors, distributors, and consumers thanks to these technologies, which promotes accountability and trust while assisting in ensuring that goods fulfill consumer expectations and legal requirements.

## 3.1. BARCODE TECHNOLOGY

In barcoding technology, a sequence of black and white lines or patterns is used to encode product information. Specialized devices can then scan and decode the information. A unique barcode is assigned to each product in the context of food traceability, signifying important information like the product's lot number, manufacturing location, production date, and expiration date. To track food products as they travel through the supply chain, from the farm to the processing facility, distribution hubs, and ultimately store shelves, this encoded data is essential.

By using barcodes, businesses can digitally store, retrieve, and distribute this data, greatly lowering the errors that come with manual record-keeping. Additionally, the barcoding procedure can be easily incorporated into current supply chain procedures. For instance, barcodes are an easy-to-use yet effective traceability tool that businesses can apply to packaging during the manufacturing or packaging phases.

### 3.1.1. Advantages of Barcoding in Food Traceability

There are many benefits for manufacturers, retailers, regulators, and consumers when barcoding technology is used in food traceability. First of all, barcodes improve food safety by making it possible to quickly access a food product's origins and journey. Barcoding enables the prompt identification of impacted batches in the event of a contamination problem or outbreak of foodborne illness, facilitating quicker recalls and preventing the distribution of potentially hazardous products to consumers. Producers and distributors are held more accountable because every product movement can be tracked back to its origin thanks to this degree of traceability. Another key benefit of barcoding is its impact on efficiency. Scanning barcodes is far faster and more accurate than manually entering product information, which means businesses can save significant time and reduce labor costs. This efficiency is especially valuable in large-scale operations where 1000s of products are processed daily. Furthermore, the ability to easily share barcoded information across different parts of the supply chain allows for better collaboration and visibility, helping companies make more informed decisions based on real-time data.

### 3.1.2. Role of Barcode Technology in Agriculture

Technological advancements are essential to improving sustainability, traceability, and efficiency in the dynamic field of agriculture. Among these developments, barcode technology has become a potent instrument that is transforming agricultural supply chains. We examine the many facets of barcodes' influence on agriculture, including their uses, advantages, and disadvantages (Figure 3.1).

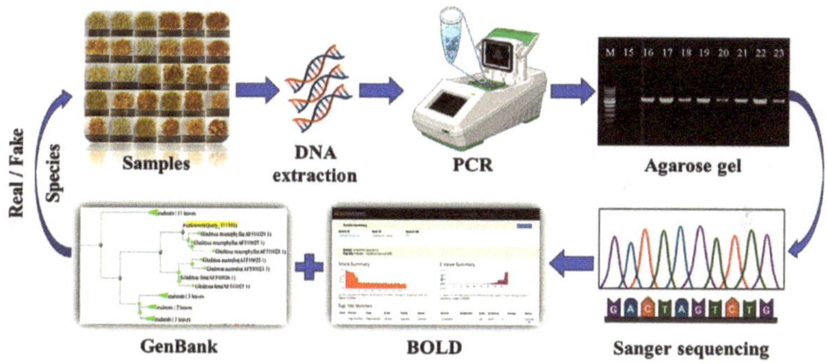


**Figure 3.1.** *Role of barcode technology in agriculture.*

*Source: Zhao G, Li L, Shen X, Zhong R, Zhong Q, Lei H. DNA Barcoding Unveils Novel Discoveries in Authenticating High-Value Snow Lotus Seed Food Products. Foods. 2024; 13(16):2580. https://doi.org/10.3390/foods13162580.*

#### *3.1.2.1. Enhanced Traceability and Transparency*

- Barcodes act as agricultural products' digital fingerprints. They facilitate smooth supply chain tracking by encoding crucial data like batch numbers, origins, and production dates.
- Imagine a situation where a customer buys an organic spinach bag from their neighborhood grocery store. The spinach was grown on a farm, and the barcode on the packaging identifies both the farm

and the particular field. Due to the ability of buyers to independently confirm claims regarding sustainability, pesticide use, and fair labor practices, this transparency promotes trust between consumers and producers.

#### 3.1.2.2. Efficient Inventory Management

- Inventory management is a problem for agricultural businesses, particularly in busy seasons. Barcodes reduce human error and automate data entry, streamlining this process.
- Consider a warehouse filled with crates of recently picked apples. A barcode corresponding to the variety, weight, and quality grade of each crate is present. These barcodes are scanned when the apples are delivered to distribution hubs or retail locations, updating inventory levels instantly. By doing this, supply chain operations are optimized and overstocking or stockouts are avoided.

#### 3.1.2.3. Quality Control and Compliance

- Barcodes make it possible for quick inspections, which supports strict quality control. For example, employees can use barcode scanning to find faulty or damaged produce when sorting fruit.
- Another crucial element is following the rules. Barcodes facilitate easy integration with certification bodies, guaranteeing the accuracy of labels pertaining to organic, non-GMO, or fair-trade products. Farmers who provide barcode-tracked data can show that they are adhering to standards.

#### 3.1.2.4. Farm-to-Table Tracing

- Customers want to know where their food comes from and how it gets there. Through their direct connection to the farm, barcodes enable this transparency.
- Think about a bottle of handmade olive oil. The grove where the olives were harvested and the pressing date can both be found by scanning the barcode. This relationship strengthens the importance of sustainable practices by fostering a sense of connection between consumers and the land.

#### 3.1.2.5. Challenges and Considerations

- **Infrastructure:** While barcodes are cost-effective, their implementation requires compatible hardware (scanners, printers) and reliable network connectivity.

- **Standardization:** There are no consistent barcode standards in the agriculture industry. It is crucial to standardize formats (like GS1 barcodes) across commodities and geographical areas.
- **Smallholder Farmers:** Ensuring that barcode technology reaches small-scale farmers is crucial. Training and support are necessary to bridge the digital divide.

## 3.1.3. Barcode Solutions for Agricultural Products

### *3.1.3.1. The Power of Barcodes in Agriculture*

Barcodes have evolved beyond their modest beginnings as simple price tags in retail establishments. They are effective instruments in agriculture for monitoring, controlling, and streamlining the flow of products from farm to table. Here's why they matter:

1. **Unique Identification:** A barcode, which serves as a digital fingerprint, is assigned to every agricultural product. The barcode gives an identity that can be tracked down the supply chain, whether it's a bag of rice or an apple crate.
2. **Efficient Data Capture:** Barcodes allow for quick data collection at multiple points of contact. From harvest to distribution, employees can use mobile apps or handheld devices to scan barcodes, instantly recording important data like batch numbers, quality, and origin.
3. **Supply Chain Visibility:** Consider the departure of a shipment of organic avocados from a Mexican farm. Barcodes allow stakeholders monitor its journey through ports, warehouses, and transportation. By predicting arrival times, retailers can minimize waste and guarantee freshness.
4. **Consumer Confidence:** The barcode on a bunch of bananas you buy at the grocery store tells you more than just the cost. Whether the fruit is fair-trade, organic, or sustainably sourced, it tells you about its journey. Barcodes meet the growing demands of consumers for transparency.

### *3.1.3.2. Traceability Challenges Addressed by Barcodes*

Let's explore how barcode solutions tackle critical traceability challenges:

1. **Product Recall Management:** Quick action is crucial in the regrettable event of a foodborne illness outbreak. Authorities can identify impacted batches thanks to barcodes, reducing the public health impact.

2. **Quality Assurance:** Temperature, humidity, and handling conditions are monitored by barcodes. This data guarantees that quality standards are fulfilled for perishable produce, such as strawberries.
3. **Origin Verification:** Customers are interested in learning the origins of their food. Barcodes provide information about the farm, processing plant, and even the particular field where those succulent tomatoes were planted.
4. **Certifications and Compliance:** Organic, non-GMO, and fair-trade certifications are vital. Barcodes link products to their certifications, building trust.

### *3.1.3.3. Examples in Action*

Let's see how barcodes are making a difference:

1. **Coffee Beans:** A Colombian small-scale coffee grower harvests beans and labels each bag with a barcode. The barcode guarantees equitable compensation and quality assurance as the beans pass through exporters, roasters, and cooperatives.
2. **Seafood:** In Norway, a fishing boat catches cod. Customers in Tokyo can scan the barcode on each fish to discover more about its journey from the icy waters of Norway to their sushi platter.
3. **Fresh Produce:** A Californian head of lettuce shows up at a grocery store in New York. Its organic certification, harvest date, and the family farm that raised it are all disclosed by the barcode.

### *3.1.3.4. Challenges and Future Directions*

While barcodes offer immense benefits, challenges remain:

1. **Standardization:** Stakeholders may become confused by the various barcode formats (UPC, QR codes, Data Matrix). Standardizing is essential.
2. **Infrastructure:** Rural areas may lack scanning infrastructure. Mobile networks and affordable devices are essential.
3. **Beyond Barcodes:** Blockchain and RFID are examples of emerging technologies that promise even more traceability. Barcodes have to change with them.

## 3.1.4. Challenges and Limitations of Barcoding in Food Traceability

While barcoding technology offers significant benefits, it is not without challenges. One of the main limitations is the reliance on accurate data input. If the information encoded in a barcode is incorrect or outdated, it can undermine the effectiveness of the traceability system. This risk is especially high in complex supply chains where multiple parties handle a product before it reaches the end consumer. Ensuring that all participants consistently update and verify product information is crucial to maintaining the integrity of the barcoding system.

Another challenge is the cost associated with implementing and maintaining barcoding systems, particularly for small and medium-sized enterprises (SMEs) with limited budgets. Although barcoding is generally less expensive than some other advanced traceability solutions, such as blockchain, the initial investment in scanners, printers, and software can be a barrier. Additionally, certain environments, such as cold storage or processing facilities, may require specialized barcode materials and equipment to ensure durability and readability, further increasing costs.

### 3.1.5. Integration with Other Traceability Technologies

To overcome some of the limitations associated with barcoding, many companies are integrating barcoding technology with other traceability solutions, such as radio frequency identification (RFID) and blockchain. RFID tags, which use radio waves to transmit product information, offer an alternative to traditional barcodes in cases where hands-free scanning is beneficial, such as in high-volume warehouses. By combining barcodes with RFID, companies can enhance their traceability capabilities, as RFID tags can store more information than standard barcodes and be read from a distance.

Blockchain technology also complements barcoding in food traceability by providing an immutable, decentralized ledger of a product's journey through the supply chain. When used in conjunction with barcoding, blockchain can offer an additional layer of transparency and security, as each stage of the supply chain can be verified and recorded on a permanent, shared ledger. This integration allows companies to create a robust traceability system that not only provides accurate data but also enhances trust by minimizing the risk of data manipulation.

## 3.2. QR CODES

Although QR codes were first developed to track tools and equipment used in manufacturing, they are now essential in many industries, including agriculture. With the growing global awareness of the value of high-quality, sustainably produced food, farmers, and other agriculturalists are exploring innovative concepts to improve their methods and achieve higher levels of productivity.

### 3.2.1. QR Codes in the Agriculture Industry

Compared to their predecessors, QR codes are made to store more data and allow for quick data decoding. 2D barcodes that are intended to be scanned by smartphones or specialized QR code readers make up Quick Response Codes or QR codes. QR codes are an excellent tool for sharing information in the agriculture sector. Farmers can use QR codes to label different seed and equipment packages with important information that benefits both workers and customers.

Advertising and marketing campaigns about farming can incorporate QR Codes since they are simple to scan. Farmers profit from marketing their goods to generate the highest returns, and QR codes can be the ideal tool for this.

### 3.2.2. Benefits of Using QR Code Technology for the Agriculture Industry

After the commercialization of QR codes, the integration of QR codes in the agriculture industry is getting various benefits:

#### *3.2.2.1. Improved Traceability and Transparency*

In the agricultural sector, enhanced traceability and transparency are two of the main advantages of QR code technology. To provide consumers with comprehensive information about their products' place of origin, the farming practices used, and even the nutritional value of the crops, farmers are adding QR codes to their packaging and goods establishing a transparent relationship and fostering trust between farmers and consumers through the use of QR codes will increase farmers' long-term loyalty.

Also, it provides consumers with knowledge that is vital to making informed food decisions. Additionally, QR codes have the potential to provide access to full supply chain data for agricultural products. With their assistance, tracking is now possible from the location of harvesting to the retail locations, with records kept at each important step of the transportation process.

#### *3.2.2.2. Enhanced Inventory Management*

Inventory management in the agricultural sector can be greatly improved by QR code technology. Each batch or individual product can have a unique QR code applied to make it simple for agribusinesses and farmers to monitor their stock. They can quickly obtain details about the product, such as its quantity, location, and expiration date, by scanning the codes. Farmers can minimize waste and overstocking by using QR codes to optimize their inventory. Additionally, it streamlines the restocking procedure, guaranteeing that thc appropriate products are restocked at the appropriate moment. While inventory management is automated by QR code technology, manual record-keeping may increase the possibility of human error. Farmers can thus save a significant amount of time and money.

### *3.2.2.3. Streamlined Supply Chain Operations*

The agriculture sector benefits from using QR codes to expedite supply chain processes. Traditionally, many parties, including farmers, distributors, processors, and retailers, are involved in the flow of agricultural products through the supply chain. To ensure efficiency and reduce delays, precise documentation and communication are necessary at every stage of the process. All parties involved in the supply chain can easily share and access information by using QR codes. Farmers can, for instance, affix QR codes to their goods so that distributors can scan them when they receive them. By giving real-time visibility into the movement of goods, they improve coordination and lower the possibility of mistakes or delays.

### *3.2.2.4. Increased Consumer Engagement*

Agribusinesses have a rare chance to interact with customers more deeply thanks to QR codes. Product packaging, promotional materials, and even physical store displays can all make use of QR codes. Customers can have an immersive and engaging experience with agribusinesses. For instance, customers may be directed to a website or mobile application that provides more details, recipes, or even promotions by scanning a QR code on a product. In addition to improving the overall shopping experience, agribusinesses can gather useful information about the preferences and actions of their customers by using QR codes. Thus, by using BLE beacons in an agri-retail store, the data can be utilized to develop personalized offers, loyalty programs, and targeted marketing campaigns.

### *3.2.2.5. Efficient Product Labeling and Packaging*

In the agricultural sector, QR code technology streamlines product labeling and packaging. Farmers can create QR codes with all the product information they need instead of relying on conventional manual-printed labels. It lowers the possibility of mistakes or out-of-date information and does away with the need for multiple labels.

QR codes are an economical and effective way to identify products because they are simple to print and affix to packaging materials. They ensure accuracy and regulatory compliance by enabling prompt updates or modifications to the product information in addition to saving time and resources.

#### *3.2.2.6. Enhanced Data Collection and Analysis*

Farmers and agribusinesses can gather useful information about their goods, clients, and operations by using QR codes. Farmers can gain insights into consumer preferences, market trends, and product performance by examining the data obtained from QR code scans. The information produced enables farmers to choose crops, farming methods, and marketing tactics with knowledge. Additionally, QR codes make it easier to gather information about environmental impact and sustainability. For instance, farmers can add QR codes to their goods to reveal details about their carbon footprint, certifications, and farming practices.

#### *3.2.2.7. Cost Savings and Sustainability in Agriculture*

Farmers and agribusinesses can save money by using QR code technology, which digitizes many tasks like product labeling, supply chain tracking, and inventory management. Farmers can cut back on manual labor, paperwork, and related expenses. Human error is less likely and overall operational efficiency is increased. QR codes to give customers comprehensive details about where their produce comes from. This promotes a deeper understanding of sustainable farming methods and locally sourced products. Additionally, a single QR code can store website links, comprehensive instructions, and promotional materials, which helps agribusinesses use less paper. To access information from various web portals, users must scan the QR code (Lv et al., 2023).

## 3.3. RFID TECHNOLOGY IN FOOD TRACEABILITY

In the food industry, RFID-based traceability systems help monitor the flow of food through the supply chain. This enables food businesses to ensure food safety, enhance supply chain management, maximize distribution, and enhance customer satisfaction.

### 3.3.1. Benefits of traceability systems with RFID in the food industry

Food traceability systems with RFID offer a series of benefits, some of them are:

- **More Precision and Reliability:** This is because RFID technology allows quick and precise reading of data, which reduces the risk of errors.
- **Best Supply Chain Control:** It is made possible by RFID traceability systems, which track a product's movement throughout the supply chain and make it simpler to identify and eliminate defective issues.

- **Enhancing Food Safety:** This is because RFID-enabled traceability technology can be used to detect contaminated or adulterated food items, protecting consumers.

### 3.3.2. Applications of Traceability Systems with RFID Applications in the Food Industry

Some sectors in which traceability systems with RFID can be used are:

- **Meat Sector:** Meat carcasses, packaged meat products, and live animals can all be identified with RFID labels. This makes it possible to keep an eye on meat products all the way from the farm to the plate.
- **Horticultural Sector:** Fresh fruits and vegetables are recognized by RFID labels. This makes it feasible to trace the products' origin and ensure their freshness.
- **Milk Sector:** Dairy products like milk, cheese, and yogurts can be tracked using RFID. Dairy product distribution and product quality can be observed.

RFID technology is an efficient tool to improve food traceability. Also, traceability systems with RFID offer a series of advantages which contribute to guaranteeing food safety and consumer protection.

## 3.4. BLOCKCHAIN FOR FOOD TRACEABILITY

Blockchain is a type of digital ledger that securely, openly, and impenetrably records transactions. Blockchain makes sure that once data is added, it can never be removed, in contrast to traditional databases where data can be changed or manipulated. This makes it revolutionary for sectors like the food supply chain that depend on openness and confidence.(Fung, F.; Wang, H.S.; Menon, et al., 2018). Traceability in the food industry refers to following a product's path from farm to fork. Food supply chains are typically intricate, involving many parties, including farmers, processors, distributors, and retailers. Vital information may be misplaced, reported incorrectly, or even tampered with while the food moves through the supply chain. Every step of the process—from the farm where it is grown to the factory where it is processed to the retail location where it is sold—is live-tracked on the blockchain. This implies that an unchangeable entry is made each time food is transferred, creating a clear and reliable record.

### 3.4.1. How Blockchain Is Used in Food Traceability

Blockchain, Verifiable Credentials, and decentralized identifiers (DIDs) are key technology components to enhance food traceability in the supply chain which defines as (Figure 3.2):

- **Blockchain:** A blockchain is a decentralized database that stores data in a way that makes it hard to tamper with or manipulate the system.
- **Verifiable Credential:** A digital record that is encrypted and secure, such as a document or certificate, and that has been validated and issued by a reliable organization or person.
- **Decentralized Identifiers (DIDs):** A product or batch of products with multiple Verifiable Credentials associated with it can be identified using a DID, which functions similarly to an identifying address registered on the blockchain.

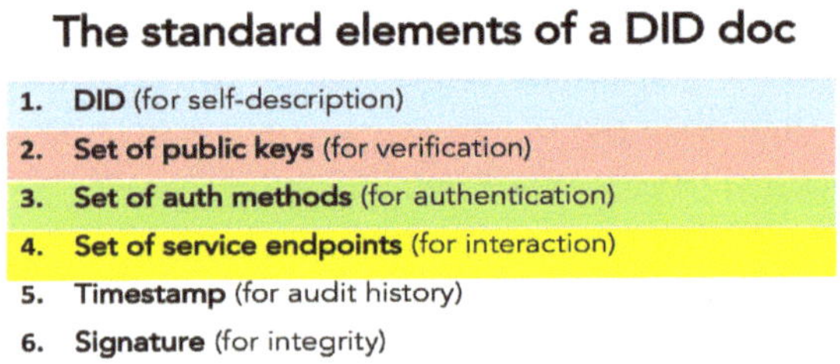


**Figure 3.2.** *Product details are added on documents issued as Verifiable Credentials by supply chain participants to a product's decentralized identifier (DID).*

*Source: By Drummond Reed - https://www.slideshare.net/SSIMeetup/decentralized-identifiers-dids-the-fundamental-building-block-of-selfsovereign-identity-ssi, CC BY-SA 4.0, https://commons.wikimedia.org/w/index.php?curid=81921111.*

Food products can all have DIDs applied to them, and as the product moves through the supply chain, various actors can add Verifiable Credentials to the product DIDs to improve food traceability.

It's crucial to understand that, for security and privacy reasons, Verifiable Credentials are not kept on the blockchain when using Dock. Instead, stakeholders securely encrypt and store them. The blockchain is only used to store public DIDs. In essence, multiple Verifiable Credential documents may be linked to each product DID, but the information on the Verifiable Credentials is not stored on the blockchain. The purpose of the blockchain is to efficiently validate data's authenticity without depending on a single party. Since each actor has access to the same source of truth, this improves transparency among them.

In contrast to days or weeks with traditional processes and systems, people can track and trace food sources in seconds when blockchain food traceability systems integrate Verifiable Credentials that are added to food product DIDs.

Food data, including its origins, movement, standards, and quality, can now be instantly verified thanks to blockchain technology.

## 3.4.2. Benefits of Using a Blockchain Food Traceability System

### *3.4.2.1. Accurate and Tamper-Resistant Food Data*

The food data is extremely accurate since Verifiable Credentials can't be altered because of cryptography's security. The verified data, which includes information about food products like quantity received, organic certification, farming community, safety standard certifications, and so forth, can be viewed upon request by any relevant party in the supply chain. A record cannot be altered once it has been entered. Anytime there is a pertinent transaction in a blockchain-based food traceability system, a Verifiable Credential can be issued. Any meaningful exchange of information about the condition of the products as well as the transfer of ownership or custodianship can be considered a transaction in a supply chain.

### *3.4.2.2. Efficiently Prevent, Contain, or Rectify Contamination Outbreaks While Reducing the Loss of Revenue*

Because blockchain food traceability systems use immutable and verifiable data gathered over time, they can quickly identify points of contamination with Verifiable Credentials. Instead of discarding the entire product inventory, a company could use blockchain technology to swiftly identify the origin and batch of a food disease pathogen—an organism that can cause disease—so that an immediate recall of the contaminated batch can occur. Customers and businesses used to have to discard the entire inventory when a food product became contaminated, which led to a lot of waste and revenue loss. Due to an *E. coli* outbreak in California, spinach was pulled from US supermarkets for two weeks, costing the state's farmers US$74 million in losses. The annual cost of food-borne illness in low- and middle-income nations is $110 billion. (Zhang, Y.; Yang, Z.; Yang, 2018).

### *3.4.2.3. Creates more Transparency and Trust in the Authenticity of Verifiable Credential Data*

Without using a middleman or getting in touch with the data's issuer, it was impossible to effectively confirm the authenticity of data using traditional supply chain recordkeeping systems. There is no need to rely on a centralized party to verify food data because blockchain-based food traceability systems can verify it quickly and effectively. Despite their lack of mutual knowledge, stakeholders can transact business with each other in a safe and efficient manner thanks to blockchain technology.

#### 3.4.2.4. Secure and Efficient Data Transfer Between Parties

Participants can exchange information quickly and effectively thanks to the digitization of data in an unchangeable format. Effective data verification reduces waste, delays, and other inefficiencies that are expensive in terms of both time and money. Data pertaining to every transaction connected to a particular food product is accessible to stakeholders, which increases system accountability and facilitates monitoring.

#### 3.4.2.5. Verifiable Credentials Help Eliminate and Prevent Fraud

The visibility of food data enables policymakers and relevant parties to monitor environmentally damaging actions like deforestation and other related activities. When food data is authenticated and issued as a Verifiable Credential, people can easily go back and track the product information.

#### 3.4.2.6. Helps Organizations Comply with Food Regulations

Systems for blockchain-based food traceability can assist businesses in adhering to food safety laws. For instance, the Food and Drug Administration (FDA) encouraged trading partners to increase the use of blockchain technology to track a product back to its source in 2020 by proposing a New Era of Smarter Food Safety Blueprint to improve food traceability.

#### 3.4.2.7. Customers Can Instantly See Information about Food Origins and Product Details Which Can Enhance the Reputation of Companies

Consumer demand for organic, slave-free, sustainably sourced food products is rising on a global scale. Additionally, consumers want assurance that the food was produced by people working in decent conditions, so they are asking companies to be more transparent about where the food comes from. According to a survey on consumer trends around the world, 79% of consumers today feel that when they purchase products, brands should be able to guarantee authenticity through certifications. Of this group, 71% are prepared to pay more for businesses that provide complete traceability and transparency.

More businesses are giving customers the ability to scan a QR code to view information verified by the blockchain, including the origin of food products, as a result of these requests. They can view videos about the food's farming community, harvesting process, and organic certification, to name a few examples of the information available to them (Figure 3.3).

**Figure 3.3.** *Scanning a QR code.*

*Source: By Hinnerk Rümenapf (HinnerkR), https://hinnerk.ruemenapf.de - Own work, CC BY-SA 4.0, https://commons.wikimedia.org/w/index.php?curid=108016734.*

## 3.4.3. Using the Dock Blockchain for Food Traceability

A DID, which is a distinct digital identity that can be stored on the blockchain, can be created by any party involved in the food supply chain that requires access to the data. To help track food products, each party can add food data in the form of Verifiable Credentials to product DIDs. Typically, farmers, cooperatives, processors, shippers, distributors, wholesalers, and retailers are common participants in the food supply chain (Figure 3.4).

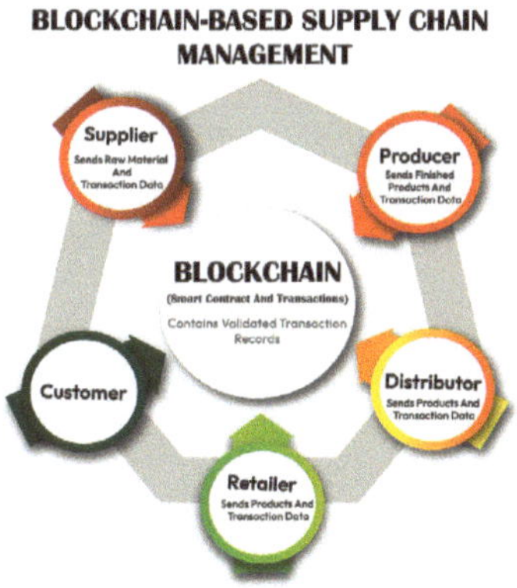


**Figure 3.4.** *Verifiable credentials are passed on to the next actor in the supply chain.*

*Source: Khan I, Ali QE, Hadi HJ, Ahmad N, Ali G, Cao Y, Alshara MA. Securing Blockchain-Based Supply Chain Management: Textual Data Encryption and Access Control. Technologies. 2024; 12(7):110. https://doi.org/10.3390/technologies12070110.*

Let's go through a step-by-step example of how a blockchain food traceability system would be integrated with a supply chain to track a product from the farm to the store. For this scenario, we will track a batch of organic peaches.

- Each participant would create a DID and they would be able to issue and add Verifiable Credentials with product details to the product's DID.
- Additionally, each batch of 500 peaches is assigned a DID. From the time the batch is harvested until it gets to the store where the customer can purchase it, its details can be tracked. Each batch is associated with a QR code that links to the product's verifiable credentials. Therefore, anyone in the supply chain can access the information by simply scanning it with their phone whenever they need to confirm the details.
- The farmer provides additional information about the growing conditions of the peach batch, such as soil temperature and weather, on a Verifiable Credential document. This document is appended to the peach's DID, enabling instantaneous verification of the product details by any party involved in the supply chain, including the final consumer.
- After a government inspector inspects a batch, the product's DID is updated with an organic certificate that certifies the product satisfies organic certification requirements and is added as a Verifiable Credential.
- A Verifiable Credential attesting to the fact that they delivered 100 peaches to the distributor is issued by the processor when they pack a box of 100 peaches.
- Upon receiving the batch of peaches, the distributor verifies the date of receipt and quantity by adding a VC to the product's DID. In addition, he provides a VC document that includes crucial information such as the date it was sent to the retailer and the warehouse's storage temperature.
- To make sure there are 100 peaches in the box, the organic store scans the peaches' code when they arrive. Since the Verifiable Credential data indicates that there were 100 peaches between the processor and distributor, if only 80 peaches are present, stakeholders can immediately identify that peaches went missing between the distributor and the store's receiver. The relevant parties can look into what happened to see if anything was stolen based on this information. The retailer issues a VC certifying that they received

100 peaches in the box if every peach is present. She quickly confirms that this batch of peaches has their organic certification by scanning the QR code on the box.

- Customers can scan the QR codes located on each product in the store to view information about the product's origin, growing conditions, farming community, and organic certification.

Dock empowers users to issue and authenticate Verifiable Credentials, while also enabling the creation of DIDs. Our technology is adaptable to collaborate with various systems, including user interfaces where data input is being carried out by stakeholders. These are just a few examples of systems that can interface with our blockchain solutions:

- Traceability systems;
- Logistics systems;
- Invoicing systems;
- Order management systems; and
- Payment systems.

Let's take an example where an Internet of Things sensor is tracking temperature and soil quality in a specific farm area on a constant basis. Verifiable Credentials can be programmed to be automatically generated and added to a product's DID based on the precise data you wish to capture using the Dock blockchain and APIs (Curto and Gaspar, 2021).

## 3.4.4. Companies Using Blockchain Food Traceability Systems

A growing number of food companies are utilizing blockchain technology to increase safety, decrease fraud, and improve product traceability.

Bumblebee Foods uses blockchain to track tuna operations from the moment fish is caught to the market. Their blockchain food traceability system is used to:

- Instantly check for food fraud and product tampering;
- Identify and classify product waste in supply chains; and
- Quickly identify food contamination issues that help with rapid product recalls.

Carrefour, the biggest retailer in Europe, uses blockchain-based food traceability systems to satisfy customers' increasing demand for information regarding the provenance and production processes of their organic products. Consumers will be able to view details about the product, such as its manufacturing location and shipping details. They intend to progressively integrate blockchain technology into every product under the Carrefour Bio brand.

Walmart has been digitizing its food supply chain and cutting down on the time it takes to find the source of food contamination by utilizing blockchain technology. To trace back their produce, the company mandates that all trading suppliers of leafy green vegetables comply with data record input into a blockchain-based platform. These days, they can quickly determine the source of tainted produce. It used to take the internal team seven days to track a package of sliced mangoes. Blockchain technology, however, allowed them to track the sliced mangoes in just two seconds, passing through every checkpoint—including a Mexican farm, a hot-water treatment plant, an importer, a processing plant, and a cold storage facility. Nestlé uses blockchain in several ways. Users can get access to key food supply chain data including the complete history and current location of any individual item, certifications, test data, and temperature data. All of this is available in seconds once it is uploaded to the blockchain. Blockchain is also used to track the origin of their luxury coffee brand Zoégas which is made from coffee beans from Brazil, Rwanda, and Colombia. Customers can scan a QR code to access blockchain data that includes information about the coffee's farmers, the transaction certificate for specific shipments, the time of harvest, and roasting period.

### 3.4.5. Comparison of Traditional Traceability System and Blockchain Traceability System

The globalization of trade has led to an increasingly complex agricultural supply chain, making it more challenging to track items through its intricate network. There is no assurance that information will flow consistently throughout the agricultural product chain with traditional traceability systems. The comparison between the blockchain traceability system and the conventional traceability system is displayed in Table 3.1.

**Table 3.1.** *Comparison of Features in Traditional System, Public Blockchain, and Hyperledger Fabric*

| Feature | Traditional System | Public Blockchain | Hyperledger Fabric |
|---|---|---|---|
| Blockchain log | No | Yes | Yes |
| Extensible event storage | Yes | No | Yes |
| Distributed ledger | No | Yes | Yes |
| Flexible transparency | No | No | Yes |
| Immutability | No | Yes | Yes |

| Feature | Traditional System | Public Blockchain | Hyperledger Fabric |
|---|---|---|---|
| Private data sharing | No | No | Yes |
| Confidentiality | No | No | Yes |
| Modularity | No | No | Yes |
| Approval process that plugs in | No | No | Yes |

*Source: Khan I, Ali QE, Hadi HJ, Ahmad N, Ali G, Cao Y, Alshara MA. Securing Blockchain-Based Supply Chain Management: Textual Data Encryption and Access Control. Technologies. 2024; 12(7):110. https://doi.org/10.3390/technologies12070110.*

Blockchain-based traceability technology is more appropriate for storing food data than traditional traceability technology, according to the analysis in Table 3.1. The specific reasons are as follows:

- **Database Operations:** Blockchain databases lack the update and delete operations found in traditional databases, which have operations like create, read, update, and delete. In the context of blockchain, no delete operation means that data uploaded to the library cannot be removed, and no update operation means that data cannot be changed as long as they are stored in the database. As a result, the openness, transparency, irreversibility, and tamper resistance of the traceable data information are essentially determined by the blockchain database, which is not possible with conventional relational databases.
- **Data Storage:** The blockchain is a distributed data storage system that simultaneously stores data on all of the decentralized system's nodes. The blockchain main chain data can be fully restored even in the event that some nodes fail, provided that there is still a functional node. This will not have an impact on the recording or updating of subsequent block data. Consequently, it is possible to solve the data loss issue. The conventional traceability system stores data centrally. The data are very vulnerable to loss or damage in the event of a server failure. As a result, data loss is common.
- **Tamper-Proof Data:** Because of its data structure and consensus mechanism, blockchain has the property that prevents data from being altered. Cryptography algorithms like Hash, Merkle Tree, SHA256, elliptic curve cryptography (ECC), etc. are used to link blocks between blockchains. As a result, once the main chain is sufficiently long, you must provide proof for each block after the

modified block using cryptography to add, remove, or change the data in one of the blocks. The cost of tampering with the data is significantly higher than the profit after tampering if the tampered block is at the front of the main chain. The nature of easily tampered-with data is determined by the single characteristic of traditional traceability data storage.

- **Consensus Mechanism:** The mechanism of consensus determines the allocation of bookkeeping rights. Only one person is permitted to keep records at a time to guarantee ledger consistency and data synchronization between nodes. With the traditional traceability system, data uploading is handled by a small group of official personnel, and no consensus mechanism is needed.
- **Trust:** Blockchain uses consensus processes and smart contracts to resolve complicated trust issues, making it possible for institutions or individuals who are strangers to come to an agreement. The food industry and its brands are the only ones that consumers can unilaterally trust under the traditional traceability system.
- **Efficiency:** Blockchain requires identity verification in addition to a consensus algorithm, the specific efficiency of which is determined by the algorithm's advantages and disadvantages. The server and database's performance affects how well the conventional traceability system works.
- **Interoperability:** Blockchain is a distributed system with good interoperability. Interoperability is the ability to exchange information between different systems. The conventional traceability systems' level of interoperability is currently fairly low.

The traditional traceability system's evasion of responsibilities, tampering with account books, accountability, private exploitation of loopholes, and information loss can all be more than compensated for by using blockchain technology for traceability. As a result, a blockchain-based traceability system has shown to be a superior option in the food supply (Sello et al., 2019).

### 3.4.6. Applications of Blockchain in the Traceability of Agricultural Products

As per the Deloitte report, an increasing proportion of consumers—51% — are willing to pay extra for food that is safe and nutritious. Customers expect verifiable traceability as a necessary component of quality and safety. A system that can offer details on the origin and life cycle of commodities is required to meet this demand. However, with the commercialization of precision agriculture technology in recent years, there has been a growing interest in agricultural data.

Agricultural data, particularly those from joint yield monitors, are unreliable as we all know. The validity of the data worries analysts, especially in light of the possibility that other factors could impact the data quality at different points along the data path. Large-scale data management also presents many difficulties, the most significant of which are caused by imprecise, fragmented, and incomplete data. Because it can confirm, audit, and transmit data and information confidentiality, blockchain technology—a decentralized ledger that enables block-encrypted data transactions—has drawn an increasing amount of attention from academic and business communities.

### *3.4.6.1. Fish*

One of the most significant primary food-based industries is seafood. But there hasn't been enough regulation of the fishery. Over 30% of the world's fish resources are overfished. In addition, it is challenging to assess the quality of seafood in real-world situations because consumers cannot see temperature records or the important certificate of origin. Therefore, it is challenging to accurately assess fish quality in practice due to a lack of trustworthy food and quality tracking data. A significant advancement for global fisheries, blockchain technology is currently being used to enhance tuna traceability to curb illicit and unsustainable fishing methods in the Pacific Island tuna industry. The authors of a study created the Blockchain-enabled Fish Provenance and Quality Tracking (BeFAQT) system in collaboration with the Sydney Fish Market (SFM). To address privacy issues raised by the use of blockchain encryption in the fish supply chain and enable the sharing of trusted and private data between supply chain participants, a multi-layer blockchain architecture built on attribute-based encryption (ABE) is suggested. Field tests demonstrate that BeFAQT can deliver accurate and thorough real-time fish provenance and quality tracking data.

The authors set out to investigate how supply chain visibility might be enhanced by blockchain technology in. Semi-structured interviews with participants and blockchain experts in the fish farming sector were used to carry out the exploratory research. The primary inference to be made from the findings is that blockchain technology can boost the value of a business by utilizing information's utility to raise industry visibility and boost operational effectiveness through improved planning and prediction via data analysis. A supply chain context-specific evaluation of blockchain technology is developed, and subsequently it is extended to a broader context. As a result, organizations or people who wish to learn more about blockchain technology or who intend to use it in the future will find the evaluation of blockchain technology helpful. By detecting issues in real time and putting solutions in place, digital technologies like blockchain can aid in the achievement of several Sustainable Development Goals pertaining to livelihoods, food security, and the environment. The study examines the architecture of a blockchain-centric food supply chain to support the Sustainable Development Goals, using the Thai fishery as a backdrop. The primary conclusions show that the supply chain has data asymmetry, which can help achieve the Sustainable Development Goals.

A blockchain-driven fish farming platform is suggested to guarantee the accuracy of agricultural data. The platform is intended to give fish farmers secure, unhackable storage for massive volumes of agricultural data. Smart contracts are used to automatically carry out the different procedures of the fish farm, lowering the possibility of error or manipulation. Tokkozhina et al., investigated the impact of adopting blockchain technology (BCT) in the context of multi-tiered supply chains (SC). Examine the real-world implementation of BCT in a multi-tiered Portuguese frozen fish product supply chain. Both qualitative and quantitative data were collected using a mixed methodology. Suppliers, transporters, and retailers from a single frozen fish SC participated in three semi-structured interviews. The findings indicated that buyers were more inclined to purchase fish that had traceability information. A private Ethereum blockchain-based system is suggested as a means of efficiently tracking and managing products in the fishery supply chain. The system efficiently oversees the operations of the fisheries supply chain in a decentralized, transparent, traceable, safe, confidential, and reliable manner. Additionally, five smart contracts and the solution architecture for automating the supply chain processes in the fishing industry are presented. To demonstrate the security and reliability of the suggested solution, a security analysis is carried out (Wang et al., 2019).

### *3.4.6.2. Fruits*

The decrease in contaminated food presents a significant challenge for the retail fruit industry. Every year, about 10% of people worldwide become ill as a result of eating tainted food, and this number is sharply rising. For this reason, a blockchain-based retail fruit chain is required to ensure society's access to food on a national and international level. Given this, agricultural product traceability necessitates the use of technology that can facilitate information flow throughout the supply chain and possibly guarantee traceability for final consumers. Shevchuk et al., to focus the study, consider the supply chain for pineapples as an example. The analysis's findings demonstrate the potential of blockchain technology as a traceability tool for pineapple supply chains. Sharif et al. studied and talked about blockchain technology and how it can be used in the retail industry. , add trust as an independent variable and apply a cross-sectional approach and the unified theory of acceptance and use of technology (UTAUT2). The research offers valuable insights into the fruit retail industry in Malaysia from various angles, including business procedures, competition, and customer-retailer collaboration. It concludes that, for the Malaysian retail sector to preserve its share of the global market going forward, it desperately needs industrial innovation, technological transformation, and new growth avenues like implementing blockchain systems. The fresh fruit supply chain is currently characterized by a wide distribution, intricate transactions, and many topics. These factors raise concerns about information tampering, the supply-demand relationship, regulatory traceability, and tamper-proofing. The fresh fruit supply chain can be improved through the application of blockchain technology in it. From the time fruit is picked until it is consumed, the blockchain system uses IoT-related technologies to gather and upload trustworthy data. The blockchain does, however, have certain drawbacks, including its reliance on the Internet of Things, its limited ability to preserve fruit, its ambiguous legal oversight, and so forth. The "double chain and double storage" blockchain model for the fruit and vegetable supply chain was created to control storage growth and enhance the blockchain's query efficiency. The Hyperledger-based double chain double storage model can greatly increase the efficiency of the blockchain storage query by accounting for the openness, security, and information privacy of businesses in all links of the transaction information (Lv et al., 2023).

### *3.4.6.3. Vegetables*

People adore vegetable-based agricultural products because they are healthy, nutritious, and environmentally friendly. However, there is a short shelf life for fruits, vegetables, and agricultural products, low temperature requirements for storage, and a high risk of food safety incidents. In the process of growing data in the blockchain traceability system, a method for on-chain and off-chain data storage based on "database + blockchain" is proposed to address the issues of high data load pressure and inadequate privacy security. The supply chain's local database houses the publicly visible data, which is then uploaded to the blockchain system via the SHA256 algorithm along with its hash value. The blockchain stores the private data that has been encrypted using the CBC encryption algorithm so that connected businesses can share it. To benefit from blockchain's crucial features of privacy and transparency, Yang et al. created a blockchain network that can address the issues with vegetable suppliers using Hyperledger Fabric. Kafka was used to create a blockchain network with nine channels. This performance suggests that a blockchain network with multiple channels could assist the vegetable supplier in enhancing their ledger capabilities to be completed in less than a day. Based on commercial-off-the-shelf (COTS) devices, VegIoT Garden is a scalable, modular, low-cost IOT/smart agriculture (SA)-oriented management platform that helps farmers improve the commodities they produce, identify critical issues, and address irrigation-related issues. It also identifies potential correlations between climatic characteristics, and this chapter presents field productivity and crop health. Massaro et al., studied blockchain technology and vegetable production technologies and outlined the whole situation and the application of blockchain technology, concluding that the tools and techniques employed in the literature are appropriate for quality traceability. Other materials can also be applied with benefit.

### *3.4.6.3. Eggs*

The use cases for deploying blockchain in food delivery are introduced, with a focus on egg supply chains in the Midwestern USA, to track the products from farm to fork using IoT-enabled technologies and blockchain technology. Customers can get the information they require and make knowledgeable food decisions by creating a transparent and traceable food supply chain. Traceability and transparency improve customer relations, boost productivity, lower costs associated with food recalls, fraud, and lost products for all parties involved in the food supply chain. It is possible to conclude that nations and supply chains, including those for chicken, tomatoes, eggs, oranges, and other food items, are boldly moving toward the large-scale use of blockchain technology by reviewing the literature from 2008–2020 that primarily examined the use of blockchain technology in supply chain management and explored two blockchain projects: Nimble and Carrefour.

**Table 3.2.** *Applications of Blockchain in the Traceability for Agricultural Products*

| Food | Goal | Method | Advantage |
|---|---|---|---|
| Fish | • Ensure agriculture data integrity.<br>• Promote sustainable development goals within the context of the fish industry.<br>• Create an assessment for blockchain technology in a specific industrial (supply chain) setting.<br>• Explore how blockchain technology could improve supply chain visibility.<br>• Accurate assessment of fish quality to ensure food safety.<br>• Explore the consequences of information availability inherent for BCT adoption pilots in the SC context.<br>• Trace the fisheries supply chain to assist stakeholders in identifying the origin and causes of product fraud and malpractice. | (1) Diverse processes of the fish farm are executed automatically by using the smart contract to reduce the risk of error or manipulation.<br>(2) Data asymmetry exists in supply chains to achieve Sustainable Development Goals.<br>(3) Using Strandhagen's control model methodology to design an AS-IS model to explore the effects of blockchain technology in Norwegian fish supply chains.<br>(4) Exploratory research was conducted through semi-structured interviews with fish industry actors and blockchain experts.<br>(5) A multi-layer blockchain architecture based on ABE is proposed to solve the privacy problems caused by the application of blockchain encryption in the fish supply chain and realize the sharing of trusted and confidential data among the parties of the supply chain.<br>(6) follows a mixed-method approach, with qualitative and quantitative data collection techniques.<br>(7) A solution based on the private Ethereum blockchain has been proposed to effectively manage fisheries supply chain operations. | (1) Providing fish farmers with secure storage for preserving the large amounts of agriculture data that cannot be tampered with.<br>(2) Contributing to the supply chain management field and ultimately impacting the resilience of fishery ecosystems and the achievement of sustainable development goals.<br>(3) Finding indicators for which types of supply chains blockchain technology could bring positive strategic implications.<br>(4) It is concluded that blockchain can use the usefulness of information to enhance the visibility of the industry, better predict and plan through data analysis, improve operational efficiency and thereby increase business value.<br>(5) The BeFAQT is able to provide trusted and comprehensive fish provenance and quality tracking information in real-time.<br>(6) The results show that consumers are more likely to buy fish products with traceability information.<br>(7) The solution effectively manages fisheries supply chain operations in a decentralized, transparent, traceable, secure, private, and trusted manner. |

| | | | |
|---|---|---|---|
| Fruits | • Research question: "Which technology should be chosen and how should it be integrated into a company's pineapple supply chain to ensure traceability?"<br>• Identifying the critical factor of fruit blockchain in the retail market and the relationship between the key factors blockchain in the retail industry.<br>• To solve the problems of anti-tampering, the supply and demand relationship and traceability of supervision in the fresh fruit supply chain.<br>• To limit storage growth and improve the query efficiency of the blockchain. | (1) Based on qualitative research, the study adopts data collection techniques such as desk research and semi-structured interview, and adopts content analysis and narrow analysis to analyze the data.<br>(2) A cross-sectional approach was used, and trust was added as an independent variable to the (UTAUT2).<br>(3) Collect and upload reliable data from fruit picking to final consumption through IoT related technologies.<br>(4) The "double chain and double storage" blockchain model for the fruit and vegetable supply chain was developed. | (1) The technology integration concept was defined. A conclusion is drawn that the hypothesis of traceability in the food industry still needs to be proven by market research.<br>(2) Providing different perspectives of Malaysian retail fruit blockchain environment, not only in terms of business processes and competition, but also cooperation between retailers and customers.<br>(3) Blockchain balances the tension between supply and demand, brings money to businesses and simplifies regulation and traceability. However, blockchain has some limitations, such as relying on the IoT, immature fruit preservation technology and so on.<br>(4) The dual-chain and dual-storage model based on Hyperledger takes into account the openness, security, and information privacy of enterprises in all links of transaction information, and can significantly improve the efficiency of blockchain storage query. |
| Vegetables | • To solve the problems of centralized management, opaque information, unreliable data and easy to generate information islands in traditional traceability systems.<br>• Discuss a case of a vegetables supplier's problems in terms of ledger.<br>• To meet the needs of agriculture 4.0 and smart agriculture for technology.<br>• Present some results of an industry research project by describing the entire vegetable scene, implementing the architecture of the blockchain, the principle of sequence diagrams, and embedding blocks and transactions in the prototype network. | (1) A dual storage structure of "database + blockchain" on-chain and off-chain traceability information is constructed, which reduces the load pressure of the chain and realizes information query.<br>(2) Use hyperledger fabric to build a blockchain network. A 9-channel blockchain network was created using kafka.<br>(3) VegIoT Garden: a modular IoT management platform for urban vegetable gardens is proposed, which is based on COTS devices, adopting short- and long-range communication protocols (IEEE 802.11 and LoRa).<br>(4) Present some results of an industry research project by describing the entire scenario, the realization of the blockchain architecture, the principle of sequence diagrams, and the prototype of network embedded blocks and transactions. | (1) The system improves the query efficiency and the security of personal information, ensures the authenticity and reliability of the data in the supply chain management, and meets the practical application requirements of agricultural product traceability.<br>(2) The results show that a multi-channel blockchain network can help vegetable suppliers increase their ledger capabilities to within a day.<br>(3) The platform can gather, monitor, and analyze sensor data collected from an urban vegetable garden, using blockchain as an enabler to satisfy security requirements.<br>(4) The preliminary results presented prove that the techniques and methods used in the literature are suitable for quality traceability. The method is useful for applying the proposed framework to other supply chains. |
| Eggs | • Track products (e.g., eggs) from farm to fork using blockchain and IoT enabled technologies.<br>• To answer: will blockchain become the norm for supply chain management? | (1) Creating traceable and transparent supply chains for food.<br>(2) Review the literature about the use of blockchain technology in supply chain management, and explore two blockchain projects: Nimble and Carrefour. | (1) Consumers can obtain the information they need and make wise choices about the food. For stakeholders in the food supply chain, traceability, and transparency can better establish relationships with customers, improve efficiency and reduce the risks and costs of food recalls, fraud, and product losses. (2) It is concluded, to the same extent, countries and supply chains are also boldly attempting to use blockchain on a large scale. |
| Pork | • Brand protection and security through transparency.<br>• To study extensively the application of blockchain-enabled food supply chains in developing countries. | (1) Using IBM's blockchain solution based on Hyperledger Fabric to realize "complete end-to -end traceability" for pork. (2) A comprehensive assessment of blockchain traceability in food supply chains is conducted through a case study of pork supply chains in Vietnam. | (1) Enhancing consumer loyalty, and farm-to-table traceability time is greatly reduced.<br>(2) Gain a deeper understanding of the potential impact on food supply chain sustainability. |

| | | | |
|---|---|---|---|
| Tea | • To know about the benefits and impact of BCT on the tea supply chain and its sustainable performance.<br>• To realize the reliable traceability technology of agricultural products.<br>• To accelerate resource deployment, reduce waste and improve the durability and sustainability of the circular tea supply chain.<br>• To achieve system automation and credibility in the field of agricultural product traceability.<br>• To achieve full chain tea counterfeiting supervision and automatic environmental management. | (1) Introduce the resource-based view (RBV) and network theory (NT) into the tea supply chain. A BCT-driven conceptual model of the tea supply chain was developed, the data using the structural equation modeling method of partial least squares regression are analyzed.<br>(2) Propose a block-chain traceability system based on smart agriculture with the integration of wireless sensor network and Ethereum. Adopt the food risk assessment and safety traceability technology based on the hazard factor to design the multi-role, multi-link, and multi-factor intelligent management system.<br>(3) A RFID technology driven by BCT practice model is proposed.<br>(4) A MBITTS is designed and implemented, in which, IoT can automate the collection of information of traceability and blockchain guarantees the security of the data entering the system.<br>(5) A blockchain–IoT-empowered decentralized framework is proposed. An incentive scheme based on blockchain technology is designed to attract tea PSC participants and encourage good behaviors of participating companies. Automation of environmental monitoring and equipment control through IoT technology. | (1) The research results show that the use of BCT has a significant positive impact on the tea supply chain; in particular, transparency and reliability are shown as sustainable performance parameters. This study is the first report to integrate BCT into the supply chain and contributes to the lack of relevant literature.<br>(2) Traceability QR code provides consumers with safe, reliable, and true traceable information for agricultural products. Create a holographic database for the entire tea industry chain from farmland to table.<br>(3) Help to manage the complexities of CTSCM and provide a deeper understanding of inventory performance, resource usage and industry process.<br>(4) The introduction of ML data verification mechanism can ensure the accuracy of information on the chain (up to 99%).<br>(5) The high throughput and the efficiency of resource-saving automation are, thus, achieved. |
| Crop | • An affordable, efficient, low cost crop insurance solution based on blockchain is proposed to solve the complexity and economic infeasibility of traditional crop insurance.<br>• By using improved blockchain technology in a breeding information management system, an architecture to store high-throughput crop breeding data efficiently and safely is proposed.<br>• Through crowdsourcing and community building to provide information to farmers and connecting them.<br>• The chapter aims to directly connect the farmers and the distributors, eliminating the middlemen using blockchain technology. • Increase crop yields and secure all valuable data related to seed type, fertilizer, and crop costs. | (1) With the proper use of blockchain based on ethereum, a blockchain-based crop insurance is proposed.<br>(2) A light blockchain is used in the storage architecture, to save key breeding data. Different types of blockchains are used to store different types of breeding data. The proxy encryption technology is used to ensure data security.<br>(3) A cost-efficient and blockchain-based secure framework for building a community of farmers is proposed. The integration of a deep neural network-based model to framework, which predicts any abnormalities in crops and gives the solutions.<br>(4) IoT-based crop monitoring and classification system is proposed, which employs LoRa for communication and blockchain for data trust and security.<br>(5) IoT sensor devices are used to collect data from farms, crop yield predictions are realized based on machine learning algorithms, all valuable data related to seed types, fertilizers, and crop costs are stored in blocks to secure the security of data and communication. | (1) The system can increase the transaction throughput and decrease the end-to-end latency.<br>(2) In the case of a large amount of data, especially when breeders query breeding data, the efficiency of GSBCP is significantly improved.<br>(3) The simulation results show that the prediction accuracy of this model is high.<br>(4) The system can automatically sense, store, and monitor real-time parameters that play an important role in determining a crop's quality and yield.<br>(5) Compared with the traditional methods, the proposed technique offers higher accuracy and profit from seed selection to transaction. Blockchain and edge computing -based transactions increase security and reduce transaction latency. The proposed system ensures the sustainability and traceability of agriculture. |

*Source: https://www.mdpi.com/2077-0472/13/9/1757.*

# REFERENCES

1. Aung, M. M., & Chang, Y. S. (2014). Traceability in a food supply chain: Safety and quality perspectives. *Food Control, 39*, 172–184.
2. Awan, S. H., Nawaz, A., Ahmed, S., Khattak, H. A., Zaman, K., & Najam, Z. (2020). Blockchain-based smart model for agricultural food supply chain. In *Proceedings of the 2020 International Conference on UK-China Emerging Technologies (UCET)* (pp. 1–5). Glasgow, UK.
3. Badia-Melis, R., Mishra, P., & Ruiz-García, L. (2015). Food traceability: New trends and recent advances. A review. *Food Control, 57*, 393–401.
4. Basnayake, B. M. A. L., & Rajapakse, C. (2019). A blockchain-based decentralized system to ensure the transparency of organic food supply chain. In *Proceedings of the 2019 International Research Conference on Smart Computing and Systems Engineering (SCSE)* (pp. 103–107). Colombo, Sri Lanka.
5. Curto, J. P., & Gaspar, P. D. (2021). Traceability in food supply chains: Review and SME-focused analysis—Part 1. *AIMS Agriculture and Food, 6*, 679–707.
6. Curto, J. P., & Gaspar, P. D. (2021). Traceability in food supply chains: SME-focused traceability framework for chain-wide quality and safety—Part 2. *AIMS Agriculture and Food, 6*, 708–736.
7. DeGroote, S. E., & Marx, T. G. (2013). The impact of IT on supply chain agility and firm performance: An empirical investigation. *International Journal of Information Management, 33*, 909–916.
8. European Union. (2002). Regulation (EC) No. 178/2002 of the European Parliament and of the Council of 28 January 2002, laying down the general principles and requirements of food law, establishing the European Food Safety Authority, and laying down procedures in matters of food safety. *Official Journal of the European Union, L 31*, 1–31.
9. Fernando, E. (2019). Success factor of implementation blockchain technology in pharmaceutical industry: A literature review. In *Proceedings of the 2019 6th International Conference on Information Technology, Computer and Electrical Engineering (ICITACEE)* (pp. 1–5). Semarang, Indonesia.
10. Karumanchi, M. D., Sheeba, J. I., & Devaneyan, S. P. (2019). Cloud-based supply chain management system using blockchain. In *Proceedings of the 2019 4th International Conference on Electrical, Electronics, Communication, Computer Technologies and Optimization Techniques (ICEECCOT)* (pp. 390–395). Mysuru, India.
11. Kumar, N., Balasubramaniam, S., Tharagesh, R. S., Kumar, P., & Janavi,

B. (2020). An autonomous food wastage control warehouse: Distributed ledger and machine learning-based approach. In *Proceedings of the 2020 11th International Conference on Computing, Communication and Networking Technologies (ICCCNT)* (pp. 1–6). Kharagpur, India.

12. Lv, G., Song, C., Xu, P., Qi, Z., Song, H., & Liu, Y. (2023). Blockchain-based traceability for agricultural products: A systematic literature review. *Agriculture, 13*(9), 1757. https://doi.org/10.3390/agriculture13091757.
13. Maouriyan, N., & Krishna, A. A. (2019). AQUACHAIN–Water supply-chain management using distributed ledger technology. In *Proceedings of the 2019 3rd International Conference on Computing and Communications Technologies (ICCCT)* (pp. 204–207). Chennai, India.
14. Perboli, G., Musso, S., & Rosano, M. (2018). Blockchain in logistics and supply chain: A lean approach for designing real-world use cases. *IEEE Access, 6*, 62018–62028.
15. Petersen, M., Yu, Z., & Lu, X. (2021). Application of Raman spectroscopic methods in food safety: A review. *Biosensors, 11*, 187.
16. Raj, R., Rai, N., & Agarwal, S. (2019). Anticounterfeiting in pharmaceutical supply chain by establishing proof of ownership. In *Proceedings of the TENCON 2019—2019 IEEE Region 10 Conference (TENCON)* (pp. 1572–1577). Kochi, India.
17. Sello, B., Tao, X., & Yong, J. (2019). A preliminary study for the risk of probabilistic blockchain protocols. In *Proceedings of the 2019 6th International Conference on Behavioral, Economic and Socio-Cultural Computing (BESC)* (pp. 1–4). Beijing, China.
18. Stranieri, S., Cavaliere, A., & Banterle, A. (2017). Do motivations affect different voluntary traceability schemes? An empirical analysis among food manufacturers. *Food Control, 80*, 187–196.
19. Stranieri, S., Riccardi, F., Meuwissen, M. P., & Soregaroli, C. (2021). Exploring the impact of blockchain on the performance of agri-food supply chains. *Food Control, 119*, 107495.
20. Swan, M. (2015). *Blockchain: Blueprint for a New Economy*. O'Reilly Media.
21. Tse, D., Zhang, B., Yang, Y., Cheng, C., & Mu, H. (2017). Blockchain application in food supply information security. In *Proceedings of the 2017 IEEE International Conference on Industrial Engineering and Engineering Management (IEEM)* (pp. 1357–1361). Singapore.
22. Wang, S., Li, D., Zhang, Y., & Chen, J. (2019). Smart contract-based product traceability system in the supply chain scenario. *IEEE Access, 7*, 115122–115133.

23. Wei, Y. (2020). Blockchain-based data traceability platform architecture for supply chain management. In *Proceedings of the 2020 IEEE 6th International Conference on Big Data Security on Cloud (BigDataSecurity), IEEE International Conference on High Performance and Smart Computing (HPSC), and IEEE International Conference on Intelligent Data and Security (IDS)* (pp. 77–85). Baltimore, MD, USA.
24. Wu, H., Cao, J., Yang, Y., Tung, C. L., Jiang, S., Tang, B., Liu, Y., Wang, X., & Deng, Y. (2019). Data management in supply chain using blockchain: Challenges and a case study. In *Proceedings of the 2019 28th International Conference on Computer Communication and Networks (ICCCN)* (pp. 1–8). Valencia, Spain.
25. Xu, X., Lu, Q., Liu, Y., Zhu, L., Yao, H., & Vasilakos, A. V. (2019). Designing blockchain-based applications: A case study for imported product traceability. *Future Generation Computer Systems, 92*, 399–406.
26. Yaga, D., Mell, P., Roby, N., & Scarfone, K. (2019). Blockchain technology overview. *arXiv*. https://arxiv.org/abs/1906.11078 (accessed on 14 December 2024).
27. Zhang, Y., Yang, Z., Yang, R., Jin, K., Lin, B., & Si, P. (2018). Traceability system of farm produce based on blockchain. *Technology and Intelligent Engineering, 3*, 4–13.

# CHAPTER 4

# TRACEABILITY IN CROP PRODUCTION

## Contents

# INTRODUCTION

Traceability in crop production refers to the ability to track and trace the journey of agricultural products from their origin on the farm to their final destination on consumers' plates. This process encompasses various stages, including planting, cultivation, harvesting, processing, and distribution. As global food systems become increasingly complex, the importance of traceability has gained significant attention for several reasons.

First and foremost, traceability enhances food safety. In the event of a foodborne illness outbreak or contamination, traceability systems allow for rapid identification of the source, enabling swift action to mitigate risks and protect public health. For example, if a specific batch of tomatoes is linked to a food safety issue, traceability allows producers and retailers to trace back through the supply chain, identify affected farms, and remove contaminated products from shelves efficiently.

Additionally, traceability supports transparency and consumer confidence. With the growing demand for sustainably produced and ethically sourced food, consumers are increasingly interested in knowing the origins of their food. Traceability systems provide essential information about agricultural practices, such as whether crops were grown using organic methods, and whether farmers adhered to environmental and social standards. This transparency can lead to more informed purchasing decisions, fostering loyalty among consumers who prioritize sustainability.

Moreover, traceability can enhance supply chain efficiency. By maintaining detailed records of production and distribution, stakeholders can optimize logistics, reduce waste, and improve resource management. This efficiency ultimately contributes to the sustainability of crop production, benefiting farmers, consumers, and the environment alike.

Traceability in crop production plays a critical role in ensuring food safety, enhancing consumer confidence, and improving supply chain efficiency. As the agricultural sector continues to evolve, the implementation of effective traceability systems will be vital for addressing the challenges of modern food production.

## 4.1. TRACEABILITY METHODS FOR PLANT PRODUCTS

The term "traceability" in plant products describes the capacity to follow and identify agricultural products from their point of origin to the final user. Ensuring food safety, quality, and sustainability requires this process. Reputable traceability programs can confirm organic and sustainable practices, help find the origin of foodborne illnesses, and build consumer confidence. This section examines the benefits, drawbacks, and uses of several plant product traceability strategies for the agricultural supply chain.

### 4.1.1. Barcoding and QR Codes

Barcoding and QR codes (Quick Response) are two of the most popular traceability techniques. These codes, which can be scanned at different stages of the supply chain, contain unique product identifiers.

- How It Works:
  - **Barcodes:** Traditional barcodes consist of vertical lines that encode information about the product, including the producer, product type, and batch number. When scanned, this information can be accessed and updated in a central database.
  - **QR Codes:** QR codes can store more information than standard barcodes and can be scanned by smartphones. This allows consumers to access detailed product information, including its origin, production methods, and nutritional data.
- **Advantages:**
  - **Efficiency:** Scanning barcodes and QR codes is quick and easy, making them suitable for high-volume environments.
  - **Consumer Engagement:** QR codes enable direct communication with consumers, allowing them to access detailed traceability information.
- **Challenges:**
  - **Infrastructure:** Effective implementation requires the necessary hardware and software infrastructure.
  - **Standardization:** Lack of standardization across different regions and systems can create discrepancies in traceability.

### 4.1.2. Electronic Data Interchange (EDI)

The electronic interchange of commercial documents between trading partners is a component of EDI. When it comes to traceability, EDI makes it easier to share information about how products are moving through the supply chain.

- **How It Works:**
  - EDI allows producers, processors, distributors, and retailers to share critical information, such as order details, shipping notices, and invoices, in real-time.

  - Data shared through EDI can include information about production practices, harvest dates, and batch numbers.
- **Advantages:**
  - **Real-Time Data Sharing:** EDI provides real-time updates, improving response times to issues such as recalls.
  - **Streamlined Processes:** Reduces paperwork and manual data entry, enhancing efficiency.
- **Challenges:**
  - **Implementation Costs:** Initial setup and ongoing maintenance can be expensive.
  - **Complexity:** Smaller producers may struggle with the technical requirements of EDI systems.

### 4.1.3. Blockchain Technology

Blockchain is a technology for decentralized digital ledgers that keeps track of transactions on several computers. An unchangeable chain of data is produced by connecting each transaction to the one before it.

- **How It Works:**
  - Every transaction involving a plant product (from planting to harvest to distribution) is recorded on the blockchain, creating a transparent and verifiable history.
  - Each participant in the supply chain has access to the same data, which ensures accuracy and reduces fraud.
- **Advantages:**
  - **Transparency:** All stakeholders have access to the same information, enhancing trust and accountability.
  - **Security:** The decentralized nature of blockchain makes it resistant to tampering.
- **Challenges:**
  - **Technical Expertise:** Implementing blockchain technology requires significant technical knowledge and resources.
  - **Scalability:** The technology can be challenging to scale, especially in complex supply chains.

### 4.1.4. GPS and Geographic Information Systems (GIS)

Agricultural products can be tracked according to their geographic location thanks to GPS and GIS technologies. This approach is especially helpful for tracking the flow of goods from the farm to the marketplace.

- **How It Works:**
  - **GPS:** Global Positioning System (GPS) devices can track the location of vehicles transporting plant products, providing real-time data on their movement.
  - **GIS:** Geographic Information Systems analyze spatial data, allowing producers and retailers to visualize the journey of products and identify potential risks in the supply chain.
- **Advantages:**
  - **Enhanced Monitoring:** GPS provides real-time tracking, improving logistics and inventory management.
  - **Risk Assessment:** GIS can identify potential risks based on geographic factors, such as weather patterns or regional outbreaks of pests.
- **Challenges:**
  - **Cost:** Implementing GPS and GIS systems can be expensive for small-scale producers.
  - **Data Management:** Managing large volumes of spatial data requires specialized skills and resources.

### 4.1.5. Supply Chain Management Software

Software for supply chain management (SCM) unifies the production, distribution, and other elements of the supply chain. Traceability features that let users follow products through each step are frequently included in these systems.

- **How It Works:**
  - SCM software can capture data from multiple sources, including EDI, barcodes, and GPS, creating a comprehensive view of product movement.
  - Users can access real-time data on inventory levels, product origins, and shipping statuses.
- **Advantages:**
  - **Centralized Information:** SCM software provides a centralized platform for managing traceability data, and improving communication among stakeholders.
  - **Improved Decision-Making:** Access to real-time data enhances decision-making related to inventory management and risk mitigation.

- **Challenges:**
  - **Complexity:** Implementing and integrating SCM software can be complex and time-consuming.
  - **Cost:** The investment required for robust SCM software may be prohibitive for some producers.

### 4.1.6. Paper-Based Documentation

Even though digital approaches are more widely used, paper-based documentation is still frequently used for traceability, particularly in areas with poor access to technology. This method entails keeping thorough records of the movements of products and agricultural practices.

- **How It Works:**
  - Producers keep logs of planting dates, harvest dates, processing methods, and distribution information in physical formats.
  - Paper-based documentation may include invoices, shipping documents, and inspection records.
- **Advantages:**
  - **Simplicity:** Easier to implement in regions with limited technology infrastructure.
  - **Accessibility:** Does not require internet access or advanced technical skills.
- **Challenges:**
  - **Prone to Errors:** Manual data entry can lead to inaccuracies and inconsistencies.
  - **Limited Accessibility:** Accessing and sharing information can be cumbersome and slow.

### 4.1.7. Certification Programs

By requiring producers to keep records and document their practices, many certification programs—including those for organic, fair trade, and sustainable agriculture—act as traceability tools.

- **How It Works:**
  - Producers must adhere to specific standards and practices to achieve certification. This often includes maintaining detailed records of inputs, production methods, and product movement.
  - Certification bodies conduct audits to verify compliance with standards.

- **Advantages:**
  - **Credibility:** Certifications enhance consumer trust by verifying that products meet specific sustainability and quality standards.
  - **Market Access:** Certified products often have access to premium markets and consumer segments.
- **Challenges:**
  - **Costs:** Achieving and maintaining certification can be expensive, especially for small producers.
  - **Complexity:** Navigating certification processes and maintaining compliance can be time-consuming.

### 4.1.8. Smart Packaging Technologies

Using technology integrated into packaging materials to deliver real-time product information is known as smart packaging. Traceability information, freshness indicators, and temperature monitoring are a few examples of this.

- **How It Works:**
  - **RFID (Radio-Frequency Identification):** RFID tags can be embedded in packaging to track products throughout the supply chain. These tags provide real-time data on product location and condition.
  - **Smart Labels:** Labels that change color or provide alerts can indicate freshness and product quality, ensuring consumers receive safe and high-quality products.
- **Advantages:**
  - **Real-Time Monitoring:** Smart packaging allows for continuous monitoring of product conditions, improving quality control.
  - **Consumer Engagement:** Interactive packaging can provide consumers with product information, enhancing transparency.
- **Challenges:**
  - **Cost:** Implementing smart packaging technologies can be expensive for producers.
  - **Consumer Acceptance:** Some consumers may be hesitant to engage with new technologies.

Plant product traceability techniques are crucial for maintaining food safety, boosting consumer confidence, and encouraging environmentally friendly farming methods. Every technique has specific benefits and drawbacks, ranging from GPS tracking and certification programs to blockchain technology and barcoding. Addressing concerns about food safety, quality, and sustainability will require the implementation of strong traceability systems as global food systems grow more intricate. Enhancing traceability in plant production will ultimately benefit producers, consumers, and the environment, and this will be made possible by ongoing technological advancements and stakeholder collaboration.

## 4.2. PESTICIDE USE INDICATORS

The pounds of active ingredient applied on an acre of a particular crop in a given year, or on all acres of a crop, or on all crops in a geographic region (a farm, county or state, nation, continent, or the world) is the most widely used indicator of "pesticide use.."

Use of pesticides in the U.S. is typically reported at the state or federal level, e. g. on all corn acres in Iowa or all corn acres nationwide. Two U.S. sources provide data on pesticide use down to the county level. S. data from the California Department of Food and Agriculture, as well as data from the U.S. government that is countrywide. S. preliminary estimates for 2013–14 and the Geological Survey for 1992–2012 (Figure 4.1).

**Figure 4.1.** *Pesticide use.*

*Source: By Roy Bateman at the English Wikipedia, CC BY-SA 3.0, https://commons.wikimedia.org/w/index.php?curid=20030511.*

Instead of pounds or kilograms of formulated products, data are reported globally in pounds or kilograms of active ingredient applied. Almost no completely pure form of pesticide active ingredient is used.

The active ingredient is mixed with adjuvants and surfactants to help keep it in the right state—solid, liquid, or gas—to ensure that it adheres to plant leaves even in the event of rain, to speed up the active ingredient's penetration into plant tissues, or to shield it from the natural degradation processes that are triggered by air, sunlight, and rain.

In prepared, ready-to-use pesticide products, the active ingredient concentration usually varies from a few percent to up to 75%, and on occasion even higher. Brand-name, concentrated products are frequently sprayed after being further combined with liquid fertilizers or water. The amount of spray material applied per acre is much less than the pounds of pesticide active ingredient used in such typical application scenarios.

## 4.2.1. Rate of Application Metrics

The first three important metrics of pesticide use are related to rates of application:

- Average "One-time Rate of Application" of active ingredient on an acre/hectare of a given crop;
- Average "Number of Applications" made with a specific active ingredient on a crop in a production cycle (usually a calendar year); and
- "Rate per Crop Year," which is the average rate of application per production cycle, is calculated as (the average one-time rate of application) multiplied by (the average number of applications).

Understanding changes in the pounds or kilograms of pesticide active ingredient applied requires an understanding of trends in the metric "Rate per Crop Year" across commonly used pesticides. This is because the pesticide industry has been concentrating on finding novel modes of action that are highly specific to target pests and efficient at low dose rates since the early 1980s. These active ingredients usually target particular processes in the target pests, such as metabolism, biochemistry, physiology, development, or reproduction.

The industry has been successful for the most part in this quest. One common outcome has been the commercialization of new active ingredients that are effective even when applied at low, or even exceptionally low rates per acre/hectare. Hence, it is useful to track over time two other metrics that are related to "Rate per Crop Year" – Number of a.i.'s Applied at < 0.1 pound/acre; and, Percent of acres treated with an a.i. applied at < 0.1 pound/acre. It is important to note that moderate, low, or very low-dose active ingredients have supplanted the majority of the commonly used, high-dose pesticides that farmers applied in the 1980s. Pounds applied, whether by crop, per acre, or for all crops, have decreased as a result. Sadly, though, this decrease does not imply that farmers are now less dependent on pesticides.

## 4.2.2. Measures of the Extent of Use

Regardless of the rate at which a pesticide is applied on a particular field, the effects of pesticides on the environment and human health are unquestionably influenced by how widely and frequently they are applied. As a result, it's crucial to consider how frequently and widely a pesticide is applied on a particular farm, to a particular crop, or to all the crops in an area.

Four additional, key metrics are needed to track the frequency and extent of use of a given pesticide:

- **"Percent Acres Treated"** is the percent of the acres of a given crop in a region that is treated with a given pesticide;
- **"Number of Acres Treated"** is the number of acres of the crop treated one or more times with a given pesticide, calculated as the (percent acres treated) multiplied by (total crop acreage);
- **"Acre-Treatments"** takes into account both the percent of acres treated and the average number of applications with each active ingredient applied, and is calculated as (acres treated) multiplied by (number of applications) made with a pesticide on the given crop; and
- **"Pounds Applied"** is the total pounds of active ingredient applied per production cycle on a given crop, or across all crops or sets of related crops (i.e., fruits or small grains).

A "Minimum Data Set" (MDS) for monitoring the application of a particular pesticide on a particular crop is represented by these seven metrics. A pesticide used on a field, all acres of a particular crop grown on a farm, or all acres of a crop grown on all farms in a county, state, province, region, country, continent, or the entire world can all be covered by this MDS.

**Figure 4.2.** *Pesticides are needed to bring a crop to harvest.*

*Source: By University of Texas at Arlington Photograph Collection - UTA Libraries Digital Gallery, CC BY 4.0, https://commons.wikimedia.org/w/index.php?curid=88996846.*

However, the total amount of pesticides used on a field, farm, and throughout all farms determines how much of an impact pesticide use has on farm production costs, the environment, the emergence and spread of resistant pests, and public health. To account for the number of different pesticides required to bring a crop to harvest in a particular production cycle, more pesticide use metrics are therefore required (Figure 4.2).

## 4.2.3. Limits of "Pounds Applied"

Although information on the pounds of active ingredient in pesticides is crucial for monitoring pesticide use, expenses, hazards, and advantages, "pounds applied" on its own may not always be a reliable indicator. This is due to the widespread belief that greater or lower application weight equates to a higher or lower risk of negative effects on the environment and public health. However, this is frequently untrue.

A few herbicides, such as pendamethalin, s-metalochlor, atrazine, and acetochlor, are applied at rates of less than 0.01, while others are applied at rates of approximately 1.0 pounds of active ingredient per acre. Most herbicides available today are applied at a rate of between 0.1 and 0.5 pounds per acre.

This also applies to insecticides. Approximately 1.0 pounds of many older insecticides, such as DDT, carbofuran, oxamyl, and endosulfan, were applied per acre. Newer insecticides with unique modes of action, as well as low-dose synthetic pyrethroids and neonicotinoids, have replaced the majority of relatively high-dose insecticides.

The main drawback of using "pounds applied" as a sole measure of pesticide use is that it ignores biological activity or potency in terms of changing the dynamics of pest populations. The term "activity" is frequently used to describe a pesticide's "potency," and highly active pesticides are used at lower concentrations than the majority of other pesticides that are authorized for use against the same crop pests.

The fact that some pesticide products are only applied once per season while others are applied several times is a second significant restriction on the number of pounds applied per acre of an active ingredient. Additionally, some farmers will divide a typical per-acre application rate into two, half-dose applications due to the necessity of re-entering fields to apply different fertilizers and pesticides.

As a result, it's critical to concentrate on pesticide use rates per crop year, accounting for all pesticide applications made on a specific crop field during that producing cycle. In contrast to a calendar year, the emphasis should be on an annual production cycle for crops such as winter wheat, which are sown in the fall of one year and harvested in the early summer or fall of the following year.

### 4.2.4. Measures of Total Use by Type of Pesticide

Data and trends on pesticide use are typically analyzed by major pesticide type and pest categories targeted by pesticides. There are three main categories of pesticides, each with a corresponding type: herbicides for weeds, insecticides for insects, and fungicides for plant diseases. It's common to refer to these three main categories of pesticides as H/I/F.

For each major type of pesticide, there are three essential metrics needed to account for aggregate pesticide use. These can be quantified per acre, on a field, across all fields on a farm, and/or in a county, state, region, nationally or globally:

- Average "Number of Active Ingredients Applied" per acre/hectare to bring a crop to harvest in a given production cycle;
- Average "Number of Acre-treatments" made with an H or I or F on an acre producing a given; and
- "Total Pounds Applied" of H/I/F active ingredient applied on the average acre in a production cycle (sum of pounds applied across all H/I/F active ingredients used on a given crop).

Farmers have used one or two active ingredients, and frequently none at all, to control weeds, insects, and plant diseases on most fields over the past few decades in the majority of years and regions. For example, until very recently, neither fungicides nor insecticides were applied to soybeans, and fungicide treatment of corn was seldom used. However, herbicides are sprayed on almost all of the traditionally managed acreage of both crops every year. On the other hand, the majority of traditionally farmed fruit and vegetable crops have been sprayed with 1–3 herbicides, 2–5 insecticides, and 2–4 fungicides for decades.

Maybe as a result of resistant pests emerging, declining pesticide efficacy in a system is almost invariably correlated with more active ingredients and modes of action being used on a given field. This is the reason it's critical to monitor changes in the quantity of active ingredients used on a particular acre.

However, farmers occasionally react to unfamiliar or more difficult-to-control pests by reapplying an active ingredient that was previously sprayed during the same production cycle. The quantity of different active ingredients applied would not account for this extra application. Metric nine above is required as a result. It considers applying a new active ingredient in addition to numerous applications of a specific pesticide.

USDA data on pesticide use makes it abundantly evident that the intensity of pesticide use has increased far more than some indicators, like the percentage of treated acres and the total number of pounds of active ingredient applied, would have us believe.

## 4.2.5. Measures of Total Pesticide Use

Farmers work hard to monitor and control pests in a way that preserves their profit margins. This calls for minimizing crop yield and/or quality losses and minimizing pest management system costs. The majority of the expenses associated with a pest management system on conventional farms go toward purchasing and using pesticides (Figure 4.3).

**Figure 4.3.** *Track and manage pests in ways of protecting crops.*

*Source: By http://www.cgpgrey.com, CC BY-SA 3.0, https://commons.wikimedia.org/w/index.php?curid=12118665.*

The entire use of pesticides, not just the use of a particular active ingredient, is what causes adverse effects on the environment, including water and soil quality, birds, and fish, earthworms, biodiversity, and non-target organisms like pollinators or other plants. Similarly, the cumulative effects of exposure to particular active ingredients and the combined, occasionally synergistic, effects of exposure to several pesticides over a period of time constitute the public health impacts of pesticide use. Therefore, total pesticide use across all H/I/F applied must also be included in a pesticide use minimum data set. The total amount of pesticide active ingredient applied must also include any additional pesticide categories that may be used, such as soil fumigants, plant growth regulators, or desiccants.Metrics that quantify the total amount of pesticides used are essentially aggregate versions of the 10 metrics listed above, adding the total amount of pesticides used. These metrics can be expressed per acre at the farm level, for all crops, across all crops of a particular type (vegetables, fruits, grains), or across all acreage of a crop grown in a region.

## 4.2.6. Measuring Other Impacts Associated With Pesticide Use

Tracking the following four significant side effects of pesticide use is necessary: potency in terms of impacting target pests, diversity in modes of action (a crucial metric associated with resilience and resistance prevention), risks to the environment and public health, and economic effects. The following is a brief introduction to each, which will be covered in subsequent sections.

### *4.2.6.1. Economics*

The economics of pesticide use has a simple cost side. By simply adding the cost of active ingredients per pound into an analysis and multiplying the number of pounds applied to a particular active ingredient by its cost per pound, the 10 indicators mentioned above can be transformed into cost estimates. It is also necessary to include the typical application costs per acre-treatment.

The economic performance of pest management systems and, consequently, the return on pesticide expenditures are also impacted by the duration and quality of control. Determining whether and how much crop yields or crop quality have declined, then converting that decline into lost farm receipts, is the process of quantifying pesticide efficacy and the "quality" of control. Profits are decreased by lower receipts, which in certain cases can be so significant as to result in overall economic losses per acre.

### *4.2.6.2. Quantifying Risks*

Projecting or monitoring the environmental and public health impacts of pesticide use requires risk assessments that take into account:

- When, how often, and how much pesticide is applied;
- The formulation of the pesticide applied and how it was applied;
- The persistence and environmental fate of applied pesticides;
- Levels reached in non-target environments (water bodies, the soil, animal, and human food); and
- Exposure levels and timing of exposures impacting non-target organisms.

### *4.2.6.3. Potency*

Monitoring changes in the total "kill" power required in a pest management system to bring a crop to harvest has received far too little attention. In this sense, "kill" refers not only to lethal methods of control but also to strategies that lower pest populations by obstructing reproduction, discouraging feeding, or removing vital habitat or food sources.

It goes without saying that the best pest management system uses an integrated set of preventive measures, commonly known as Integrated Pest Management, or IPM, to control pests and needs little to no outside assistance. For instance, most land under organic management is not sprayed with pesticides. A portion of small grain crops and conventional hay are also not treated with pesticides in the majority of years and areas.

When farmers must use stronger herbicides, insecticides, or fungicides overall during a crop production cycle, it indicates that pest infestations are becoming harder to control and therefore more likely to cause financial losses as well as hazards to the environment and public health. The three most common causes of slipping pesticide efficacy are:

- The emergence and/or spread of less sensitive, or fully resistant pest phenotypes;
- A reduction in the health of the crop because of changes in the genetics of the seed, changes in soil health, or increases in seeding rates; and
- A new, exotic pest becomes established that was not previously in a crop production region.

### *4.2.6.4. Diversity in Modes of Action*

Relying primarily on one or a small number of pesticides with a common mechanism of action is the most effective way to encourage resistance in target pest populations. Applying active ingredients in secession, or together, that have multiple modes of action, is the best way to prevent resistance for farmers who have decided to rely primarily on pesticides in their IPM systems.

Spreading out the burden of control over multiple families of pesticide chemistry is a sensible strategy that will help avoid major impacts on vulnerable, non-target organisms since the majority of the environmental and human health risks associated with pesticide use are directly linked to how widely and how often a particular pesticide is used. It's a strategy to disperse the risk of negative effects so that, in the unlikely event that a single pesticide application proves harmful in some way, collateral damage is contained and can be more easily avoided going forward.

These are the main arguments for why it is preferable to use a variety of pesticides and why it is crucial to measure the homogeneity of pesticide use overall when it comes to a specific kind of pest. As with crops genetically modified to withstand glyphosate applications, problems are usually lurking around the corner when farmers rely solely on one method of controlling a class of pests.

## 4.3. TRACKING FERTILIZERS

To increase crop productivity and guarantee food security, fertilizers are necessary. On the other hand, misuse of them can result in water body contamination, soil health problems, and environmental degradation. Thus, in contemporary agriculture, monitoring fertilizer use has taken on greater significance.

There are many ways to track fertilizers in crop production, including:

- **Software:** Farmers can monitor the total amount of nutrients applied as well as the amount of fertilizer applied to each crop and field using software such as AGRIVI.
- **Plant Tissue Testing:** The elements, such as potassium, phosphorus, and nitrogen, that are present in plant tissue are measured in this analysis.
- **Soil Testing:** This analysis examines the texture, organic matter, and nutrient levels in the soil.
- **Irrigation Water Testing:** This analysis measures the pH, electrical conductivity, nitrate levels, phosphorus, chlorine, and sulfates in the irrigation water.
- **Satellite Imagery:** Satellite imagery analytics can help farmers track vegetation development, identify issues, and apply fertilizers precisely.
- **Optical Biosensors:** These sensors use fluorescent dye or molecules to check how much nutrients crops and soil are taking up.
- **Flame Atomic Absorption Spectroscopy:** This method measures how much of the required nutrients are present in the fertilizer.
- **GPS and GIS:** Farmers can use GPS to track the location of farmed equipment and soil samples, and then integrate that data with GIS mapping.
- **Drones:** Farmers can use drones to collect data and high-resolution photos to assess field conditions and monitor crop health.
- **Mobile Apps:** Farmers can use mobile apps to remotely monitor and manage their fields.

### 4.3.1. Increase Productivity by Simple Fertilization Tracking

Every crop production process starts with soil tillage because only well-prepared soil can create the ideal environment for crop growth. Still, other factors must also be considered to achieve successful crop production, including soil tillage. There is something far more significant that, when combined with soil, air, and water, will result in healthy crops and a bountiful yield. These are nutrients for plants.

#### *4.3.1.1. Food for Plant Life*

Plant nutrients, which supply every element required for growth, are the foundation of plant life. They take the shape of fertilizers, either liquid or solid (Figure 4.4).

**Figure 4.4.** *Applying plant nutrients through fertilizers will provide high and quality yield.*

*Source: By Alandmanson - Own work, CC BY-SA 4.0, https://commons.wikimedia.org/w/index.php?curid=49622141.*

A farmer must supply plants with nutrients throughout the crop production cycle to guarantee a good and high-quality yield. A planned yield can only be produced by well-nourished plants, and more significantly, they will be more resilient to diseases, insect pests, and adverse weather and soil conditions.

#### *4.3.1.2. Fertilization Plan Is a Smart Start for Achieving Higher Productivity*

While discussing fertilizers, it's crucial to maintain a balanced approach. It's true that plants benefit from fertilizers, yet they require them in a precise quantity. Excessive use of fertilizers does not guarantee increased yields. Quite the opposite, it could lead to significantly lower yields. To prevent such outcomes, farmers must administer fertilizers according to a predetermined amount, which varies based on the specific crop and any prior fertilizers added (Figure 4.5).

**Figure 4.5.** *The right amount of fertilizers given at the right time will provide abundant and quality crop yield.*

*Source: By SuSanA Secretariat - https://www.flickr.com/photos/gtzecosan/3617543234/, CC BY 2.0, https://commons.wikimedia.org/w/index.php?curid=36540726.*

Planning all fertilization activities is crucial before the vegetation season starts. For good reason, a farmer ought to take that action. A farmer can more easily budget for this expense when they have a plan in place that specifies precisely when and how much fertilizer to apply to a given crop.

Numerous fertilization processes take place throughout the crop production cycle. A farmer can prevent a delay in fertilization application, which can significantly affect the final yield, by planning each one in advance and preparing all stocks on time. Task planning enables a farmer to get ready on time, whether he rents farm equipment like a tractor or fertilizer spreader or perhaps hires more workers for this task. In this manner, he can oversee everything as planned.

### *4.3.1.3. Fertilization Dashboard for Smart Maintenance of High Productivity*

Fertilizers enrich the soil with a variety of nutrients, creating an ideal environment for plant growth. As a result, it is very difficult to monitor the entire amount of soil reserve nutrients because of the multiple fertilizations that occur during the vegetation season. When creating a fertilization plan for the upcoming season, this is taken into consideration.

## 4.4. TRACKING CROP TREATMENTS

To achieve maximum agricultural productivity while maintaining environmental sustainability and food safety, crop treatments must be managed effectively. For many reasons, it is essential to monitor crop treatments, such as the application of fertilizers, growth regulators, pesticides, and other agrochemicals. This section addresses the significance of monitoring crop treatments, the different approaches used to do so, the difficulties encountered, and possible ways to improve monitoring procedures.

### 4.4.1. Importance of Tracking Crop Treatments

- **Food Safety and Consumer Health:** Monitoring crop treatments helps ensure that the products reaching consumers are safe and meet regulatory standards for pesticide residues. This is critical for maintaining public health and consumer trust.
- **Environmental Protection:** Proper tracking of crop treatments minimizes the risk of chemical runoff into water bodies and helps protect biodiversity. By tracking applications, farmers can optimize their use of inputs and reduce negative environmental impacts.
- **Integrated Pest Management (IPM):** Tracking crop treatments is a fundamental component of IPM, which promotes the use of various strategies to manage pests sustainably. Accurate records enable farmers to assess the effectiveness of treatments and make data-driven decisions.
- **Regulatory Compliance:** Many countries have regulations governing the use of agrochemicals. Keeping detailed records of crop treatments helps farmers comply with these regulations and avoid penalties.
- **Data for Research and Development:** Collecting data on crop treatments provides valuable insights for agricultural research. Understanding treatment efficacy and crop response can inform future practices and innovations in crop management.

### 4.4.2. Methods for Tracking Crop Treatments

- **Manual Record Keeping:**
  - Traditional record-keeping involves farmers maintaining written logs of their crop treatments, including product details, application rates, dates, and locations.
  - **Advantages:** Simple to implement and cost-effective.
  - **Challenges:** Prone to human error and may lack consistency over time.

- **Precision Agriculture:**
  - Precision agriculture utilizes technology such as GPS, sensors, and data analytics to monitor and manage crop treatments more effectively.
  - **Advantages:** Provides real-time data, enabling precise application and reducing waste.
  - **Challenges:** Requires significant investment in technology and training.
- **Mobile Applications and Software:**
  - Various mobile apps and software platforms are available for farmers to track crop treatments. These tools often feature user-friendly interfaces, reminders, and reporting capabilities.
  - **Advantages:** Easily accessible and allows for real-time updates.
  - **Challenges:** Accessibility issues, especially in rural areas with limited internet connectivity.
- **Geographic Information Systems (GIS):**
  - GIS technology can map and analyze spatial data related to crop treatments, providing farmers with visual insights into application patterns across fields.
  - **Advantages:** Facilitates informed decision-making and resource management.
  - **Challenges:** Complexity in operation and the need for specialized training.
- **Remote Sensing Technologies:**
  - Remote sensing techniques, such as satellite imagery and drones, can assess crop health and monitor the application of treatments.
  - **Advantages:** Allows for large-scale monitoring without disrupting farming operations.
  - **Challenges:** High initial costs and the need for expertise to interpret data.
- **Blockchain Technology:**
  - Blockchain provides a secure and transparent way to track crop treatments throughout the supply chain.
  - **Advantages:** Enhances traceability and accountability, fostering consumer confidence in agricultural products.

- **Challenges:** Complexity of implementation and the need for widespread adoption.

## 4.5. EMERGING CROP TRACEABILITY SYSTEMS IN SMART FARMING

The agricultural sector has grown rapidly in recent years and is expected to keep changing, particularly in this IR4.0 era. According to projections, the world's food production needs will rise by 70% by 2050 to feed one billion people (Kc et al., 2018). Food security will suffer if traditional farming practices are still widely employed (Hamadani et al., 2021). Malnutrition, food shortages, and price inflation will result from an inability to balance the supply and demand for food. Farmers and agribusiness owners must therefore take the necessary steps to become more productive and efficient to meet the constantly rising demand for food. As a result, smart farming is starting to replace conventional farming. By collecting data and analyzing it with ICT-based technology, smart farming revolutionized traditional farming and increased farm operations' productivity, sustainability, and efficiency. However, the quality of crops and agri-products should never be compromised because smart farming can maximize crop quantity. As a result, there is a growing need for a way to monitor crop quality and food safety throughout the supply chain. A crop traceability system is therefore essential to any smart farm. Crop traceability, by definition, is the process of tracking crops from the farms to the marketplace to ensure food safety throughout the supply chain. The following factors make crop traceability crucial for farmers and agribusiness owners: (i) guarantees the quality of agricultural products and attains disease-free crop production (Mykhailenko et al., 2020), (ii) permits export to numerous nations by adhering to regulations and guidelines (Miatton and Amado, 2020) and (iii), which distinguish genuine agricultural products from fakes, also emphasized how the traceability system enables farmers to communicate with final consumers directly, without the involvement of a third party. This will increase consumer confidence in the product and fortify its brand. As a result, this chapter examines the crop traceability systems that are currently in use in smart farming and talks about potential ways to enhance them from an ICT standpoint.

### 4.5.1. Agriculture Farming

#### *4.5.1.1. Farming: Conventional Versus Smart Farming*

Since ancient times, conventional farming has been used to produce crops and food. As a result, this farming method primarily relies on the expertise and experience of the farmers to handle pest control, fertilization, and crop growth requirements. Since agriculture is one of the most significant industries in Malaysia and the world at large, conventional farming is starting to give way to smart farming, which is a longer-term, more sustainable method. El Chami et al. (2020) claim that an agricultural practice must fulfill these requirements to be considered sustainable.

(i) Fulfill the demand of people through its production; (ii) It is economically viable; and (iii) Preserve the environment in the long run. Table 4.1 compared the characteristics of conventional and smart farming to highlight how smart farming can revolutionize the agriculture sector.

## 4.5.2. Roles of ICT in Smart Farming

### *4.5.2.1. Automate Tasks*

Automating tasks that are frequently completed by hand in traditional farming methods is one of the key functions of ICT in smart farming. Drones and autonomous robots, for instance, can now automate tasks like applying fertilizer to crops and watering them. A study conducted by Wongpatikaseree et al. (2018) stated that smart farms equipped with automatic fogging or dosing systems guarantee that ideal farming conditions can be automatically maintained, minimizing human labor that is more prone to mistakes.

**Table 4.1.** *The Main Parameters Measured by IoT Systems Across the Agri-Food Supply Chain*

| Supply Chain Part | Parameter Type | Measured Parameter |
|---|---|---|
| Farm | Weather | Ambient temperature<br>Relative humidity<br>Atmospheric pressure<br>Wind velocity<br>Precipitation |
| | Soil | Soil temperature<br>Moisture content<br>Electrical conductivity |
| | Environment | Greenhouse gas emissions |
| | Location | Tractor's location tracking |
| Transportation and storage | Transportation and storage environment conditions | Ambient temperature<br>Relative humidity<br>$CO_2$ concentration |

**Table 4.2.** *Comparison Between Conventional Farming and Smart Farming*

| Category | Conventional Farming | Smart Farming | References |
|---|---|---|---|
| **Definition** | Farming that practices uniform resource inputs such as fertilizers, pesticides, and excessive irrigation without much consideration of the conditions between and within the fields. It commonly focuses on frequent, single-crop planting to grow food. | Farming focuses on capturing data and interpreting them using ICT technologies to make farm operations more predictable and efficient. | [K. Wongpatikaseree,et al. 2018, D. L. Corwin and E. Scudiero 2019, Q. A. Panhwa et al. 2019] |
| **Aim** | - Growing food as economically as possible<br>- Ensuring that food is affordable for all<br>- Allowing farmers to make a living through minimum earnings | - Reducing the need for human resources and efforts<br>- Making maximum utilization of the available resources<br>- Improving monitoring capabilities and farming efficiency<br>- Ensuring maximum profitability through increased quantity and quality of agricultural production<br>- Allowing farmers to obtain maximum earnings | [Q. A. Panhwa et al. 2019, F. Tian et al. 2016] |
| **Emphasis** | Localization, biodiversity, shared genetic resources, and cultural appreciation for many different crops | Emphasis on production, capital gain, input intensity, and crop consistencies | [C. Elliott 2011] |

*Source: Tagarakis AC, Benos L, Kateris D, Tsotsolas N, Bochtis D. Bridging the Gaps in Traceability Systems for Fresh Produce Supply Chains: Overview and Development of an Integrated IoT-Based System. Applied Sciences. 2021; 11(16):7596.* https://doi.org/10.3390/app11167596.

#### *4.5.2.2. Allow for Real-Time Monitoring*

Data collection in traditional farming is frequently done by hand. For instance, farmers document crop production data and environmental parameters on paper records, which agribusiness employees must subsequently enter into spreadsheet software (Wongpatikaseree et al., 2018). This manual procedure limits data analysis and increases data inaccuracy in addition to delaying data summaries. In the meantime, real-time data collection is made possible by ICT in smart farming by means of IoT sensors (Prathibha et al., 2017). Therefore, with all the information from environmental conditions parameters (such as soil pH, fertilizer, temperature, humidity, and water irrigation) available. Farming conditions can be swiftly modified in real-time to ensure the quantity and quality of their crop production (Othman et al., 2021).

#### *4.5.2.3. Curate Meaningful Data Visualizations*

Images and data cannot be integrated into one data collection method in conventional farming. However, to convert data into meaningful information, these data form variations are crucial. Thus, curating insightful data visualizations is another way that ICT is used in smart farming (Kuppusamy et al., 2021) using the information gathered. This can be accomplished using a variety of platforms for data visualization, including dashboards, mobile apps, and online platforms. These can show users relevant data visuals based on their needs and support their decision-making (G. D. Tikas and K. B. Akhilesh, 2020). For instance, a smart farm dashboard can offer data visualizations to help farmers monitor the conditions of their smart farms, let agribusiness owners monitor the entire crop production from several farms, and let customers learn more about an agriproduct by simply scanning its QR code.

#### *4.5.2.4. Enables Crop Data Traceability*

The potential for crop traceability features is one of the benefits of smart farming. Once a crop is on the market, it can be challenging to track down its information in conventional farming.

However, as data is collected from planting, harvesting, processing, distribution, and delivery, crop traceability is receiving increased attention in smart farming (M. Hate et al., 2018). The smart farming system incorporates many ICT-based technologies, including RFID, QR codes, artificial intelligence, and the Internet of Things, to enhance crop traceability. Crop traceability is significant primarily because it can be used as a tool for brand awareness, export compliance, and quality control.

## 4.6. CROP TRACEABILITY IN SMART FARMING

Traceability is, by definition, the capacity to obtain information about a registered identification (I) about its history, use, location, or activities. Tzoulis et al., 2013). Thus, crop traceability refers to tracking crops as they move from the farm to the market, via wholesalers and distributors, and finally to final consumers. As stated by Yang et al. (2018), product identification, data to be traced, product touting, and traceability tools are the four pillars of an agri-product traceability system. Crop traceability is crucial for guaranteeing product quality and safety as well as streamlining the supply chain, all of which have an indirect impact on labor and resource consumption. Because of this, the perfect traceability system should include the following: (i) the full history of a specific crop, from seed to finished product on the shelf; (ii) environmental data for every farm, from planting to harvesting; and (iii) details on agricultural treatments and inputs (e.g., seedling, growth, fertilization, pesticide application, harvesting, and (iv) digital records for every step of the supply chain process, including delivery and distribution.

According to González-Amarillo et al. (2018), traceability can be either internal or external. While external traceability monitors the crop from harvest, agri-product production, and market distribution, internal traceability tracks the crop from the seedling stage to harvest to ensure its quality and safety. An outline of the crop traceability system in smart farming can be found in the next subsection.

### 4.6.1. Crop Traceability Systems in Smart Farming

Given the significance of crop traceability in the agricultural sector, it's critical to comprehend the extent to which smart farming systems have advanced. The crop traceability systems currently used in smart farming are listed in Table 4.3, while An overview of the technologies used in the crop traceability system in smart farming is shown in Figure 4.6.

Generally speaking, the majority of crop traceability systems combine several different technologies. But the majority of the systems are frequently centralized. Centralized systems are often opaque, asymmetric, and monopolistic (Tian, 2017). This prompted a more thorough investigation into the possibilities of blockchain-based crop traceability systems to shift the environment in favor of a decentralized system to address the problems of data validity and accuracy. However, a deeper and more comprehensive exploration of blockchain technology is required before it can be fully integrated into smart farms.

Concerns regarding food safety and the caliber of agri-products that reach the market have grown as a result of the constant recalls of agricultural products because of safety and sanitation problems. This problem will weaken consumer confidence in the food markets if it is not resolved (Mishra et al., 2020). Furthermore, the needs and demands of the food market are not adequately satisfied due to the expanding global population. As a result, a trustworthy traceability system is now required. The process of following a specific crop through its whole life cycle, from planting to harvest, is known as crop supply chain tracing. In smart farming, crop traceability is an important quality management tool. Smart farm owners can efficiently stop the circulation of unqualified agricultural products before they reach the market. Unqualified agricultural products can be easily identified in any part of the supply chain. A crop traceability system can be used to quickly recall unqualified products and identify the causative agent in the event that a safety or sanitation issue arises after the agriproducts are put on the market. According to consumers, the crop traceability system makes it simple for them to track down information about the quality of agriproducts (Wang et al., 2018).

As stated by González-Amarillo et al. (2018), given the enormous advantages it offers, a traceability system is highly likely to be implemented, particularly in smart farming. The ability to predict a product's post-harvest shelf life is one advantage. Variables related to farm conditions, such as temperature, humidity, water irrigation use, crop growth patterns, seedling behavior, and crop harvest schedule, can be precisely monitored from the system. This translates into the highest yield and quality of smart farm agriproducts.

**Table 4.3.** *Comparison of the Available IoT-Based Traceability Systems on the Market with the Proposed Solution (AgroTRACE)*

| IoT System | GS1 Standard | Supply Chain Range | Minimum Reference Level | Data Transfer | Modular Structure | Waste Treatment | Open Access |
|---|---|---|---|---|---|---|---|
| GR-LIVE | x | Field-Retailer | Product | RFID [1] | √ | x | x |
| iApp | x | Packaging plant-Retailer | Container | RFID, Satellite, BLE [2] | √ | x | x |
| AutoSense | x | Packaging plant-Retailer | Container | RFID, GSM [3], NFC [4], | √ | x | x |
| Harvest Mark | x | Packaging plant-Consumer | Product | RFID | x | x | x |

| **IoT System** | **GS1 Standard** | **Supply Chain Range** | **Minimum Reference Level** | **Data Transfer** | **Modular Structure** | **Waste Treatment** | **Open Access** |
|---|---|---|---|---|---|---|---|
| Food-LogiQ's | √ | Field-Retailer | Product | x | x | x | x |
| iris | √ | Packaging plant-Consumer | Product | x | √ | x | x |
| Agro-Trace | √ | Field-Consum-er-Field | Product | RFID, Lo-RaWAN, BLE | √ | √ | √ |

[1] RFID: Radio-Frequency IDentification; [2] BLE: Bluetooth Low Energy; [3] GSM: Global System for Mobile communications; [4] NFC: Near Field Communication.

**Table 4.4.** *Comparison of Technologies in Supply Chain*

| **Technology** | **Cost** | **Sector of Use** | **Efficiency** | **Features** | |
|---|---|---|---|---|---|
| **Barcode** | Low cost | Fresh products, pesticide administration | Less efficient | 2-dimensional code image, limited data traceability | [M. Hate,et al . 2018, Q. Zhang 2016, T. Nanseki 2005] |
| **Quick Response (QR) Code** | Low cost | Track across the supply chain | Fairly efficient in all sectors | Simple with the use of smart-phone devices and in combination with IoT | [K. Wongpatikaseree et al. 2018] |
| **Radio Frequency Identification (RFID)** | Slightly higher than the barcode/QR code. High investment limits usage to specific processes | Bigger production scale | Fairly efficient in all sectors | Simple and versatile, sensor-based data transmission | [F. Yang et al. 2018, R. Rayhana et al. 2021] |

*Source: Tagarakis AC, Benos L, Kateris D, Tsotsolas N, Bochtis D. Bridging the Gaps in Traceability Systems for Fresh Produce Supply Chains: Overview and Development of an Integrated IoT-Based System. Applied Sciences. 2021; 11(16):7596. https://doi.org/10.3390/app11167596.*

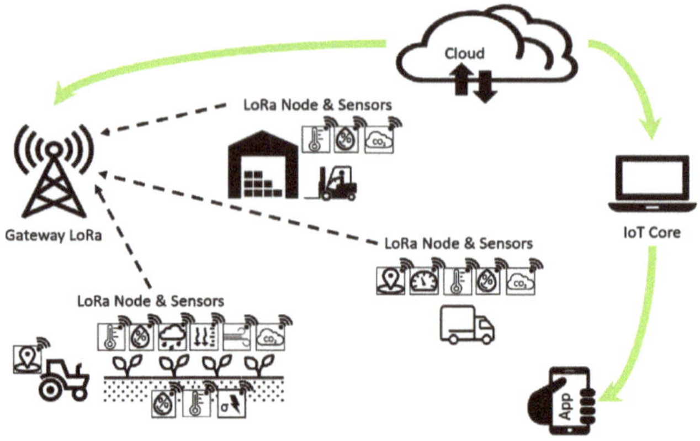


**Figure 4.6.** *Overview of the technologies involved in the crop traceability system in smart farming. Note ADV = advantage, DIS = disadvantage.*

*Source: Tagarakis AC, Benos L, Kateris D, Tsotsolas N, Bochtis D. Bridging the Gaps in Traceability Systems for Fresh Produce Supply Chains: Overview and Development of an Integrated IoT-Based System. Applied Sciences. 2021; 11(16):7596. https://doi.org/10.3390/app11167596.*

In terms of ICT technology, barcodes have been used for more than 30 years in the crop traceability system. It is inexpensive and easy to use. However, barcodes only store a limited amount of information that can be retrieved when the scanner is used. Therefore, data loss and inconsistent data storage may be problems for the barcode-based traceability system. Subsequently, barcode technology is transformed into QR codes. Among their many uses, QR codes are particularly popular for recording data throughout the supply chain. QR codes are frequently used to communicate information to the supply chain's end users because of their adaptability in providing precise information and details.

In addition, RFID technology uses wireless sensors or tags to automatically collect and exchange data, which helps the crop traceability system. Because of its enormous data storage capacity, fast enough data reading speeds, and degree of reusability, RFID can lower production and distribution costs in smart farming. Nevertheless, barcodes are less expensive than RFID tags. As a result, smart farm RFID adopters frequently limit the application of RFID to specific supply chain operations. Due to its ability to collect real-time data about the crop life-cycle and supply chain with less human labor, this technology has become popular in smart farm crop traceability systems as a result of the Internet of Things' rapid development. However, data fragmentation and centralized controls make the current crop traceability practice in smart farming—or agriculture in general—vulnerable to fraud, tampering, falsifying, and altering data. Therefore, it is essential to make sure that the information in crop traceability is reliable, accurate, and valid. As a result, there has been a lot of interest in the study of blockchain technology in crop traceability systems for smart farming.

Sneri et al. (2021) stated that blockchain technology aids in the establishment of a consensus mechanism whereby the participants in the peer-to-peer network maintain a decentralized system by openly sequentially recording data. Because crop-related information is transparently provided, the use of this disruptive technology fosters trust between consumers and agricultural producers. The customer realized that the information is legitimate and cannot be altered without the consent of both the seller and the customer thanks to blockchain technology (Alobid et al., 2022).

Therefore, a crop traceability system built on blockchain technology produces a highly reliable, fraud-proof system with a high degree of data integrity (Casino et al., 2019). To put it briefly, this will help with regulation and adherence to the crop export policy (T. Surasak et al., 2019). The application of smart contracts is another way that blockchain can transform the crop traceability system. With the use of script technology, smart contracts enable the development of reliable applications that complement current blockchain technologies (H. Xiong et al., 2020). For instance, to enable instantaneous transactions devoid of middlemen and to guarantee genuine consensus between farmers and entrepreneurs.

## 4.7. THE FARM-TO-TABLE MOVEMENT

Farm-to-table refers to both an agricultural practice and a food concept. It links customers with wholesome, fresh produce and proteins. Though it's commonly perceived as a direct connection between nearby farms and customers, a variety of suppliers are included in the movement.

Food that is delivered straight from farms to commercial or residential kitchens is the most basic definition of farm-to-table. It isn't provided by conventional channels like wholesale or retail sellers. The phrase isn't regulated, though.

The term "farm-to-table" can apply to food that is obtained locally from either large or small farms. It can also be used on meats and vegetables obtained from community agricultural projects and farmers markets. A restaurant advertises that its menu is built around farm-to-table ingredients, but how can you be certain that the food on your plate was bought straight from the grower? How do you know the vegetables you buy are sourced from farm-to-table operations?

Asking direct questions and listening for direct answers is the easiest way to be certain. Information sharing should be a pleasure for a farmers market vendor. Find out which suppliers are used by the front and back of house staff of restaurants. Farm-to-table is made possible by the connections that exist between growers and sellers. Your questions and comments may be a very helpful component of the procedure.

### 4.7.1. A History of the Farm-To-Table Movement

In the early days of our country, supply, and demand served as the fundamental link between farmers and nearby towns. Usually, there was a farmer nearby who sold meats, eggs, milk, and vegetables.

The population moved into urban areas as cities expanded. Naturally, growers and markets were farther apart. Food was frequently transported by truck and train from farms to markets by the early 1900s. This is when the idea of "farm-to-table" started to take shape. Restoring direct distribution from grower to consumer was the idea. These are some of the historic highlights:

- **1915:** The U.S. Postal Service initiated food shipments from farm to city via parcel post.
- **1940s:** The majority of the country's population settled in urban centers away from farming operations.
- **1950s:** Commercially processed, canned, and frozen foods became widely available.
- **1960s:** Consumers began looking for alternatives to traditional supermarket vegetables and proteins.
- **1971:** Chef Alice Waters opened Chez Panisse, one of California's first restaurants featuring farm-to-table fare.
- **2007:** The National Farm to School Network was established to connect schools with farm-to-table food sources.

### 4.7.2. Why is Farm-To-Table Important?

Once thought to be merely a fad, the farm-to-table idea is now acknowledged as a crucial operational element of agricultural sustainability. Fertile fields and clean water sources are preserved thanks to the production that goes into farm-to-table operations.

Additionally contributing to community and economic sustainability is farm-to-table dining. It improves ties between growers, customers, and the foodservice sector. We are all involved in a process that encourages best growing practices and produces a healthier food supply, from farmers to restaurant employees to home cooks.

### 4.7.3. The Benefits of Farm to Table

There are four principle benefits that anchor the farm-to-table concept which include:

- **Enhanced Food Security:** An improved supply chain between growers and communities;
- **Less Physical Distance:** Reduced environmental impact from food transportation operations;
- **More Self-Reliance:** Increased community ability to source a dependable food supply; and
- **Food System Sustainability:** Best growing practices that foster environmental conservation.

There are other benefits associated with farm-to-table, but these four fundamentals stand out as the most significant. This specialized foodservice model has extra advantages for caterers, large commercial kitchens, and small eateries. These are just a few examples.

- Cooperative and profitable relationships between growers and food service professionals
- Reliable sources for a variety of fresh, quality produce and proteins available from established growers
- Greater flexibility for designing healthy menus in all types of restaurant business models
- Access to seasonal items for planning specialty dishes that bring in customers through promotional LTOs.
- Reduced environmental impact from carbon footprints, especially in farm-to-table transportation
- Increased agricultural sustainability developed through innovations, such as regenerative agriculture.

### 4.7.4. Challenges of the Farm-to-table Process

As we've seen, farm-to-table is now a strong and influential force in the foodservice sector, rather than just a passing fad. However, as the process develops, it encounters unique difficulties.

- **False Authenticity:** Restaurant operations that falsely claim to serve farm-to-table eventually create consumer skepticism.
- **Lack of Regulatory Certification:** There are currently no industry-certified standards that can discourage false claims of farm-to-table authenticity.

- **Seasonal Challenges:** Restaurants constantly adapt menus to incorporate seasonal changes; however large operations can supply most fresh produce year-round.
- **Higher Food Costs**: Small growers in the farm-to-table system deal with higher production costs, and that can impact a restaurant's bottom line.
- **Consumer Misconceptions:** Most consumers assume farm-to-table starts with small operations, but large growers and shippers are also a part of the process.

Growers, suppliers, vendors, and restaurants are constantly coming up with new ways to solve existing issues as farm-to-table operations continue to grow. The industry is steadfast in its commitment to a future where foodservice workers from all backgrounds are included.

## 4.8. LINKING FARM-TO-TABLE FOR CROP-BASED PRODUCTS

As consumers grow more conscious of the sources of their food and its effects on local economies, the environment, and health, the farm-to-table movement has gained significant traction in recent years. To create a transparent food supply chain that supports sustainable agricultural practices, this concept places an emphasis on direct connections between farmers, consumers, and food suppliers. For crop-based products, connecting the farm to the table entails combining many elements, such as community involvement, local distribution channels, sustainable farming methods, and consumer education. This section will discuss the fundamentals and strategies for successfully bringing crop-based products from farm to table, as well as the advantages of this strategy and the problems and solutions involved in setting up a successful farm-to-table system.

### 4.8.1. Principles of Farm-to-Table

#### *4.8.1.1. Sustainability*

Sustainability is the cornerstone of the farm-to-table movement. This idea includes sustainable farming methods that enhance soil health, biodiversity, and ecosystem resilience by reducing the usage of chemical pesticides, fertilizers, and herbicides. Crop rotation, cover crops, and integrated pest management are the main components of sustainable agriculture, which aims to minimize negative environmental effects and increase long-term productivity. Customers can help farmers who prioritize sustainable practices and select crops that are grown in harmony with the environment by supporting the farm-to-table movement.

#### *4.8.1.2. Local Sourcing*

By bringing customers and local farmers and food producers together, farm-to-table highlights the value of local sourcing. Local sourcing ensures fresher products, reduces packaging waste, and lowers transportation-related carbon emissions by shortening the distance food travels from the farm to the plate. By assisting small-scale farmers and food businesses, generating employment, and building community resilience, local sourcing also boosts regional economies. Farm-to-school programs, farmers' markets, and community-supported agriculture (CSA) programs are all successful models for encouraging the local sourcing of agricultural products.

#### *4.8.1.3. Transparency*

An essential component of the farm-to-table movement is transparency. Customers are becoming more and more inquisitive about the origins, production methods, and contents of the food they buy. Farmers and food suppliers can gain the trust of consumers by giving comprehensive information about the origin, cultivation methods, and handling of crop-based products. This openness promotes consumers making knowledgeable decisions about what to buy and creates a bond between them and the food they eat. Enhancing accountability and transparency in the food supply chain can be achieved through the implementation of traceability systems, such as certifications and labeling.

#### *4.8.1.4. Community Engagement*

Linking farm-to-table initiatives requires community engagement. A thriving food culture that honors regional crops, customary cooking methods, and seasonal eating can be developed through cooperation between farmers, consumers, and neighborhood organizations. Farm tours, cooking classes, and educational workshops that instruct customers on sustainable agriculture, nutrition, and food preparation are just a few of the various ways that the community can be engaged. Farm-to-table programs can strengthen ties between farmers and consumers while advancing a common understanding of the food system by creating a sense of community around food.

#### *4.8.1.5. Consumer Education*

Promoting involvement in the movement requires educating consumers about the advantages of farm-to-table practices. Customers can make well-informed decisions that are consistent with their values if they are given information about the nutritional value of fresh, locally grown crops, the effects that food choices have on the environment, and the significance of supporting local farmers. Numerous platforms, such as school programs, cooking demonstrations, and social media campaigns, can be used to spread this knowledge. Farm-to-table programs can increase demand for crop-based products and encourage healthier eating habits by educating consumers.

## 4.8.2. Methods for Linking Farm-to-Table

### *4.8.2.1. Farmers' Markets*

A key component of the farm-to-table movement, farmers' markets facilitate direct communication between nearby farmers and customers. These markets give farmers a place to sell their seasonal, fresh produce directly to customers, fostering community involvement and fostering relationships. Customers can learn about various crops, ask farmers about their growing methods, and find new varieties by going to farmers' markets. Farmers' markets also frequently feature regional food producers, artists, and educational materials, which improves the whole farm-to-table experience.

### *4.8.2.2. Community-Supported Agriculture (CSA)*

Customers can sign up for CSA programs to receive regular shares of fresh produce straight from nearby farms. Farmers are able to plan their production according to the expected demand because members pay in advance for the entire crop season. In return for receiving fresh, in-season produce all growing season long, members of this model financially support farmers, fostering a symbiotic relationship between them and the food supply. As CSAs frequently place a high value on sustainable farming methods, they are a great choice for customers looking to get involved with local agriculture.

### *4.8.2.3. Farm-to-School Programs*

Through farm-to-school programs, nearby farms collaborate with educational institutions to supply students with educational opportunities and fresh produce for school lunches.

Garden-based education is one of these programs; students grow their own crops while learning about sustainability, nutrition, and food production. Farm-to-school initiatives increase students' access to fresh produce, encourage healthier eating habits, and assist local farmers by incorporating locally sourced food into school meals. This method instills values about agriculture and nutrition that students will carry with them throughout their lives by creating a sense of connection between them and their food sources.

#### *4.8.2.4. Restaurants and Local Chefs*

The farm-to-table movement can be strengthened by working with nearby eateries and chefs to include locally grown crops on menus. Restaurants that put a high priority on local sourcing can design seasonal menus that highlight the variety and tastes of regional produce. Chefs and farmers can collaborate by showcasing the farmers' produce and exchanging anecdotes about their farming methods. This partnership not only improves the dining experience but also spreads knowledge about the value of encouraging regional farming. Restaurants can also host farm-to-table gatherings where patrons can talk with farmers face-to-face and discover more about where their food comes from.

#### *4.8.2.5. Online Platforms and Delivery Services*

Farm-to-table connections are made easier by the emergence of online platforms and delivery services brought about by technological advancements. Through these services, customers can place direct orders for fresh produce, meats, and other products from local farmers and food producers. Through the use of technology, people can support local agriculture and obtain seasonal crops without having to leave their homes. These platforms, which support the fundamental ideas of the farm-to-table movement, frequently place an emphasis on transparency by offering details about the farms and growing techniques that produce the goods.

### 4.8.3. Benefits of Linking Farm-to-Table

#### *4.8.3.1. Healthier Food Choices*

Connecting the farm to the table encourages people to eat seasonal, fresh produce, which is often more nutrient-dense than processed foods. Consuming fruits and vegetables that are grown nearby has been shown to enhance dietary quality and general health. Farm-to-table programs lower the risk of diet-related diseases and promote healthier eating habits by promoting the consumption of fresh crops.

#### *4.8.3.2. Economic Support for Local Farmers*

By keeping food dollars in the community, direct consumer-farmer connections boost regional economies.

Customers support farmers and promote the sustainability of local agriculture when they decide to buy crop-based products from nearby suppliers. This financial assistance protects farmland, sustains rural livelihoods, and promotes the expansion of small-scale farming enterprises.

#### *4.8.3.3. Reduced Environmental Impact*

Linking farm-to-table lessens the environmental impact of food production and distribution by encouraging sustainable farming methods and local sourcing. Transportation emissions can be reduced by shortening the supply chain, and biodiversity and soil health can be enhanced by sustainable practices. This strategy supports the food system's overall sustainability and satisfies the growing demand for ecologically friendly farming practices.

#### *4.8.3.4. Strengthened Community Connections*

Farm-to-table programs help to strengthen the bond and sense of community that exists between farmers and their customers. Engaging in local food systems encourages consumers to take an active role in shaping their communities through agricultural practices and food choices. This link strengthens social cohesion, fosters neighborhood cooperation, and advances a common knowledge of the significance of sustainable agriculture.

#### *4.8.3.5. Cultural Preservation*

Connecting farm to table can aid in maintaining regional culinary customs and food traditions. Through encouraging the utilization of locally sourced and seasonal produce, customers can become involved in their community's culinary legacy. This focus on regional produce enhances cultural diversity and promotes the exploration of customary recipes, tastes, and cooking methods, enhancing a community's culinary scene.

### 4.8.4. Challenges in Linking Farm-to-Table

#### *4.8.4.1. Supply Chain Complexity*

Since the food supply chain is so varied and dispersed, connecting farm to table can be challenging. Farmers may encounter difficulties with pricing, distribution, and logistics, especially when engaging directly with customers. The establishment of effective supply chains that give local sourcing and transparency priority calls for cooperation between multiple stakeholders.

#### *4.8.4.2. Consumer Awareness and Education*

Even though the farm-to-table movement is becoming more well-known, many consumers might not be familiar with its guiding ideals and advantages. To educate consumers about the value of supporting local agriculture and making informed food choices, effective education and outreach programs are required.

#### *4.8.4.3. Seasonal Limitations*

The seasonality of crop-based products can make it difficult to connect farm to table throughout the year. In a market that values consistency and convenience, consumers may need to modify their eating habits to accommodate what is readily available locally, which can be challenging. These difficulties can be lessened by encouraging seasonal eating and preservation techniques like freezing and canning.

#### *4.8.4.4. Market Competition*

The farm-to-table movement faces competition from industrial food systems that place an emphasis on affordability and efficiency. Customers might be hesitant to pay more for crops that are sourced locally, especially in areas where the economy is struggling. To promote fair farm-to-table relationships, pricing models that accurately account for the costs associated with sustainable agriculture must be developed. Additionally, food access concerns must be addressed.

#### *4.8.4.5. Regulatory Hurdles*

Regulatory obstacles could stand in the way of farmers and food producers trying to sell directly to consumers. Local sourcing opportunities can be restricted and direct sales can become more difficult due to health codes, food safety regulations, and licensing requirements. Farm-to-table connections can be facilitated by advocating for laws that are supportive and for simplified regulations (Balafoutis et al., 2020).

## REFERENCES

1. Abdul-Hamid, M. L. T. A., Ali, M. H., & Osman, L. H. (2021). Data information via blockchain transparency in poultry supply chain: Empirical evidence using fuzzy Delphi method. In *International Management & Accounting Conference (IMAC) 10* (pp. 136–139).
2. Ali, M. H., Chung, L., Kumar, A., Zailani, S., & Tan, K. H. (2021). A sustainable blockchain framework for the halal food supply chain: Lessons from Malaysia. *Technological Forecasting and Social Change, 170*, 120870.
3. Alobid, M., Abujudeh, S., & Szűcs, I. (2022). The role of blockchain in revolutionizing the agricultural sector. *Sustainability, 14*(7), 4313.
4. Balafoutis, A. T., Van Evert, F. K., & Fountas, S. (2020). Smart farming technology trends: Economic and environmental effects, labor impact, and adoption readiness. *Agronomy, 10*(5), 743.
5. Caro, M. P., Ali, M. S., Vecchio, M., & Giaffreda, R. (2018). Blockchain-based traceability in AgriFood supply chain management: A practical implementation. In *2018 IoT Vertical and Topical Summit on Agriculture-Tuscany (IOT Tuscany)* (pp. 1–4).
6. Casino, F., Dasaklis, T. K., & Patsakis, C. (2019). A systematic literature review of blockchain-based applications: Current status, classification and open issues. *Telematics and Informatics, 36*, 55–81.
7. Corwin, D. L., & Scudiero, E. (2019). Review of soil salinity assessment for agriculture across multiple scales using proximal and/or remote sensors. *Advances in Agronomy, 158*, 1–130.
8. el Chami, D., Daccache, A., & el Moujabber, M. (2020). How can sustainable agriculture increase climate resilience? A systematic review. *Sustainability, 12*(8), 3119.
9. Elliott, C. (2011). A pressure to modernize: A comparison between traditional and industrial agricultural practices.
10. González-Amarillo, C. A., Corrales-Muñoz, J. C., Mendoza-Moreno, M. Á., Hussein, A. F., Arunkumar, N., & Ramirez-González, G. (2018). An IoT-based traceability system for greenhouse seedling crops. *IEEE Access, 6*, 67528–67535.
11. Hamadani, H., et al. (2021). Traditional farming practices and its consequences. In *Microbiota and Biofertilizers, Vol 2* (pp. 119–128). Springer.
12. Hate, M., Jadhav, S., & Patil, H. (2018). Vegetable traceability with smart irrigation. In *2018 International Conference on Smart City and Emerging Technology (ICSCET)* (pp. 1–4).

13. Hong, W., Cai, Y., Yu, Z., & Yu, X. (2018). An agri-product traceability system based on IoT and blockchain technology. In *2018 1st IEEE International Conference on Hot Information-Centric Networking (HotICN)* (pp. 254–255).
14. Katyal, N., & Pandian, B. J. (2020). A comparative study of conventional and smart farming. In *Emerging Technologies for Agriculture and Environment* (pp. 1–8). Springer.
15. Kc, K. B., et al. (2018). When too much isn't enough: Does current food production meet global nutritional needs? *PLoS One, 13*(10), e0205683.
16. Kumar, A., & Rao, N. S. (2021). Artificial intelligence technologies driven smart agriculture. *Agricultural Science: Research and Reviews*, 23–34.
17. Kuppusamy, P., Shanmugananthan, S., & Tomar, P. (2021). Emerging technological model to sustainable agriculture. In *Artificial Intelligence and IoT-Based Technologies for Sustainable Farming and Smart Agriculture* (pp. 101–122). IGI Global.
18. Lin, J., Shen, Z., Zhang, A., & Chai, Y. (2018). Blockchain and IoT based food traceability for smart agriculture. In *Proceedings of the 3rd International Conference on Crowd Science and Engineering* (pp. 1–6).
19. Mat, I., Kassim, M. R. M., Harun, A. N., & Yusoff, I. M. (2018). Smart agriculture using internet of things. In *2018 IEEE Conference on Open Systems (ICOS)* (pp. 54–59).
20. Miatton, F., & Amado, L. (2020). Fairness, transparency and traceability in the coffee value chain through blockchain innovation. In *2020 International Conference on Technology and Entrepreneurship- Virtual (ICTE-V)* (pp. 1–6).
21. Mishra, N., Mistry, S., Choudhary, S., Kudu, S., & Mishra, R. (2020). Food traceability system using blockchain and QR code. In *IC-BCT 2019* (pp. 33–43). Springer.
22. Mohamood, N., Zainal, N., Kadmin, A. F., Gani, S. F. A., & Wook, T. (2019). Integrated smart farming based on Internet of Things (IoT) system for figs cultivation. *International Journal of Innovative Technology and Exploring Engineering, 9*(2), 1429–1436.
23. Mykhailenko, O., Desenko, V., Ivanauskas, L., & Georgiyants, V. (2020). Standard operating procedure of Ukrainian saffron cultivation according to good agricultural and collection practices to assure quality and traceability. *Industrial Crops and Products, 151*, 112376.
24. Nanseki, T. (2005). A navigation system for appropriate pesticide use and food safety. In *Proceedings of International Seminar on Technology Development for Good Agriculture Practice in Asia and Oceania* (pp. 134–152).

25. Othman, M. M., Ishwarya, K. R., & Ganesan, M. (2021). A study on data analysis and electronic application for the growth of smart farming. *Alinteri Journal of Agriculture Sciences, 36*(1), 209–218.
26. Panhwar, Q. A., Ali, A., Naher, U. A., & Memon, M. Y. (2019). Fertilizer management strategies for enhancing nutrient use efficiency and sustainable wheat production. In *Organic Farming* (pp. 17–39). Elsevier.
27. Prathibha, S. R., Hongal, A., & Jyothi, M. P. (2017). IoT based monitoring system in smart agriculture. In *2017 International Conference on Recent Advances in Electronics and Communication Technology (ICRAECT)* (pp. 81–84).
28. Rayhana, R., Xiao, G., & Liu, Z. (2021). RFID sensing technologies for smart agriculture. *IEEE Instrumentation & Measurement Magazine, 24*(3), 50–60.
29. Said, H. B., Nordin, R., & Abdullah, N. F. (2019). Energy models of ZigBee-based wireless sensor networks for smart-farm. *Jurnal Kejuruteraan, 31*(1), 77–83.
30. Sneri, S. S., Chaware, S. R., Bodele, P. S., Thakre, P. R., & Taiwade, H. (2021). A proposed agriculture trade and supply chain traceability system using blockchain. *Journal of Agricultural Research and Technology, 3*(6), 392–395.
31. Surasak, T., Wattanavichean, N., Preuksakarn, C., & Huang, S. C. H. (2019). Thai agriculture products traceability system using blockchain and Internet of Things. *System, 14*, 15.
32. Tian, F. (2016). An agri-food supply chain traceability system for China based on RFID & blockchain technology. In *2016 13th International Conference on Service Systems and Service Management (ICSSSM)* (pp. 1–6).
33. Tian, F. (2017). A supply chain traceability system for food safety based on HACCP, blockchain & Internet of things. In *2017 International Conference on Service Systems and Service Management* (pp. 1–6).
34. Tikas, G. D., & Akhilesh, K. B. (2020). Smart agriculture: A tango between modern IoT-based technologies and traditional agriculture techniques. In *Smart Technologies* (pp. 387–394). Springer.
35. Tzoulis, I., & Andreopoulou, Z. (2013). Emerging traceability technologies as a tool for quality wood trade. *Procedia Technology, 8*, 606–611.
36. Wang, J., Wang, J. M., & Zhang, Y. J. (2018). Agricultural product quality traceability system based on the hybrid mode. In *2018 4th Annual International Conference on Network and Information Systems for Computers (ICNISC)* (pp. 392–395).

37. Wongpatikaseree, K., Kanka, P., & Ratikan, A. (2018). Developing smart farm and traceability system for agricultural products using IoT technology. In *2018 IEEE/ACIS 17th International Conference on Computer and Information Science (ICIS)* (pp. 180–184).
38. Xiong, H., Dalhaus, T., Wang, P., & Huang, J. (2020). Blockchain technology for agriculture: Applications and rationale. *Frontiers in Blockchain, 3*, 7.
39. Yang, F., Wang, K., Han, Y., & Qiao, Z. (2018). A cloud-based digital farm management system for vegetable production process management and quality traceability. *Sustainability, 10*(11), 4007.
40. Zainal, N., Mohamood, N., Norman, M. F., & Sanmutham, D. (2019). Design and implementation of smart farming system for fig using connected-argonomics. *International Journal of Electrical & Computer Engineering, 9*(6).
41. Zhang, Q. (2016). *Precision Agriculture Technology for Crop Farming*. Taylor & Francis.
42. K. Wongpatikaseree, P. Kanka, and A. Ratikan, "Developing smart farm and traceability system for agricultural products using IoT technology," in 2018 IEEE/ACIS 17th International Conference on Computer and Information Science (ICIS), 2018, pp. 180–184.
43. D. L. Corwin and E. Scudiero, "Review of soil salinity assessment for agriculture across multiple scales using proximal and/or remote sensors," Advances in agronomy, vol. 158, pp. 1–130, 2019.
44. Q. A. Panhwar, A. Ali, U. A. Naher, and M. Y. Memon, "Fertilizer management strategies for enhancing nutrient use efficiency and sustainable wheat production," in Organic Farming, Elsevier, 2019, pp. 17–39.
45. F. Tian, "An agri-food supply chain traceability system for China based on RFID & blockchain technology," in 2016 13th international conference on service systems and service management (ICSSSM), 2016, pp. 1–6.
46. C. Elliott, "A pressure to modernize: a comparison between traditional and industrial agricultural practices.," 2011.
47. M. Hate, S. Jadhav, and H. Patil, "Vegetable traceability with smart irrigation," in 2018 International Conference on Smart City and Emerging Technology (ICSCET), 2018, pp. 1–4.
48. Q. Zhang, Precision agriculture technology for crop farming. Taylor & Francis, 2016.

49. T. Nanseki, "A navigation system for appropriate pesticide use and food safety," inProceedings of International Seminar on Technology Development for Good Agriculture Practice in Asia and Oceania, 2005, pp. 134–152.
50. R. Rayhana, G. Xiao, and Z. Liu, "Rfid sensing technologies for smart agriculture," IEEE Instrumentation & Measurement Magazine, vol. 24, no. 3, pp. 50–60, 2021.
51. F. Yang, K. Wang, Y. Han, and Z. Qiao, "A cloud- based digital farm management system for vegetable production process management and quality traceability," Sustainability, vol. 10, no. 11, p. 4007, 2018.

# CHAPTER 5

# TRACEABILITY IN LIVESTOCK PRODUCTION

## Contents

# INTRODUCTION

One essential element of Precision Livestock Farming (PLF) in the livestock industry is traceability. Precision livestock farming is a technological framework designed to increase production efficiency while improving welfare for both humans and animals. It uses advanced information, communication technologies, and computerization to obtain precise control over the production process. PLF offers the potential to improve the health and welfare of the animals used in the production cycle in addition to increasing productivity and cutting expenses. PLF is bringing fresh perspectives to bear on sectors such as feedlots and cow-calf operations. It is possible to record and analyze enormous volumes of data. New understanding of the entire production cycle and its effects on the agricultural production process is being made possible by big data (Figure 5.1).

**Figure 5.1.** *Traceability in the livestock industry.*

*Source: By Cattle_Feedlot_near_Rocky_Ford,_CO_IMG_5651.JPG: Billy Hathornderivative work: AlanAdler - Own work (Billy Hathorn)This file was derived from: Cattle Feedlot near Rocky Ford, CO IMG 5651.JPG:, CC BY-SA 3.0, https://commons.wikimedia.org/w/index.php?curid=21608441.*

Identification and tracking of livestock from farm to table is known as traceability. The first step is to identify the animals. Cattle are identified individually by means of RFID ear tags. Geographical locations are identified to identify the premises on which an animal is housed, such as a feedlot or ranch. Animal movement is the last step. From the moment of birth through the completion of the animal's life cycle, records are maintained.

The PLF framework includes traceability via RFID ear tags and a database. To guarantee health, welfare, and productivity, each individual cow can be individually observed in many ways. PLF is the tracking of a single animal wearing an ear tag from one location to another and the analysis of this tracking information. With the use of software, databases, reporting, and RFID communication capabilities, management is able to make decisions regarding a single animal or a group of animals.

Traceability is met with resistance in the cattle sector. Parties must exchange information in order for traceability to function. This covers rivals or the government. This data includes the number of animals sold each month and the size of operations. Competitors are reluctant to divulge details. Certain businesses choose not to file all of their taxes and do not wish to disclose their activities to the government by storing herd data in a database.

A disease outbreak that spreads across feedlots may also be disastrous. Larger businesses can survive an outbreak, but smaller businesses risk closure if an outbreak is linked to them. Smaller operators are not as well protected. They are more vulnerable if they disclose their production information in a database.

Participation in national traceability programs is mandatory in certain countries. For instance, since 2001, the Canadian Cattle Identification Agency has been in charge of Canada's traceability system. A national database that is maintained from the time of birth until the point of export or carcass inspection includes an animal's RFID ear tag and unique identification. In Australia, cattle, sheep, and goats can be traced using the National Livestock Identification System (NLIS). The US Department of Agriculture is working to implement an RFID ear tag system in the near future, despite the lack of a national traceability system in the US.

## 5.1. LIVESTOCK PRODUCTION

Both directly through grazing and indirectly through the consumption of feed grains and fodder, livestock production is the largest land user in the world. Approximately 40% of the gross value of agricultural production worldwide is currently derived from the production of livestock. This percentage is greater than half in industrialized nations. Its percentage is rapidly increasing in developing nations, where it makes up one-third of the total; changes in lifestyles and dietary habits, as well as increases in population and income, are all contributing to the rapid rise in livestock production.

The livestock industry has continuously grown faster than the crop industry. By 2030, it is anticipated that the overall demand for animal products in developing nations will have more than doubled. In contrast, the industrial world's demand for animal products has been increasing slowly, and during the projection period, livestock production in this group of nations is only anticipated to grow slowly (*see* Table 5.1).

One of the main issues facing global agriculture today is meeting the growing and shifting demand for animal products while maintaining the foundation of natural resources (soil, water, air, and biodiversity). Many of the trends in the livestock subsector are already evident, and they will continue to drive global agriculture as a whole such as:

- The production of livestock will increasingly come from warm, humid, and disease-prone environments.
- Practices for raising livestock will shift from being a multipurpose, local endeavor to becoming a more intensive, market-driven, and increasingly integrated process.
- Competition for and pressure on common property resources like water and grazing areas will grow.
- There will be an increase in large-scale industrial production near urban areas, which carries risks to the environment and public health.
- The production of pigs and poultry will increase. There will be a substantial rise in the use of cereal-based feeds.

Addressing these challenges brings up important national and international public policy issues. In general, these include public health and food safety, the environment and natural resource management, and equity and poverty reduction.

### 5.1.1. Consumption of Livestock Products

Demand for more diverse food grows as incomes rise. When compared to foods originating from plants, like cereals, the demand for higher-quality, higher-value foods is rising, such as meat, eggs, and milk. The overall demand for animal products has increased significantly in many developing nations as a result of these changes in consumption and significant population growth, and this trend is expected to continue.

In developing nations, there is a noticeable increase in the consumption of animal products. Cereal calories have increased in absolute terms, but they have decreased as a percentage of total calories, going from 60% in 1961–1963 to an anticipated 50% in 2030. Similar to this, by 1997/99, the contribution of other traditional staples (potatoes, sweet potatoes, cassava, plantains, and other roots) to dietary calories dropped to the lowest level (6.2%) from the second-largest (10%) in 1961/63. At that point, animal products accounted for 10.6% of calories in developing nations, making them the second most important source. In developed nations, cereals make up about 34% of total calories consumed, a much smaller percentage than animal products, which continue to make up roughly 23% of calories. Significant potential exists to increase the percentage and absolute amount of animal products in the diet in developing nations, where the per capita consumption is still less than one-third that of industrialized nations.

Animal protein consumption rose from 44–55 grams per capita per day in industrialized nations during the 1960s and 1970s. Consumption of animal protein stayed relatively constant after this. However, even though the amount of animal protein consumed in developing nations increased gradually from 9 g/capita/day in 1961–1963–20 g/capita/day in 1997–1999, there is still a lot of room for growth.

It is predicted that between 1997–1999 and 2030, the amount of meat consumed annually in developing nations will rise from 25.5–37 kg per person, while in industrialized nations, it will rise from 88–100 kg. In developing nations, milk, and dairy product consumption will increase from 45 kg per person per year to 66 kg, while in industrialized nations, it will increase from 212–221 kg. Consumption of eggs will increase from 13.5–13.8 kg in industrialized nations and from 6.5–8.9 kg in developing nations.

Large regional and national variations in the amount and kind of animal products consumed also show traditional preferences based on availability, relative costs, and religious and personal preferences. Among the more crucial elements are the following:

- Over the past three decades, there has been minimal change in the low levels of animal product consumption in sub-Saharan Africa. These account for roughly 5% of the calories consumed per person, which is roughly half that of developing nations overall and 5% of industrial nations. In recent years, milk's share of total calories and protein per person has stayed steady, suggesting that overall milk availability has increased in tandem with population growth. Consumption is only expected to rise slightly.
- Only 8.7% of the relatively high calorie consumption in the Near East and North Africa in 1997–1999 came from animal products. By 2030, animal products (mainly milk and poultry meat) will account for 11.4% of total caloric intake.
- The consumption of animal products, such as meat, has historically been higher in Latin America and the Caribbean (a region that does not include Brazil), and this trend is expected to continue. Animal products currently account for 16% of dietary energy, but meat consumption per person is still only 60% of what it is in industrialized nations, suggesting room for growth.

- Brazil stands out due to the significant and growing amount of animal products it consumes (18.8% in 1997–99). It is anticipated that Brazil will become more isolated from the rest of Latin America. By 2030, it is anticipated that meat consumption per capita, which is currently more than three-quarters that of industrialized nations, will catch up to that level. It is anticipated that by 2030, per capita milk consumption will have increased to three-quarters of that of industrialized nations, from just over half at present.
- The consumption of animal products has grown gradually but steadily throughout South Asia (apart from India). This increase is primarily the result of rising milk contributions, which are already high at a per capita level 50% higher than the average for developing nations, combined with rising poultry meat contributions. Egg contribution is significantly less than the average for developing nations.
- Up until 2030, it is anticipated that the proportion of animal products in Indian diets will rise, mostly due to rising milk and milk product consumption.
- The amount of animal products in the diet is also steadily rising in East Asia (apart from China). However, meat—mostly pork—has contributed to this increase, in contrast to South Asia.
- Pork and poultry will account for a significant portion of China's anticipated sharp increase in the percentage of animal products in diet energy, from 15%–20% between 1997/99 and 2030. Milk consumption per person is currently very low and is expected to stay that way (growing from 7 kg pa in 1997/99–14 kg pa in 2030). In comparison, China consumes a lot of eggs—more than twice as much as the average for developing nations and even higher than the average for industrialized nations—and is expected to increase from 15 kg per person per year in 1997–1999 to 20 kg in 2030.

### 5.1.2. Production

More on the implications is covered in this section. An increasing number of factors, including the cost, quality, and availability of inputs as well as the proximity to markets, influence the location of production and processing operations. It is anticipated that the market for animal products will grow, making traditional mixed farming methods insufficient to meet demand.

Increasing the number of animals, as well as their productivity, processing, and marketing effectiveness, as well as utilizing different combinations of these factors, can all contribute to an increase in the supply of animal products. Since land availability in most regions restricts the growth of livestock numbers in extensive production systems, the majority of the increase in production will come from productivity gains achieved through intensification and broader adoption of improved production and marketing technologies.

The development of intensive, vertically integrated, intensive establishments near major urban centers has been the strongest structural trend in livestock production. This is especially true for the production of broiler chicken in South Asia and East Asia, as well as pig and poultry meat in Latin America and East Asia. Though to a lesser extent, similar trends can be seen in the production of dairy and beef. Increased imports of feed grains have effectively replaced livestock product imports in East Asia as a result of the growing demand for feed grains linked to these production systems.

Governments and other stakeholders must make significant trade-offs when drafting public policies. For instance, industrial livestock production is preferred in many developing nations to supply urban populations with inexpensive animal protein (benefits to public health and nutrition). However, this might come at the price of small rural producers' ability to compete and access markets. Stricter food safety laws aimed at improving public health also act as obstacles that keep impoverished farmers from participating in official markets.

### 5.1.2.1. Livestock Production and Productivity

The estimates of the total demand for livestock products were calculated by taking the estimates of the demand for food for direct human consumption and adding other demand components, like the use of livestock products for industrial purposes other than food, the milk that children consume, and the hatching and waste of eggs.

A thorough examination of historical livestock product trade, including trade in inputs like cereals and oilseeds used to make animal feed, was done to determine production estimates for each nation.

Table 5.1 presents the overall results. The primary characteristic is a slowdown in the increase in livestock production worldwide. This will consist of a marked slowdown from rapid to moderate growth rates in the developing countries, a recovery in the transition countries, and slow growth in the industrialized countries. The anticipated slowdown in China and Brazil, two significant nations that previously saw rapid growth in livestock production, is largely responsible for the latter trend. With these two nations excluded, the slowdown occurs much more gradually. Developments are far more varied among individual nations, with several seeing faster growth in livestock production.

**Table 5.1.** *Annual Growth Rates of Total Livestock Production*

**Annual Growth Rates of Total Livestock Production**

| | **1969–99** | **1979–99** | **1989–99** | **1997/99 -2015** | **2015 -2030** | **1997/99 -2030** |
|---|---|---|---|---|---|---|
| | **Percentage** | | | | | |
| World | 2.2 | 2.1 | 2.0 | 1.7 | 1.5 | 1.6 |
| excl. China | 1.7 | 1.3 | 0.8 | 1.6 | 1.5 | 1.5 |
| Developing countries | 4.6 | 5.0 | 5.5 | 2.6 | 2.1 | 2.4 |
| excl. China | 3.5 | 3.5 | 3.6 | 2.8 | 2.5 | 2.7 |
| excl. China and Brazil | 3.3 | 3.3 | 3.3 | 2.9 | 2.6 | 2.8 |
| Sub-Saharan Africa | 2.4 | 2.0 | 2.1 | 3.2 | 3.3 | 3.2 |
| Latin America and the Carib-bean | 3.1 | 3.0 | 3.7 | 2.4 | 1.9 | 2.1 |
| excl. Brazil | 2.3 | 2.1 | 2.7 | 2.4 | 2.1 | 2.3 |
| Near East/ North Africa | 3.4 | 3.4 | 3.4 | 2.9 | 2.6 | 2.7 |
| South Asia | 4.2 | 4.5 | 4.1 | 3.3 | 2.8 | 3.1 |
| East Asia | 7.2 | 8.0 | 8.2 | 2.3 | 1.6 | 2.0 |
| excl. China | 4.8 | 4.7 | 3.7 | 3.0 | 2.7 | 2.8 |
| Industrial countries | 1.2 | 1.0 | 1.2 | 0.7 | 0.4 | 0.6 |
| Transition countries | -0.1 | -1.8 | -5.7 | 0.5 | 0.6 | 0.5 |

Note: Total livestock production was derived by aggregating four meats, milk, and eggs at 1989/91 international commodity prices used to construct the FAO indices of agricultural production.

*Source: https://www.fao.org/4/y4252e/y4252e07.htm#:~:text=Annual%20growth%20of%20meat%20production,percent%20(484%20million%20tonnes).*

The data and projections for the six livestock products covered in this study are detailed in Table 5.2, which provides context for the overall findings shown in Table 5.1. Production has been steadily increasing in developing nations. Between 1989 and 1999, the six commodities' yearly growth rates varied from 3.7%–9.4%. In contrast, production in the industrialized countries increased only marginally during this same period, while it actually decreased in the transitional countries.

**Table 5.2.** *Livestock Production by Commodity: Past and Projected*

**Livestock Production by Commodity: Past and Projected**

| | 1967/69 | 1987/89 | 1997/99 | 2015 | 2030 | 1969 -1999 | 1989 -1999 | 1995/97 -2015 | 2015 -2030 |
|---|---|---|---|---|---|---|---|---|---|
| | Million Tonnes | | | | | % p.a. | | | |
| *Total meat* | | | | | | | | | |
| World | 92 | 166 | 218 | 300 | 376 | 2.9 | 2.7 | 1.9 | 1.5 |
| excl. China | 84 | 142 | 162 | 218 | 277 | 2.1 | 1.3 | 1.8 | 1.6 |
| Developing countries | 28 | 66 | 116 | 181 | 247 | 5.2 | 5.9 | 2.7 | 2.1 |
| excl. China | 21 | 41 | 60 | 98 | 147 | 3.8 | 3.9 | 3.0 | 2.7 |
| excl. China and Brazil | 18 | 34 | 47 | 79 | 123 | 3.5 | 3.3 | 3.1 | 2.9 |
| Sub-Saharan Africa | 3 | 4 | 5 | 9 | 16 | 2.3 | 2.2 | 3.3 | 3.5 |
| Latin America and the Carib-bean | 10 | 19 | 28 | 43 | 58 | 3.5 | 4.5 | 2.6 | 2.1 |
| excl. Brazil | 7 | 11 | 15 | 24 | 33 | 2.5 | 3.1 | 2.7 | 2.3 |
| Near East/ North Africa | 2 | 5 | 7 | 13 | 19 | 4.4 | 3.8 | 3.5 | 2.9 |
| South Asia | 3 | 5 | 7 | 13 | 23 | 3.7 | 2.8 | 3.6 | 3.9 |
| East Asia | 10 | 33 | 69 | 103 | 131 | 7.1 | 7.6 | 2.4 | 1.6 |
| excl. China | 3 | 8 | 13 | 21 | 32 | 5.1 | 4.1 | 3.0 | 2.8 |
| Industrial countries | 46 | 71 | 85 | 99 | 107 | 1.9 | 1.8 | 0.9 | 0.5 |
| Transition countries | 17 | 29 | 17 | 20 | 22 | 0.0 | -6.4 | 0.8 | 0.8 |
| *Bovine meat* | | | | | | | | | |
| World | 38.0 | 53.7 | 58.7 | 74.0 | 88.4 | 1.4 | 0.8 | 1.4 | 1.2 |
| Developing countries | 11.8 | 19.3 | 28.0 | 41.2 | 55.0 | 3.0 | 3.8 | 2.3 | 2.0 |
| excl. China | 11.7 | 18.4 | 23.2 | 33.5 | 44.1 | 2.5 | 2.2 | 2.2 | 1.8 |
| excl. China and Brazil | 10.0 | 14.4 | 17.3 | 25.2 | 34.1 | 2.0 | 1.5 | 2.3 | 2.0 |

| | | | | | | | | | |
|---|---|---|---|---|---|---|---|---|---|
| Sub-Saharan Africa | 1.6 | 2.2 | 2.6 | 4.3 | 6.7 | 1.5 | 1.7 | 3.0 | 3.0 |
| Latin America and the Carib-bean | 6.8 | 10.4 | 13.1 | 18.2 | 22.5 | 2.5 | 2.1 | 1.9 | 1.4 |
| excl. Brazil | 5.1 | 6.5 | 7.2 | 9.9 | 12.5 | 1.4 | 0.4 | 1.9 | 1.6 |
| Near East/ North Africa | 0.7 | 1.3 | 1.8 | 2.8 | 4.1 | 3.2 | 3.4 | 2.4 | 2.6 |
| South Asia | 1.7 | 3.1 | 4.0 | 5.7 | 7.4 | 3.1 | 2.3 | 2.1 | 1.7 |
| East Asia | 1.0 | 2.3 | 6.4 | 10.1 | 14.4 | 6.4 | 11.5 | 2.7 | 2.4 |
| excl. China | 0.8 | 1.4 | 1.6 | 2.5 | 3.5 | 2.1 | 2.3 | 2.6 | 2.2 |
| Industrial countries | 19.1 | 23.8 | 25.0 | 26.6 | 26.5 | 0.6 | 0.6 | 0.4 | 0.0 |
| Transition countries | 7.0 | 10.6 | 5.7 | 6.3 | 6.9 | -0.3 | -7.5 | 0.5 | 0.6 |
| *Ovine meat* | | | | | | | | | |
| World | 6.6 | 9.1 | 10.8 | 15.3 | 20.1 | 1.9 | 1.4 | 2.1 | 1.8 |
| Developing countries | 3.0 | 5.0 | 7.4 | 11.2 | 15.4 | 3.4 | 3.7 | 2.5 | 2.1 |
| Sub-Saharan Africa | 0.6 | 0.9 | 1.3 | 2.2 | 3.4 | 2.8 | 3.5 | 3.1 | 3.0 |
| Near East/ North Africa | 0.9 | 1.5 | 1.8 | 2.6 | 3.5 | 2.3 | 1.9 | 2.2 | 2.0 |
| South Asia | 0.6 | 1.1 | 1.3 | 2.1 | 3.1 | 3.5 | 1.4 | 2.6 | 2.6 |
| East Asia | 0.4 | 1.1 | 2.5 | 3.8 | 4.8 | 7.0 | 8.1 | 2.6 | 1.5 |
| Industrial countries | 2.4 | 2.8 | 2.7 | 3.1 | 3.5 | 0.6 | -0.8 | 0.9 | 0.8 |
| Transition countries | 1.3 | 1.3 | 0.8 | 0.9 | 1.1 | -1.0 | -6.4 | 1.3 | 1.1 |
| *Pig meat* | | | | | | | | | |
| World | 34.1 | 66.3 | 86.5 | 110.2 | 124.5 | 3.2 | 2.7 | 1.4 | 0.8 |
| excl. China | 28.1 | 46.2 | 48.1 | 57.9 | 66.2 | 1.7 | 0.4 | 1.1 | 0.9 |
| Developing countries | 9.7 | 28.0 | 49.3 | 69.5 | 82.8 | 6.1 | 5.7 | 2.0 | 1.2 |
| excl. China | 3.8 | 7.9 | 10.9 | 17.2 | 24.5 | 3.7 | 3.4 | 2.7 | 2.4 |
| Latin America and the Carib-bean | 1.8 | 3.0 | 3.9 | 6.0 | 7.8 | 2.1 | 3.9 | 2.5 | 1.8 |
| excl. Brazil | 1.1 | 1.9 | 2.3 | 3.4 | 4.4 | 1.7 | 2.8 | 2.3 | 1.8 |
| East Asia | 7.6 | 24.2 | 44.3 | 61.6 | 71.9 | 6.8 | 6.0 | 2.0 | 1.0 |
| excl. China | 1.6 | 4.0 | 5.9 | 9.3 | 13.6 | 5.1 | 3.3 | 2.8 | 2.5 |
| Industrial countries | 16.6 | 26.0 | 29.3 | 32.3 | 33.1 | 1.8 | 1.4 | 0.6 | 0.2 |
| Transition countries | 7.7 | 12.3 | 7.9 | 8.4 | 8.6 | -0.1 | -5.3 | 0.4 | 0.1 |

| | | | | | | | | | |
|---|---|---|---|---|---|---|---|---|---|
| *Poultry meat* | | | | | | | | | |
| World | 12.9 | 37.2 | 61.8 | 100.6 | 143.3 | 5.2 | 5.4 | 2.9 | 2.4 |
| excl. China | 12.1 | 34.6 | 51.2 | 81.4 | 117.5 | 4.8 | 4.1 | 2.8 | 2.5 |
| Developing countries | 3.3 | 13.2 | 31.3 | 59.1 | 93.5 | 7.9 | 9.4 | 3.8 | 3.1 |
| excl. China | 2.5 | 10.6 | 20.7 | 39.9 | 67.7 | 7.4 | 7.2 | 4.0 | 3.6 |
| excl. China and Brazil | 2.2 | 8.6 | 15.6 | 31.9 | 56.4 | 6.9 | 6.4 | 4.3 | 3.9 |
| Sub-Saharan Africa | 0.3 | 0.7 | 0.9 | 1.9 | 4.1 | 3.8 | 2.6 | 4.3 | 5.1 |
| Latin America and the Caribbean | 1.0 | 4.7 | 10.5 | 18.2 | 27.3 | 7.8 | 9.0 | 3.3 | 2.7 |
| excl. Brazil | 0.7 | 2.7 | 5.4 | 10.2 | 16.0 | 6.7 | 8.4 | 3.8 | 3.0 |
| Near East/ North Africa | 0.4 | 2.1 | 3.2 | 7.1 | 11.6 | 7.7 | 5.2 | 4.7 | 3.3 |
| South Asia | 0.2 | 0.5 | 1.1 | 3.9 | 10.6 | 7.7 | 7.2 | 7.9 | 6.9 |
| East Asia | 1.5 | 5.3 | 15.5 | 27.9 | 39.9 | 8.5 | 11.7 | 3.5 | 2.4 |
| excl. China | 0.7 | 2.6 | 4.9 | 8.7 | 14.1 | 7.3 | 6.1 | 3.4 | 3.2 |
| Industrial countries | 8.1 | 18.8 | 27.7 | 37.5 | 44.1 | 4.0 | 3.9 | 1.8 | 1.1 |
| Transition countries | 1.5 | 5.2 | 2.9 | 4.1 | 5.7 | 1.6 | -6.7 | 2.0 | 2.3 |
| *Milk (whole milk eq.)* | | | | | | | | | |
| World | 387 | 528 | 562 | 715 | 874 | 1.3 | 0.6 | 1.4 | 1.3 |
| Developing countries | 78 | 149 | 219 | 346 | 484 | 3.6 | 4.1 | 2.7 | 2.3 |
| excl. China and Brazil | 69 | 128 | 189 | 301 | 425 | 3.5 | 4.1 | 2.8 | 2.3 |
| Sub-Saharan Africa | 8 | 13 | 16 | 26 | 39 | 2.7 | 1.9 | 3.0 | 2.8 |
| Latin America and the Caribbean | 24 | 40 | 57 | 81 | 105 | 2.6 | 3.9 | 2.1 | 1.8 |
| excl. Brazil | 17 | 26 | 36 | 52 | 69 | 2.2 | 4.0 | 2.1 | 1.9 |
| Near East/ North Africa | 14 | 21 | 28 | 41 | 56 | 2.3 | 3.1 | 2.2 | 2.1 |
| South Asia | 30 | 65 | 104 | 174 | 250 | 4.5 | 4.9 | 3.1 | 2.4 |
| East Asia | 3 | 10 | 15 | 25 | 34 | 6.9 | 4.5 | 2.9 | 2.2 |
| excl. China | 1 | 4 | 5 | 8 | 12 | 7.3 | 3.2 | 3.0 | 2.4 |
| Industrial countries | 199 | 236 | 246 | 269 | 286 | 0.7 | 0.5 | 0.5 | 0.4 |
| Transition countries | 110 | 144 | 97 | 100 | 104 | -0.3 | -4.6 | 0.2 | 0.2 |
| *Eggs* | | | | | | | | | |

| | | | | | | | | | |
|---|---|---|---|---|---|---|---|---|---|
| World | 18.7 | 35.6 | 51.7 | 70.4 | 89.9 | 3.4 | 4.2 | 1.8 | 1.6 |
| Developing countries | 4.9 | 16.2 | 33.7 | 50.7 | 69.0 | 7.0 | 8.0 | 2.4 | 2.1 |
| excl. China | 3.2 | 9.5 | 13.5 | 24.6 | 37.8 | 5.0 | 3.4 | 3.6 | 2.9 |
| Sub-Saharan Africa | 0.3 | 0.7 | 0.9 | 1.8 | 3.4 | 3.7 | 2.6 | 4.0 | 4.1 |
| Latin America and the Carib-bean | 1.2 | 3.6 | 4.6 | 7.3 | 10.4 | 4.5 | 2.5 | 2.8 | 2.3 |
| Near East/ North Africa | 0.4 | 1.5 | 2.2 | 3.6 | 5.3 | 6.0 | 4.1 | 3.0 | 2.6 |
| South Asia | 0.3 | 1.4 | 2.2 | 5.7 | 9.9 | 6.3 | 4.7 | 5.8 | 3.7 |
| East Asia | 2.6 | 9.1 | 23.8 | 32.1 | 40.0 | 8.3 | 10.7 | 1.8 | 1.5 |
| excl. China | 0.9 | 2.4 | 3.6 | 6.0 | 8.8 | 5.0 | 3.5 | 3.0 | 2.6 |
| Industrial countries | 10.7 | 12.8 | 13.7 | 14.8 | 15.5 | 0.6 | 0.9 | 0.5 | 0.3 |
| Transition countries | 3.1 | 6.5 | 4.3 | 5.0 | 5.5 | 0.7 | -4.7 | 0.8 | 0.7 |

*Source: https://www.fao.org/4/y4252e/y4252e07.htm#:~:text=Annual%20growth%20of%20meat%20production,percent%20(484%20million%20tonnes).*

By 1997/99, the developing countries produced 39% of the world's milk and 53% of its meat, respectively, compared to only 28% and 40% 10 years earlier. Although the production collapse in the transition countries played a role, this trend persisted even in the absence of this phenomenon. Meat production in developing nations is expected to grow by 2.4% annually until 2030, while milk production is predicted to grow by 2.5% annually. By 2030, this would increase the proportion of developing nations producing milk to 55% (484 million tonnes) and meat to 66% (247 million tonnes) globally.

Between 1989 and 1999, the production of white meat (pork and poultry) increased dramatically in developing nations; it was more than twice as fast as that of red meat (cattle, sheep, and goats). Nonetheless, there are significant regional variations. East Asia has seen phenomenal growth in poultry production (11.7% p. a.) as well as South Asia (7.2% p. a.) and illustrates how quickly the chicken business has become more intense in these areas. Annual growth rates in Latin America were 9%. Nonetheless, the annual growth rate in sub-Saharan Africa was a mere 2.6%. In the late 1980s, red meat made up approximately 37% of all meat produced in developing nations; however, by 1997–1999, that percentage had dropped to 31%, and it is anticipated that it would continue to fall.

Over the past 10 years (1989–1999), developing nations have seen increases in egg production with comparable regional variations. East Asia, South Asia, and sub-Saharan Africa had annual growth rates of 10.7%, 4.7%, and 2.6%, respectively. While production in the industrialized countries increased by 0.9% per year and in Latin America by 2.5% per year, it decreased by 4.7% per year in the transition countries. Over the same time period, milk production in developing nations increased at a rate of 4.1% per annum, with South Asia experiencing the highest annual growth (4.9%) and sub-Saharan Africa the lowest (1.9%). In both industrialized and developing nations, milk production followed the same pattern as egg production.

Meat or milk production per animal annually can be used to gauge productivity. More complex productivity analyses that are based on financial flows or on the unit of output per unit of biomass or feed input are far more challenging to carry out. For instance, despite lower carcass weight and lower offtake rates, farmers may still be able to turn a profit because of the rising demand for premium, free-range meat in high-income nations.

A combination of increased productivity and animal population growth can lead to higher production. Higher offtake rates (shorter production cycles by, for example, faster fattening) and higher carcass weight or milk or egg yields combine to produce higher productivity. According to the projections, the increase in livestock numbers will continue to be substantial, albeit not as much as it has historically been. In the production of beef, higher carcass weights will be more significant, whereas in the production of pig and poultry meat, higher offtake rates (shorter production cycles) will be more significant.

Reliable data on carcass weights and offtake rates are difficult to obtain. To get around these, herd sizes and meat production can be directly compared. For instance, while cattle numbers increased by only 1.3% (Table 5.3), beef production in developing nations increased by 3.8% per annum during the last 10 years (1989–1999), suggesting a 2.5% increase in productivity annually. While flock size increased by only 1.5%, small ruminant production increased by 3.7% per year, indicating a 1.2% increase in productivity annually.

**Table 5.3.** *Meat Production*

**Meat Production: Number of Animals and Carcass Weight**

| | **Number of Animals (Millions)** | | | | **Number of Animals (% p.a.)** | | | **Carcass Weight (Kg/Animal)** | | |
|---|---|---|---|---|---|---|---|---|---|---|
| | **1967/69** | **1987/89** | **1997/99** | **2030** | **1969 -1999** | **1989 -1999** | **1997/99 -2030** | **1967 /69** | **1997 /99** | **2030** |
| **World** | | | | | | | | | | |
| Cattle and buffaloes | 1189 | 1418 | 1497 | 1858 | 0.8 | 0.5 | 0.7 | 174 | 198 | 211 |
| Sheep and goats | 1444 | 1708 | 1749 | 2 309 | 0.9 | -0.1 | 0.9 | 14 | 14 | 17 |

| | | | | | | | | | | |
|---|---|---|---|---|---|---|---|---|---|---|
| Pigs | 566 | 838 | 873 | 1062 | 1.4 | 0.3 | 0.6 | 65 | 78 | 84 |
| Poultry | 5585 | 10731 | 15067 | 24804 | 3.8 | 3.4 | 1.6 | 1.3 | 1.6 | 1.8 |
| **Developing Countries** | | | | | | | | | | |
| Cattle and buffaloes | 799 | 1013 | 1156 | 1522 | 1.3 | 1.3 | 0.9 | 150 | 163 | 188 |
| Sheep and goats | 862 | 1121 | 1323 | 1856 | 1.6 | 1.5 | 1.1 | 13 | 13 | 16 |
| Pigs | 297 | 493 | 581 | 761 | 2.2 | 1.6 | 0.8 | 49 | 73 | 82 |
| Poultry | 2512 | 6168 | 10544 | 19193 | 5.6 | 5.5 | 1.9 | 1.2 | 1.4 | 1.8 |
| **Sub-Saharan Africa** | | | | | | | | | | |
| Cattle and buffaloes | 130 | 159 | 200 | 285 | 1.5 | 2.4 | 1.1 | 137 | 130 | 157 |
| Sheep and goats | 182 | 269 | 346 | 501 | 2.4 | 2.6 | 1.2 | 12 | 12 | 17 |
| Pigs | 6 | 13 | 18 | 27 | 4.5 | 2.3 | 1.4 | 45 | 47 | 63 |
| Poultry | 313 | 555 | 720 | 1459 | 3.1 | 2.4 | 2.2 | 0.9 | 1.0 | 1.4 |
| **Latin America and the Caribbean** | | | | | | | | | | |
| Cattle and buffaloes | 219 | 317 | 350 | 483 | 1.6 | 0.9 | 1.0 | 191 | 211 | 230 |
| Sheep and goats | 152 | 145 | 119 | 145 | -0.5 | -2.5 | 0.6 | 15 | 13 | 16 |
| Pigs | 63 | 74 | 76 | 108 | 0.6 | 0.1 | 1.1 | 65 | 72 | 83 |
| Poultry | 558 | 1248 | 2075 | 3815 | 4.5 | 5.7 | 1.9 | 1.2 | 1.5 | 1.9 |
| **Near East/North Africa** | | | | | | | | | | |
| Cattle and buffaloes | 37 | 37 | 39 | 62 | 0.0 | 0.7 | 1.5 | 107 | 158 | 194 |
| Sheep and goats | 205 | 241 | 256 | 350 | 0.9 | 0.5 | 1.0 | 14 | 16 | 20 |
| Poultry | 215 | 722 | 1101 | 2135 | 6.3 | 4.9 | 2.1 | 1.1 | 1.1 | 1.6 |
| **South Asia** | | | | | | | | | | |
| Cattle and buffaloes | 293 | 348 | 384 | 424 | 1.0 | 1.0 | 0.3 | 95 | 121 | 151 |
| Sheep and goats | 148 | 241 | 289 | 405 | 2.5 | 1.7 | 1.1 | 11 | 12 | 15 |
| Pigs | 6 | 12 | 17 | 23 | 3.4 | 3.6 | 1.0 | 35 | 35 | 55 |
| Poultry | 232 | 472 | 717 | 2256 | 4.4 | 4.7 | 3.6 | 0.9 | 0.9 | 1.6 |
| East Asia | | | | | | | | | | |
| Cattle and buffaloes | 121 | 153 | 183 | 268 | 1.8 | 2.0 | 1.2 | 147 | 144 | 176 |
| Sheep and goats | 174 | 226 | 312 | 455 | 2.0 | 3.1 | 1.2 | 12 | 13 | 15 |
| Pigs | 221 | 393 | 470 | 602 | 2.4 | 1.7 | 0.8 | 47 | 75 | 83 |
| Poultry | 1195 | 3171 | 5930 | 9529 | 6.5 | 6.1 | 1.5 | 1.3 | 1.6 | 1.8 |
| **Industrial Countries** | | | | | | | | | | |
| Cattle and buffaloes | 263 | 253 | 254 | 243 | -0.5 | 0.2 | -0.1 | 212 | 284 | 308 |

| | | | | | | | | | | |
|---|---|---|---|---|---|---|---|---|---|---|
| Sheep and goats | 397 | 394 | 341 | 358 | -0.1 | -2.2 | 0.2 | 16 | 17 | 20 |
| Pigs | 172 | 206 | 210 | 220 | 0.7 | 0.4 | 0.1 | 75 | 85 | 89 |
| Poultry | 2167 | 2941 | 3612 | 4325 | 1.8 | 2.2 | 0.6 | 1.4 | 1.8 | 2.1 |
| **Transition Countries** | | | | | | | | | | |
| Cattle and buffaloes | 127 | 152 | 87 | 94 | -1.0 | -6.4 | 0.2 | 144 | 155 | 170 |
| Sheep and goats | 185 | 193 | 85 | 95 | -1.9 | -9.3 | 0.3 | 14 | 15 | 18 |
| Pigs | 97 | 139 | 81 | 82 | -0.5 | -6.2 | 0.0 | 77 | 82 | 84 |
| Poultry | 906 | 1622 | 920 | 1287 | 0.4 | -6.9 | 1.1 | 1.3 | 1.4 | 1.6 |

*Source: https://www.fao.org/4/y4252e/y4252e07.htm#:~:text=Annual%20growth%20of%20meat%20production,percent%20(484%20million%20tonnes).*

Nonetheless, there are significant variations between nations and regions. There was a fall in meat productivity in sub-Saharan Africa, where the increase in cattle numbers outpaced the growth in production. The growth in cattle and buffalo herd sizes in Asia, a region with limited land, was substantially less than the growth in production, suggesting that intensification and higher productivity were comparatively more significant.

The rises in the production of eggs and white meat were also brought about by increases in productivity. Compared to developing nations, industrialized nations continue to produce more meat or milk per animal. For instance, the average milk yields in industrialized and developing nations in 1997–1999 were 1.1 and 5.9 tonnes pa, respectively, while the yield of beef per animal (carcass weight) in developing nations was 163 kg versus 284 kg. Because production techniques are easier to transfer and adopt, pork and poultry productivity levels are more similar across regions.

### *5.1.2.2. Production Systems*

The primary production systems found in developing nations are significantly impacted by shifts in the global market for animal products as well as growing resource constraints.

### *5.1.2.3. Grazing Systems*

Roughly 25% of the world's land is dedicated to grazing, with extensive pasture accounting for 23% of mutton production and 30% of total beef production (FAO, 1996e). Extensive grazing systems have generally enhanced production in developing nations by herd expansion as opposed to significant productivity gains. On the other hand, in comparison to other production systems, the market share of these large systems is decreasing globally. Rangelands are becoming less available due to arable land encroachment, land degradation, conflict, and other factors. Thus, in these systems, there is still little room for herd numbers to grow further.

### *5.1.2.4. Crop-livestock Production Systems*

The majority of ruminant livestock are found in mixed farming systems in developing nations. According to estimates, they supply more than 65% of beef, 69% of mutton, and 92% of cow milk (FAO, 1996e). It is common knowledge that the production of crops and livestock complement each other. While livestock provide animal traction, manure, food, a form of savings or collateral, income diversification, and risk reduction, crops and crop residues supply feed. While short-cycle species, like pigs and chickens, are frequently crucial for meeting immediate financial needs and ensuring the security of household food supplies, only ruminants possess the ability to transform highly fibrous materials and forages with minimal or no alternative use into valuable products. About half of all cropland in developing nations is under cultivation thanks to the work of an estimated 250 million work animals.

### *5.1.2.5. Intensive Industrial Livestock Production Systems*

Asia is seeing the biggest intensification trend due to a lack of available land and a surplus of labor at a low cost. Small-scale intensive systems that require less land but require more labor, like "cut and carry" and stall feeding, have been encouraged as a result. A greater availability of capital has made it possible to invest in equipment, buildings, and inputs like better breeds, concentrate feeds, and veterinary medications. This has increased productivity and explained why the production of monogastric animals—like pigs and chickens—has grown more quickly than that of ruminants. As a result, the value of livestock's alternative uses has decreased as the value of its food products has risen in proportion.

Semi-intensive and intensive dairying have grown in sub-Saharan Africa in close proximity to urban areas and, where agro-ecological conditions allow, on the basis of farmed fodder and agro-industrial byproducts. In response to high levels of urbanization and a return to economic growth in the 1990s, intensive poultry production and, to a lesser extent, dairying have emerged in Latin America.

In the developing world, especially in East Asia, the number of large-scale, vertically integrated, intensive industrial systems for the production of pigs and poultry has dramatically increased. They require extremely competent technical and business management, and they employ enhanced genetic material and complex feeding systems. They also require high-energy, high-protein diets as well as preventive measures for animal health, and they use a significant amount of fossil fuels both directly and indirectly. The ease and speed at which the necessary equipment and infrastructure can be operationalized in so-called "turnkey" operations has made the wholesale transfer of these production systems possible. In recent years, the annual growth rate of industrial livestock production has exceeded six times that of production based on grazing (0.7%; FAO, 1996e) and doubled that of more traditional, mixed farming systems (4.3% against 2.2%). Because they have shorter reproductive cycles than ruminants and can turn feed concentrates (cereals) into meat more effectively, pigs, and poultry production has seen the biggest expansion in industrial systems. Currently, 74% of the world's poultry production, 40% of pig meat, and 68% of eggs are produced by industrial enterprises (FAO, 1996e).

## 5.2. LIVESTOCK IDENTIFICATION AND TRACEABILITY

In all 47 counties in the nation, animal identification and traceability will be applied to animals in all production systems and ecological zones. All domestic animal species and livestock species will gradually undergo this.

Considerations such as risk assessment results, the state of animal and public health, including zoonoses and related programs, animal population parameters like species and breeds, numbers, and distribution, production types, animal movement patterns, available technologies, trade in animals and animal products, cost-benefit analysis, and other economic indices, as well as geographical, environmental, and cultural factors, will all be taken into account when designing and developing the system.

Affordability, negative effects on the food value chain, animal welfare, dependability, sustainability, utilizing existing technological innovations, forming partnerships with the private sector for sustainability, and the kind of animal are additional factors that were taken into account during the system's design.

The ability to track an individual animal or a group of animals from one location in the supply chain to another is known as livestock traceability.

Three fundamental components form the basis of livestock traceability systems are:

- Animal identification;
- Premises identification; and
- Animal movement.

Traceability systems are important, effective tools that can be used for many things, including:

- Protection of animal health;
- Public health; and
- Food safety.

They can lessen the amount of time it takes to respond to emergencies or outbreaks, which can lessen the effects that disease outbreaks, problems with food safety, and natural disasters have on the economy, the environment, and society.

## 5.2.1. Animal Identification

Animal identification is the process of assigning a unique identity to a single animal, a group of animals, or a combination of two or more identifiers using characters, signs, symbols, letters, or naming. This involves identifying animals according to their breed, species, age, sex, and color and connecting them to their owners and the locations where they are raised.

Identification of livestock makes it easier to track down animals and animal products for disease prevention, animal health services, livestock trade, ownership, animal welfare, certification, food safety, production, breeding, and genetics, and security. Important information for planning is provided by livestock identification. At the pre-export finishing and processing stage, all live animals for export that have not yet been identified will be tagged or branded appropriately using authorized identification devices and registered in the central database. For use in Kenya, identification devices must have color and alpha-numeric codes prescribed by the Director of Veterinary Services (DVS). The DVS will establish distinct county color codes for livestock identification in areas where cattle rustling is likely to occur, in consultation with participating county governments and the ministry in charge of internal security.

### *5.2.1.1. Identification Methods*

A few of the current techniques for identification are RFID ear tags, visually printed ear tags, hot-iron branding, paint marks, photographs, biometrics, rumen bolus, microchips, ear notching, and naming.

### *5.2.1.2. Prescribed Technologies for Kenya*

The following are the recommended identification methods in Kenya:

- **Large Food Animals:** Electronic ear tags and/or printed visual ear tags.

- **Small Food Animals:** Electronic ear tags and/or printed visual ear tags.
- **Companion and Sport Animals:** Microchips and/or collars
- **Others:** The DVS will prescribe appropriate identification methods for other animal species from time to time.
- **Wildlife:** As determined by the Kenya Wildlife Service in consultation with the DVS.

To serve the nation and the county, the DVS will approve and authorize frequencies and identification coding in accordance with ICAR and ISO guidelines.

## 5.2.2. Animal Registration

Under the supervision of the appropriate authority, registration is the process by which data about animals, including identification, health, movement, certification, breed, sex, production, and family lineage, is gathered, documented, safely stored, and made suitably available for use by producers, service providers, regulators, and industry. Additionally, ANITRAC will work to register important facilities for animal holding, marketing, and production, including quarantine stations.

## 5.2.3. Traceability

The ability to track an animal or group of animals through all phases of their lives is known as animal traceability, whereas product traceability is the ability to determine who supplies or receives an animal product. To achieve traceability, ANITRAC will collect, preserve, and strategically access data to key nodes along the animal value chains, such as feed lots, markets, processing facilities, slaughter facilities, and quarantine facilities. Information about animal registration, the location, the history of the animal's health and husbandry, the animal's movements, and the ownership and termination trail are some examples of this data. Using bar codes, QR codes, or other appropriate technologies, products will be graded at termination and connected to the animal records.

## 5.2.4. Tracking

Animal tracking is the ability to determine an animal's location at any given time. It makes use of active radio signal detection and satellites.

Globally, animal identification is predicated on traceability and food safety regulations; however, due to Kenya's distinct livestock production methods and high rate of livestock theft, there is a significant need for real-time tracking technologies to combat cattle rustling.

Battery power, device size, and cost limitations make it impractical to implement this system on a large scale in public programs. Real-time remote tracking that is both universal and reasonably priced is currently not possible with any off-the-shelf technologies.

The size, covertness, and powering of devices may be addressed by current and upcoming technological advancements as well as satellite communication. It is imperative that any new technology undergo field testing and validation by the Director of Veterinary Services to determine its suitability and alignment with regional systems prior to being put into practice.

## 5.3. MAJOR PERSPECTIVES, ISSUES, AND POSSIBLE POLICY RESPONSES

Several chosen livestock-related topics that are anticipated to gain significance over the projection period will be covered in this section.

### 5.3.1. Livestock, Economic Development and Poverty Alleviation

The identification and promotion of broad-based income opportunities that could result in significant pro-poor growth is a major challenge for development and poverty alleviation. A significant chance to improve the lives of many of the world's impoverished people seems to exist in the livestock industry. Approximately 675 million rural poor people currently depend on livestock ownership for their livelihoods (Livestock in Development [LID], 1999). These people rely entirely or in part on livestock for their livelihood. In addition to increasing overall farm productivity, livestock can offer a consistent supply of food and revenue. Because they permit the exploitation of resources belonging to common property for private benefit, they are frequently the only source of income available to those without land. Furthermore, the only way to diversify risks and build assets that can keep the rural poor in marginal areas from falling into poverty is frequently through livestock. According to recent statistics, women make up an estimated 70% of the impoverished, and for them, livestock is one of their most valuable assets and sources of income (DFID, 2000a). Additionally, owning livestock tends to boost employment opportunities and the consumption of animal products. The majority of the food consumed in developing nations is still produced by semi-subsistence farmers, despite the trend toward larger production scales and vertical integration. Since the rural poor already have a substantial stake in livestock production, the anticipated increase in demand for animal products presents opportunities for them. Regretfully, the vast majority of rural impoverished people have not been able to benefit from these opportunities up to this point. Processors and traders, middle-class urban consumers, and a small number of large producers in high-potential areas with easy access to markets have so far benefited the most.

The impoverished are prevented from utilizing the potential for development by a variety of obstacles. Below is a summary of these.

### *5.3.1.1. Financial and Technical Barriers*

Small farmers are prevented from increasing their output by financial obstacles. Their capital wealth is frequently exceeded by the necessary investment. To ensure that the poor are included in the future expansion of livestock production and processing, policies and institutions must support targeted small- to medium-sized credit programs that are based, among other things, on the strengthening of property rights. Small producers are unable to provide the market with a product that is both safe and reasonably uniform due to technical obstacles. The negotiating power of small production units is impacted by the lack of infrastructure necessary to preserve perishable goods, especially if those units are far from centers of consumption. As a condition of trade, sanitary regulations (which include animal welfare) create technical barriers. Real or perceived incidence of livestock diseases may keep certain farmer groups or entire nations out of local, regional, and global markets. Institutions and policies need to support the development of product standards and safety norms that protect public health without discriminating against smaller producers and while facilitating access to technologies, goods, and services.

Higher production and transaction costs together have the potential to reduce small producers' access to markets and render them uncompetitive. Their production costs are typically higher than any cost benefits from the discounted value of family labor because they are not able to take advantage of the economies of scale that are available to large-scale units. There is a dearth of objective data to evaluate the potential effects of covert and overt subsidies on small-scale producers, public health, and the environment as they relate to the easy access to inexpensive animal products in urban areas.

Due to the limited amounts of marketable product and the lack of suitable physical and market infrastructures in remote locations, transaction costs may be unaffordable for small-scale producers. When producers are dependent on middlemen and lack access to market data or negotiating power, transaction costs are also raised. The development of market infrastructures, such as suitable information systems that empower small-scale producers to make well-informed marketing decisions, requires public policies. By lowering transaction costs, producers' associations or cooperatives allow them to take advantage of economies of scale.

A sustainable reduction in poverty requires lowering risks and lessening their impact on those who depend on livestock but are otherwise impoverished. Production and market risks are linked to small-scale production. Market risks are frequently linked to a weak negotiating position and include changes in the prices of both products and inputs. Even though subsistence farming frequently has effective risk-reduction techniques, many small-scale farmers lack the resources or know-how to continue taking full risks associated with the market. Safety nets are necessary to help the impoverished deal with the inevitable economic shocks that come with free markets if they are to fully engage in the economy.

Disease outbreaks, extreme weather events like droughts and floods, and resource degradation are the main causes of production risks. Droughts and epidemic diseases can devastate both intensive and small-scale livestock production systems, but the impoverished are especially susceptible to these kinds of shocks due to their lack of resources and insurance policies. In poor countries that are prone to disasters, both public and private services are almost always ill-equipped to prepare for such risks and to react quickly enough. Enhancing the ability of communities and institutions to respond to emergencies is crucial, and public policies should incorporate plans for preparedness against diseases and droughts.

Development initiatives in the livestock industry haven't always been very effective. Inappropriate technology and a failure to provide services to underprivileged farmers have contributed to the failure of numerous livestock development projects. Even in situations where the focus was clearly pro-poor and the technologies were appropriately targeted, many technological projects have fallen short of significantly improving the lives of the impoverished. Undoubtedly, for interventions and strategies aimed at the poor to be sustained, a supportive institutional and political environment is necessary (LID, 1999; IFAD, 2001).

## 5.3.2. Livestock Health, Welfare, and Nutrition

### *5.3.2.1. Animal Health*

In numerous developing regions, infectious and parasitic diseases of livestock continue to pose significant obstacles to the production of more profitable and productive livestock. Diseases have two main effects on farm incomes: first, they cause significant losses in productivity and stock, which requires farmers to spend money and labor controlling them; second, they result in export restrictions.

In developing nations, infectious diseases like rinderpest, foot-and-mouth disease, contagious bovine pleuropneumonia, classical and African swine fever, and peste des petits ruminants continue to pose serious risks to livestock production. In industrialized nations, they also pose a threat to production through the increased movement of people, livestock, and livestock products.

The goal of eliminating rinderpest worldwide by 2010 is still attainable. Only through intensive, globally coordinated animal health programs can the other epizootic diseases be brought under control gradually. Countrywide eradication programs have largely given way to more adaptable control strategies that concentrate interventions on regions with the highest potential returns. Animal health economics and risk analysis are used to identify areas where investments in disease control will be most beneficial. Because of the significant externalities associated with these disease outbreaks, the public sector is still in charge of managing their control. But public funding for disease control has been declining over the past few decades, and many developing nations lack adequate veterinary facilities.

Trypanosomosis, also known as sleeping sickness, is one of the parasitic diseases that severely limits cattle production in the majority of Africa's humid and subhumid regions. The gradual recovery of infested areas for mixed farming is possible with the help of trypanocidal medications and adaptable combinations of adhesive pyrethroid insecticides, impregnated screens and traps, sterile insect techniques (SIT), and aerial spraying. The advantages of these tactics will far outweigh those of the increased livestock production alone. They will improve human welfare by preventing sleeping sickness, generating higher and more stable incomes, and enhancing nutrition. They will also increase crop output and permit the introduction of mixed farming.

Tick-borne illnesses and helminthosis are especially significant from a production perspective. Although they are rarely lethal, helminths (worms) can have a significant impact on output and profitability. The improper use of anthelmintics can lead to parasite resistance to medications, despite the fact that helminths can be effectively controlled. Diseases, particularly East Coast fever in eastern and southern Africa, can be spread by ticks. However, traditional acaricide dipping for tick control is getting too expensive and causes environmental issues with the way the waste chemical is disposed of. Nutritional imbalances and diseases that impair reproductive performance will also become more significant as production systems become more intensive.

Thus, both endemic and epidemic disease agents will present increasing challenges to intensifying livestock production. For instance, in many regions of China, high densities of pigs and poultry near high human population concentrations serve as breeding grounds for both established and newly emerging diseases, such as avian influenza and pig Nipah virus. India has an extremely high density of cattle, making foot-and-mouth and other diseases endemic. Rift Valley fever and bluetongue are among the illnesses that affect the vast numbers of sheep in the Near East. Additionally, certain zoonoses (livestock diseases) can spread from animals to people directly or via a middle host (e.g., new strains of influenza via mosquitoes).

Some livestock diseases have been eradicated and the incidence of others has been significantly decreased in industrialized nations due to centuries of methodical eradication efforts. With the increasing intensity of international travel and the expanding long-distance trade in animals and livestock products, maintaining this situation is getting harder and harder. The response includes veterinary barriers, such as the foot-and-mouth disease control program that is safeguarding Australia and Indonesia or the screwworm control program that began in the United States and Mexico and is currently safeguarding nearly all of Central America.

Through better animal disease prevention, diagnosis, and treatment, biotechnology holds the potential to address some technological limitations. For instance, the development of novel vaccines employing recombinant antigens to pathogenic agents may very well benefit from the application of genomics. Future developments should bring about a far greater variety of affordable, efficient, simple-to-use vaccines that do not require a cold chain. Veterinary services provided in developing nations will be improved by the development of reliable, affordable pen-side diagnostics. However, to keep up with the rapid advancements in technology, institutions involved in the control of animal diseases at all levels—from local to international—must greatly improve their coordination, as well as their epidemiological and logistical capabilities.

### *5.3.2.2. Animal Welfare*

Uncontrolled intensification of livestock production is linked to animal management techniques that prevent the animals from expressing their natural behavior, such as restricting their movement, space, and light. In wealthier societies, such practices are becoming less and less popular. Similar concerns are raised regarding specific feeding and medication procedures, as well as the long-distance transportation of animals.

The skeletal and circulatory systems of broilers are strained by the rate of muscle growth, which increases the risk of heart failure and broken limbs. Genetic selection for increased weight in broilers has also been connected to health issues in the animals. Over the anticipated period, such concerns are probably going to have a bigger impact on production systems, especially in the industrial world. EU regulations pertaining to battery hens' minimum cage sizes already reflect this trend.

#### *5.3.2.3. Feed Quantity and Quality*

The vast majority of animals worldwide experience either seasonal or permanent nutritional stress, especially ruminants in low-input mixed farming systems and pastoral environments. Optimizing the utilization of fibrous plant material is a top priority. Proven methods for handling crop residues and other low-quality roughage have been made possible by a better understanding of how the rumen works, and more advancements are likely to come. Enhancing the rumen's ability to break down diets rich in fiber has the potential to significantly enhance ruminant productivity in regions where roughage of poor quality is easily accessible.

Fertilizer shortages during the dry season can be mitigated by cultivating, gathering, storing, and conserving fodder, forage, and residues. Considerable opportunity exists for combining a variety of superior products (e.g., legume feeds) into the diets of animals. Feeding strategically and adding essential ingredients (e.g., nitrogen or minerals) can enhance the overall uptake of poor diets. A variety of enzyme additives have been introduced by the feed industry to improve the quality and nutritional value of feeds. For instance, it has been demonstrated that silage additives containing bacterial inoculants and the enzymes cellulase and hemi-cellulase enhance preservation and increase digestibility. Antinutritional factors (ANFs), like phytic acid in maize, have been identified in plants by plant genomics and phytochemistry. Under test conditions, it has been possible to successfully transfer a detoxifying gene (dehalogenase) from the soil bacterium Morexella to rumen bacteria. This suggests that ruminant ecosystems may benefit from the addition of microbes that have a better capacity to detoxify ANFs. Feed additives that are microbial and ionophoresic have been demonstrated to boost feed conversion efficiency by as much as 7.5%; however, the exact mechanisms and intricate interactions between breed, microorganism, and animal remain unclear, and the results are inconsistent. Profitability for pigs and poultry will depend on increasing feed conversion efficiency. When necessary, it will also lessen reliance on feed grains. Feed conversion rates have increased by 30%–50% for pigs and poultry over the last 10 years, partly due to improved breeding practices and partially due to the addition of enzymes to feeds. Nevertheless, only 25%–35% of the nutrients ingested are retained in the finished products in monogastrics. It is anticipated that increased knowledge of the biochemistry and physiology of the digestive system will enhance feed utilization.

### 5.3.3. Livestock and Trade

The past 10 years have seen a sharp increase in the trade of meat products due to the growing demand for livestock products, especially in developing nations. A shifting policy environment that has lowered trade barriers has enabled this dynamic growth. Technical aspects like growing specialization in production and processing processes, as well as developments in cold-chain and transportation technologies, have also supported it. Trade patterns will increasingly be determined by the complexity of the world's meat markets, which is being fueled by rising consumer demands regarding product type, quality, and safety. The ability of meat exporters to adapt to quickly shifting consumer preferences and the numerous international laws pertaining to food safety and animal health standards will become more and more important to their competitiveness.

The ongoing specialization in production and processing is one of the elements influencing the direction of the global meat trade. Different kinds of meat cuts will be traded between nations as a result of growing production specialization brought on by consumer preferences for particular meat types and cuts. For instance, China purchases premium grain-fed beef from the United States while the former imports beef fit for manufacturing. Japan receives China's imports of chicken feet and wings as well as higher-value processed cuts from China, which occasionally use cheaper imports of foreign cuts as an input. Future developments in consumer sophistication and demand will accelerate the process of increasing product differentiation to cater to diverse preferences across markets, resulting in an increasingly complex and diverse meat economy.

It is anticipated that consumers' concerns about the production and distribution methods of meat will further complicate the global meat product trading system. The increased risk of food-borne infections, zoonoses, and animal diseases spreading internationally due to increased trade in livestock and livestock products is another factor contributing to the complexity. Divergent food standards, sanitary assurances, and certification processes could proliferate as a result of increased scrutiny of meat production systems and the "hoof to plate" approach to food safety and quality. Developing nations that export meat would be at a disadvantage because it would be more expensive for them to comply with these evolving regulations.

Countries have supported efforts to establish science-based food safety regulations in terms of the sanitary and phytosanitary standards (SPS) under the Uruguay Round global trade agreements in response to concerns about animal disease and food safety. The goal of the SPS Agreement is to do away with the use of irrational, non-scientific regulations to impede trade. Countries are still free to determine their own "acceptable risk" or desired level of protection for food safety, animal health, and plant health under the new regulations, but they must either follow international standards or, if they choose to set their own, support them with scientific evidence. Industrialized nations have primarily used and tested this right. However, in the future, challenging and contentious trade disputes may center on national standards as well as certain international standards. Additionally, they will put more pressure on both industrial and developing nations to improve the way veterinary and public health services are delivered.

Along with efforts to harmonize measures based on international standards of the International Office of Epizootics (OIE) and the Codex Alimentarius Commission, the SPS notification process should offer a mechanism for introducing increased transparency into the system of notification, consultation, and resolution of trade disputes. The fact that the formal dispute settlement body has only been involved in one livestock/meat case to date may be a sign of the SPS Committee's effectiveness. With Canada's help, the World Trade Organization Hormone Case settled a protracted conflict between the US and the EU in 1999. Imports of red meat from animals given six different growth promotants have been prohibited by the EU since 1989. In retaliation for the EU's refusal to allow hormone-treated beef on its market, WTO arbitrators ruled that the US and Canada could halt tariff concessions for specific EU goods.

Reduced construction, operating, and maintenance costs as well as the expansion of cold-chain facilities have aided in the trade of livestock products. This has significantly decreased transportation costs and opened up the possibility of transporting fresh livestock products to a far wider range of nations. Previously, these kinds of advancements might have encouraged agribusinesses to relocate their operations to nations with laxer labor, environmental, and public health laws. But these changes are no longer supported by global regulations and consumer concerns. More controls and infrastructure investment are required by stricter food sanitation laws, labeling, and certification requirements, and other requirements that may be challenging for most developing nations to meet in the near future. The expenses and dangers associated with transporting animal feed as opposed to livestock products are other considerations. Even when measured in livestock product equivalent units, shipping animal feeds is still far less expensive than shipping livestock products. Furthermore, compared to fresh livestock products, livestock feeds are much less perishable and pose less of a risk during transportation.

Technological developments in information, transportation, and packaging will remain important factors in meat trading trends. They will make it possible for meat shippers to transport fresh, frozen, and chilled meat products to customers 1000s of kilometers away without experiencing a significant decline in quality. A sizable portion of the overall cost of transportation has always gone toward loading and unloading. However, these front-end and back-end expenses have drastically decreased with the use of containers. Trade transaction costs will continue to decline as a result of shorter delivery times, better pre-

and post-shipment handling procedures, and technological advancements that prolong shelf life, including packaging.

Enhancements to internal transportation networks in developing nations will impact their trade and consumption habits. It will be simpler to transport imported perishable meat products within the nation as infrastructure develops and intermodality—the linking of two or more modes of transportation—improves.

The types of meat that can be transported in refrigerated containers without spoiling will increase as a result of improvements in cold-chain facilities, additional refinements, and controlled atmosphere (CA) technologies.

Attention will be drawn to each nation's comparative advantage and global competitiveness as a result of the ongoing liberalization of meat markets and increased regional integration, particularly the EU's accession of Eastern European nations.

The patterns of the meat trade will increasingly be determined by the cost of labor and capital as well as the capacity of industries to adapt to consumer demands in meat processing. Cross-border investments in meat production and processing, combined with the quick speed of technology transfer, could hasten trade changes and boost product diversity.

Growing environmental regulations that target animal feeding operations and waste management have put pressure on the competitiveness of traditional developed country meat exporters. Strict phosphate quota laws, waste treatment laws, storage, and field application limitations, and, more recently, direct output controls are thought to cost the Netherlands up to US$4.05/hog (Metcalfe, 2001).

According to estimates, waste management expenses in the US range from US$0.40 to US$3.20 per hog, or 1%–8% of the total costs associated with producing hogs. The implementation of these regulations leads to increased expenses for producers and a decline in competitiveness for particular sectors, especially those within the European Union.

Animal production-related social, health, and environmental concerns are increasingly being brought up in WTO negotiations. If these concerns are officially recognized as "trade issues" and associated international regulations are harmonized (as has been done with animal diseases, zoonoses, and food safety), the competitive landscape for the animal industries will shift. Exporter competitiveness will also be impacted by choices about how to address traceability, meat labeling, and animal welfare issues. In particular, it will hurt exporters from developing nations who, due to their smaller export operations, will incur higher compliance costs.

## 5.3.4. Intensification: Public Health Risks and Consumer Choices

### *5.3.4.1. Food Safety and Emerging Zoonoses*

Having access to wholesome food products is a significant public benefit. There are risks to human health associated with animal products, particularly animal fat, but some of these risks are limited to overconsumption. Products made from meat, milk, and eggs are healthy when consumed in small to moderate amounts and offer vital vitamins, minerals, and amino acids. In fact, increasing the consumption of animal products in developing nations will aid in the fight against certain types of undernutrition.

Concerns regarding diseases, microbial contamination of food, and overall food safety have increased as a result of changing dietary habits, the importation of feedstuffs from far-off places, and growing livestock densities, especially in peri-urban and urban areas.

The risk and pattern of disease transmission may change as a result of modifications to production systems, feeding procedures, and animal feed safety. When production systems change without corresponding changes in veterinary regulations and enforcement, a zoonotic disease can become a serious risk, as demonstrated by the recent increase in human cysticercosis in eastern and southern Africa after the expansion of pork production.

Humans can contract some livestock diseases (zoonoses), such as brucellosis and tuberculosis, and livestock populations may give rise to novel zoonoses (e.g., bird flu and nipah). The development of bovine spongiform encephalopathy (BSE) in cattle and its consequences for human health (variant Creutzfeld-Jakob disease) and the livestock industry amply illustrate the possible risks. Over the past 30 years, an estimated one new disease has been discovered annually.

Eggs, milk, and meat are perishable goods that can get contaminated by microorganisms. A few of these, like *Escherichia coli* and *salmonella*, live in the digestive tracts of animals. Thus, throughout the food chain, improper handling, slaughter hygiene, processing, or preservation can lead to microbial contamination and propagation, posing a serious risk to consumers' health. Microbial contamination is the main cause of the ever-increasing sanitary standards placed on the processing and transportation of livestock, even though many of these contaminants have little to no effect if products are prepared correctly. More affluent consumers willing to pay more for organic or free-range livestock products, along with stricter regulations, will probably have a long-lasting impact on production practices, especially in industrialized nations. Changes will include the implementation of pathogen reduction or elimination programs along the production chain, such as the removal of Salmonella enteritidis from breeder flocks, and pre-harvest food safety programs using the Hazard Analysis Critical Control Point (HACCP) concept at the farm level, from breeding to the slaughterhouse.

Chemical and biological pollutants are also a concern for food safety. For example, aflatoxins are very important in warm, humid environments, and drug residues are another important class of contaminants.

Basic food quality control systems are developing into quality assurance systems, which are then progressing towards total quality management (TQM) systems, in an effort to enhance food safety from feed production to the store shelf. The food chain will become even more integrated and concentrated as a result of the high costs of complying with these systems.

### *5.3.4.2. Antimicrobials and Hormones*

Antimicrobials are used extensively in the livestock industry for reasons other than medical treatment. Supplementing feed with subtherapeutic doses of antimicrobials to boost growth rates is a common practice in high animal density modern farming systems. However, persistent antimicrobial exposure encourages the growth of resistant bacteria. Through the exchange of genetic material between microorganisms, farm animal pathogens that are resistant to antibiotics can spread to human bacteria. Due to the need for more costly medications and longer hospital stays, this may raise the expense of public health care. The fact that no truly novel antibiotics have been developed in the last 10 years adds to the problem of antibiotic resistance. Although improper use of these medications in human medicine is the primary cause of most antimicrobial resistance in human pathogens, the use of antimicrobials in the livestock industry also contributes to this problem.

The EU has imposed a partial ban on six antibiotics that are also used to treat humans, and the World Health Organization (WHO) has recently called for a ban on the practice of giving healthy animals low doses of antibiotics to increase their productivity. Livestock producers are vehemently against these measures, but they are likely to gain traction given consumer demands. In 1998, the Danish poultry industry made the voluntary decision to stop using any antimicrobial growth promoters, even though there were worries that this would lead to higher mortality and lower productivity. Against expectations, the feed-conversion ratio increased only slightly and mortality was unaffected. The biggest chicken producer in the United Kingdom has made similar moves.

Around the world, hormones that boost feed conversion efficiency are widely used, especially in the cattle and pig industries. Science has not demonstrated any harmful effects on human health as a result of their proper application. Nonetheless, the EU has adopted a strong stance against the use of hormones in livestock production, in part due to consumer pressure. Significant trade disputes have resulted from this with the United States. These illustrations demonstrate how consumer concerns will progressively impact both the final product's quality and its manufacturing process.

### *5.3.4.3. Livestock Biotechnology*

Biotechnology is defined as "any technological application that uses biological systems, living organisms, or derivatives thereof, to make or modify products or processes for specific use" by the Convention on Biological Diversity (CBD). Applications of biotechnology to crops have dominated much of the discussion to date, but animals can benefit from many aspects of the debate as well, such as biodiversity and intellectual property rights.

In terms of animal biotechnology, there are two main categories: DNA-based genetic interventions, which include marker-assisted selection, DNA vaccines, GM marker vaccines, recombinant vaccines, and the creation of genetically modified animals; and reproductive interventions, which include artificial insemination (AI), embryo transfer (ET), *in vitro* maturation, fertilization, and sexing, and cloning.

AI is a well-established technology, especially in the industrialized world's commercial dairy industry. The more favored production environments of developing nations, where the need for milk provides an economic incentive for its introduction, are likely to see the widespread use of AI in the future. High genetic potential cows can have significantly more calves thanks to ET than they could through regular reproduction. However, in certain industrialized nations, this technology is currently restricted to a tiny portion of commercial herds and breed improvement initiatives. In the near future, it is unlikely that this and other cutting-edge reproductive technologies will become extensively used in developing nations.

Recent developments in mammalian cell cloning have the potential to have a significant impact, especially for dairy cattle in industrialized nations. Nonetheless, there is a risk of further reducing disease resilience and the genetic base. Only 2%–5% of animal cloning attempts currently succeed, and the resulting animals frequently experience severe health issues. Furthermore, cloning remains inefficient. Thus, there are still a lot of difficult ethical and scientific questions surrounding cloning.

Animal breeding is probably going to be significantly impacted by the quick rise in knowledge of animal genetics. The use of neutral markers in conjunction with the direct identification of genes influencing critical traits for disease resistance or economic performance could expedite genetic improvement.

This technology is being developed for a wide range of traits and may offer developing nations a shortcut to genetic development for both productive traits and adaptation to unfavorable environmental conditions like diseases and climate.

Many developing nations underutilize current technologies, including the more fundamental biotechnologies mentioned above. Numerous factors contribute to this, such as a lack of investment in infrastructure related to education, health, finance, communication, and transportation, which would lower the expenses and risks associated with product marketing and input supply (Thompson, 1997). These restrictions are probably going to make it even harder to use the new, advanced technologies, which could lead to a wider rather than narrower technological divide between industrialized and developing nations.

Large private companies carry out more than 80% of biotechnology research activities for commercial exploitation to satisfy the demands of developed markets and large-scale commercial producers (Persley and Lantin, 2000). As a result, they are probably not going to be very appropriate for the circumstances of small-scale farmers in tropical areas, which could result in a rise in income and wealth inequality within nations. Furthermore, the introduction of commercial products may be impeded by legal registration requirements, which favor large, primarily multinational corporations with the financial and institutional capacity to comply with stringent registration requirements.

### *5.3.4.4. GM Livestock and Feeds*

It's challenging to determine how many genetically modified animals are currently being produced. The great majority have been modified to produce various products, such as human proteins in their milk, and have been employed in biomedical research. Currently, scientists are working to alter chickens so that their eggs express specific compounds that could be used medicinally.

Also being produced for potential direct human consumption are genetically modified cattle, sheep, pigs, and chickens, though these are still mostly in the experimental stage. To accelerate growth rates, for instance, genes are being injected into cattle in an effort to increase endogenous growth hormone production. But thus far, the welfare of the animals has suffered as a result of this procedure.

The primary hazards posed by genetically modified livestock to the environment, human health, and animals are pleiotropy—unexpected changes in form and function brought on by gene insertion—particularly when these occur without adequate testing prior to broad release.

Humans do not currently consume genetically modified animals or their products. However, products from livestock fed genetically modified crops are available on the market (e.g., corn and soybean feeds used in poultry meat and egg production and cottonseed cake in dairy production). The Indian government ordered transgenic cotton field tests in 2001 in many agroclimatic zones to evaluate the product's environmental safety. Transgenic cottonseed and oil were tested for potential effects on animal health, milk production, and quality in relation to human health and toxicity through parallel nutritional studies conducted in cows and buffaloes.

Although GM food and feed have not yet had a significant impact on global markets, some biotechnology-related changes to international trade are anticipated. Currently, importers have the option of importing feed from GM-free regions or segregated non-GM products from GM-producing regions to meet the high demand for GM-free food. Food product market segmentation is occurring, especially in the EU and Japan, where GM-free soy products are more expensive than those made from a combination of GM and GM-free soy. Since GM-free products can be acquired for a relatively small additional cost, this market development is said to have had little effect thus far on the pattern of international food trade. Consumer demand for goods derived from animals fed GM-free diets is just starting to emerge, though, and this trend could have a big effect on global bulk-based feed handling and commodity trade flows.

## 5.3.5. Livestock, Environment, and Animal Genetic Diversity

The practice of keeping livestock in mixed farming systems can have significant positive effects on the environment. Agricultural intensification has historically been based on mixed crop-livestock systems. Livestock in these systems serve as a means of importing and concentrating nutrients, which is essential for intensification, in addition to speeding up nutrient turnover. Additionally, livestock can play a significant role in maintaining ecosystems and landscapes.

Many controversial issues still need to be resolved, though. In semi-arid and humid regions, widespread poverty-driven land degradation and related biodiversity losses are taking place as a result of growing population pressure, unclear access to land and other resources, and limited access to financial services and markets. By decreasing the productivity of shared resources and raising vulnerability, degradation perpetuates poverty.

Awareness of the related environmental issues has increased as a result of intensive livestock production and processing in both industrial and developing nations, which pollutes the land, water, and air. In addition to causing direct harm to the environment, this pollution has the potential to be a major source of disease transmission. Global warming is also linked to animals. Domesticated animals both directly and indirectly produce the greenhouse gases carbon dioxide ($CO_2$) and methane ($CH_4$), along with trace amounts of nitrous oxide

($N_2O$) and ozone ($O_3$). Below is a summary of additional livestock-related environmental effects include:

- **Land Degradation:** Land tenure, crop encroachment, fuelwood collection, and livestock movement and density all play a complex role in land degradation. The semi-arid regions of Africa and the Indian subcontinent are where it is most noticeable. The situation has frequently gotten worse due to shifting land tenure, conflicts, and settlement and incentive policies that have weakened traditional land use practices. Although the scope of the issue may have been overstated in previous reports of extensive irreversible degradation, there is no excuse for complacency and coordinated action is required.
- **Deforestation:** A significant loss of biodiversity has resulted from the destruction of large tracts of rain forest due to commercial ranching. Policies that encouraged improper ranching were the root cause of the issue, which was mostly, though not entirely, limited to Central and South America.
- **Animal Genetic Diversity:** Intensification is linked to the possibility of reducing the gene pool and thereby losing the genetic diversity of farm animals. There are currently close to 5,000 different breeds and strains of farm animals. There are currently about 600 breeds in danger of going extinct, and many traditional and regionally adapted breeds are predicted to continue to erode.

Numerous new breeds of farm animals have been successfully created in recent decades. As long as they are fed a lot of high-quality feed and are shielded from environmental stressors like diseases, pests, and severe weather, these animals will produce more meat, milk, or eggs. However, as a result of biotechnologies like artificial intelligence (AI), which make it simple to transfer genetic material across international borders, there is now a greater reliance on a smaller pool of genetic resources.

One of the biggest threats to domestic animal diversity is the mass importation of breeds that are suited to high-input production systems from industrial to developing nations. Such introductions are encouraged by development policies, which typically hinder the survival of native breeds. For instance, it is estimated that only about 400 of the world's remaining breeds—nearly all of which are in industrialized nations—are the focus of genetic improvement programs, despite the fact that 4,000 of them are still preferred by farmers.

Farmers can choose stock or create breeds in response to changing market and societal demands, disease threats, environmental changes, and consumer demands by preserving animal genetic diversity. Thus, genetic diversity is a treasure trove of mostly unexplored possibilities. Breeds that can thrive in challenging conditions, use inexpensive feeds, or are resistant to or tolerant of particular diseases could be very beneficial in the long run. Particularly, wild relatives of popular breeds might have valuable but as-yet-unknown traits that could prove advantageous in the present or the future. To adapt livestock production to changing global production environments, tropical animal breeds must be identified, conserved, and used. Adapted livestock can withstand disease and environmental obstacles better, allowing them to continue producing without requiring expensive inputs. This helps farms make more money and reduce poverty.

Given that the majority of indigenous populations have not experienced significant selection pressure for specific traits, there is a great deal of potential for the within-breed improvement of locally adapted breeds. Although these programs are long-term in nature and demand a large amount of dedication and funding, they promise substantial returns (Delgado et al., 1999).

## REFERENCES


1. Backmann, P. (1997). Pesticide inputs now and into the twenty-first century. In B. White & L. Chuang (Eds.), *Crop and Livestock Technologies* (RCS III Symposium). Iowa University Press.
2. Battaglia, R. A. (2001). *Handbook of Livestock Management* (3rd ed.). Prentice-Hall, Inc.
3. De Haan, C., Steinfeld, H., & Blackburn, H. (1998). *Livestock and The Environment: Finding a Balance*. CEC/FAO/WB.
4. Delgado, C., Rosegrant, M., & Meijer, S. (2001). Livestock to 2020: The revolution continues. IFPRI. Paper presented at the Annual Meeting of the International Agricultural Trade Research Consortium (IATRC), Auckland, New Zealand, 18–19 January.
5. Delgado, C., Rosegrant, M., Steinfeld, H., Ehui, S., & Courbois, C. (1999). *Livestock to 2020: The Next Food Revolution*. Food, Agriculture and the Environment Discussion Paper No. 28. IFPRI.
6. DFID. (2000a). *Halving World Poverty by 2015: Economic Growth, Equity and Security*. DFID Strategy Paper. London.
7. DFID. (2000b). *Eliminating World Poverty: Making Globalization Work for the Poor*. White Paper on International Development. London.
8. FAO. (1996a). *Food, Agriculture and Food Security: Developments Since the World Food Conference and Prospects*. Technical Background Document No. 1 for the World Food Summit. Rome.
9. FAO. (1996b). *The Sixth World Food Survey*. Rome.
10. FAO. (1996c). *Mapping Undernutrition: An Ongoing Process*. Poster for the World Food Summit. Rome.
11. FAO. (1996d). *Environment, Sustainability and Trade Linkages for Basic Foodstuffs*. Commodities and Trade Division. Rome.
12. FAO. (1996e). *World Livestock Production Systems: Current Status, Issues and Trends*. FAO Animal Production and Health Paper No. 127. Rome.
13. Fleischauer, E., Bayer, W., & Lossau, A. (1998). Assessing and monitoring environmental impact and sustainability. In *Proceedings of the International Conference on Livestock and the Environment* (June 1997, Wageningen). IAC for FAO and the World Bank.
14. Hendy, C., Nolan, J., & Leng, R. (1995). Interactions between livestock production systems and the environment: Impact domain: Concentrate feed demand. FAO consultancy report for livestock and the environment study, Rome.

15. IFAD. (2001). *Rural Poverty Report 2001: The Challenge of Ending Rural Poverty*. Oxford University Press.
16. LID. (1999). *Livestock in Poverty-Focused Development*. Livestock in Development.
17. Metcalfe, M. (2001). Environmental regulation and implications for competitiveness in international pork trade. Paper presented at the IATRC Symposium on International Trade in Livestock Products, Auckland, New Zealand.
18. Persley, G., & Lantin, M. (2000). *Agricultural Biotechnology and the Poor*. Proceedings of an International Conference, 21–22 October 1999, CGIAR, Washington DC.
19. Pitcairn, C., Leith, I., Sheppard, L., Sutton, M., Fowler, D., Munro, R., Tang, S., & Wilson, D. (1998). The relationship between nitrogen deposition, species composition and foliar nitrogen concentrations in woodland flora in the vicinity of livestock farms. *Environmental Pollution, 102*, 41–50.
20. Thompson, R. (1997). Technology, policy and trade: The keys to food security and environmental protection. Presidential Address of the 23rd International Conference of Agricultural Economists.

# CHAPTER 6

# SUPPLY CHAIN MANAGEMENT AND TRACEABILITY

## Contents

# INTRODUCTION

The ability to follow food items from their farm of origin to their final destination—typically the consumer—is known as supply chain traceability in the context of agricultural products. Documenting and recording each stage of the supply chain, including harvesting, processing, transportation, and distribution, is part of this process. Assuring food safety, quality, and transparency as well as adhering to legal requirements depend on traceability. Companies can minimize the risk of widespread foodborne illnesses and enable targeted recalls by implementing traceability systems, which enable them to promptly identify the source of any contamination or defect.

The growing consumer demand for transparency regarding the origins of their food, its production process, and its compliance with ethical and environmental standards has made traceability in agriculture more and more crucial. Real-time information about the product's journey is provided by technologies like blockchain, RFID (radio-frequency identification), and barcoding, which are frequently used to improve traceability. In particular, blockchain provides a decentralized, unchangeable ledger that documents each transaction, facilitating information verification and trust-building for all parties involved.

Furthermore, by empowering consumers to make knowledgeable decisions and motivating producers to use eco-friendly practices, traceability promotes sustainable agricultural practices. Since many nations now require comprehensive information regarding the origin and handling of agricultural products, it also helps to meet international trade requirements. All things considered, supply chain traceability promotes increases in agricultural production's sustainability and efficiency while also boosting food safety and consumer confidence.

## 6.1. TRACEABILITY IN SUPPLY CHAIN MANAGEMENT

Traceability in supply chain management is the capacity to monitor the flow of materials and products from raw materials to final goods that are delivered to the final customer. Every step of the product's journey, including sourcing, manufacturing, distribution, and transportation, must be meticulously documented and kept on file. By giving businesses insight into the supply chain, traceability aims to enable them to confirm the movement, quality, and origin of goods at any stage of the process. Enhancing supply chain transparency, risk management, and regulatory compliance all depend on traceability. It assists companies in reacting promptly to problems like product recalls, contamination, or flaws by locating the issue and separating impacted products. In sectors like food and pharmaceuticals, for instance, where product safety is crucial, traceability guarantees that businesses can adhere to strict safety standards and reduce the possibility of consumer injury.

To improve traceability, technologies like barcodes, RFID, GPS tracking, and blockchain are frequently employed. Particularly, blockchain is becoming more and more well-liked because it offers a safe and decentralized method of transaction recording, enabling real-time information access and product authenticity verification for all supply chain participants. By allowing businesses to keep an eye on the ethical and environmental effects of their supply chain operations, traceability not only helps with risk management but also advances sustainability. It enables companies to guarantee that their goods are sourced ethically and in accordance with environmental and labor regulations. All things considered, supply chain management traceability boosts operational effectiveness, increases consumer trust, and guarantees that companies can satisfy the rising demand for accountability and transparency.

### 6.1.1. Necessity of Traceability

The EU's food safety policy is based on traceability. As a risk-management tool, traceability gives food business owners or authorities the ability to respond to possible hazards that may occur in food and feed, isolate the issue by recalling or withdrawing products, and subsequently stop tainted or dangerous goods from getting to consumers. Numerous researchers have noted that effective traceability in food supply chains can eliminate food safety hazards and lower the risks and expenses related to food-borne illnesses. For instance, traceability lowers labor productivity losses and medical expenses (Kelepouris et al., 2007; Lee and Özer, 2007; Chryssochoidis et al., 2009; Veronneau and Roy, 2009; Mehrjerdi, 2010), reduces recall scope and time (Hobbs et al., 2005; Banterle and Stranieri, 2008; Bechini et al., 2008; Bevilacqua et al., 2009), and ensures the consistent safety of food.

Another way that traceability helps customers is by giving them targeted and accurate information about a particular product and allowing them to learn about pertinent information about food safety and quality issues. As a result, customers are willing to pay more for products that meet certain guarantees or have a desired origin.

As a result, businesses who are able to offer such efficient traceability systems for their goods not only boost operational safety measures but also build client confidence and trust by guaranteeing both quality and safety. Efficient traceability systems can also increase the products' value, which will increase overall profits. One could see this as a win-win scenario. Effective traceability systems can also reduce the likelihood that a supplier or operator will be held accountable for a product safety issue by offering thoroughly documented traceable data to demonstrate their compliance with legal requirements and lack of risk. Liability claims and lawsuits will be avoided in that situation, and the reputation of the business will not be impacted.

### *6.1.2. Tracking and Tracing*

Tracing is the capacity to determine the origin of an item or set of items upstream in the supply chain using records. Tracing is the ability to track an item as it travels downstream through the chain from start to finish. Information flows for supply chain tracking and tracing are depicted in Figure 6.1.

A tracking and tracing system may still have advantages in other areas even if the quality of the transportation process is extremely high. It could be used administratively, for instance, as the foundation for hauler payments. To verify whether the process quality is maintained at a satisfactory level, the gathered data could also be statistically processed and integrated into an information system. Everything pertaining to the products prior to, during, and following the manufacturing, packaging, and distribution process—including ingredients, procedures, tests, and test results, the environment, resources used, and modes of transportation—is also covered by traceability.

**Figure 6.1.** *Information flows of tracking and tracing in supply chains.*

*Source: Lee Z-Y, Kang Y-H, Chang Y-T, Lin S-H, Chien C-C, Lee S-C, Ko W-C. Carbon Footprint Assessment Based on Agricultural Traceability System Records: A Case Study of Onion Production in Southern Taiwan. Sustainability. 2024; 16(20):8817. https://doi.org/10.3390/su16208817.*

Two categories of tracking and tracing definitions—tracking and tracing in a limited sense and tracking and tracing in an extensive sense—could be established to account for the quality variation on tactical and operational production levels. Table 6.1 contains the attributes of each type. The primary distinction between tracking and tracing in a limited sense and tracking and tracing in an extensive sense is that the latter includes the former and allows the traceability data to be utilized in multiple supply chain dimensions rather than just product tracing.

**Table 6.1.** *Two Types of Tracking and Tracing*

| Tracking and tracing in a restricted sense | Tracking and tracing in an extensive sense |
|---|---|
| • Provides real-time visibility and disposition<br>• Creates historical record for the traceability of components and usage of each end product<br>• Provides forward and backward traceability | • Encompasses tracking and tracing in a restricted sense<br>• Information is used in the control and management of successive stages of production<br>• Provides dynamic lot allocation<br>• Optimizes and controls of processes in and between separate links of the supply chain |

*Source: Prashar D, Jha N, Jha S, Lee Y, Joshi GP. Blockchain-Based Traceability and Visibility for Agricultural Products: A Decentralized Way of Ensuring Food Safety in India. Sustainability. 2020; 12(8):3497. https://doi.org/10.3390/su12083497.*

### 6.1.3. Principle of Tracking and Tracing

The interface between an information system and a physical transportation system should be a part of a tracking and tracing system, according to Stefansson and Tilanus (2000). Eight characteristics were used to categorize tracking and tracing systems are:

- **Goods Identification Technology:** For example, alphanumerical codes, bar codes, RFID etc.
- **Scope of the Tracking and Tracing System:** The three dimensions of transformation—transportation (change of place); storage (change of time); and chain-wide conversion operations (change of form)—define the scope of the tracking and tracing system.
- **Registration Timing and Placing:** There are instances when tracking is done at specific times and locations, in which case the time and location must be recorded. For instance, time and location must be noted to maintain the integrity of transportation data when the shipment is transferred from one hauler to another, which may involve several modes of transportation.
- **Hierarchical Level:** Discrete registration instance may refer to different hierarchical packaging levels and different hierarchical assembly levels.
- **Attributes Recorded:** The identity of the entity, the current location, and the current time are the three attributes that a tracking

and tracing system may record. Additional entity attributes include quantity (if the shipment is complete, even if it contains multiple units) and quality (if there is any noticeable damage).

- **Organization of the Information System:** The information system that stores tracking and tracing data may be centralized or shared by multiple participants.
- **Accessibility of the Information System:** Interested parties should be able to follow and locate entities traveling from point A to point B using a tracking and tracing system. The information system has two different kinds of accessibility. Automated queries can be made automatically regardless of whether the information is centralized or decentralized, via EDI or the Internet; non-automated queries must be made and answered manually.
- **Activity Level of the Tracking and Tracing System:** Entities in fixed locations are registered by a passive tracking and tracing system as soon as they arrive or depart. An active tracking and tracing system notifies the user if something unexpected is registered and tracks the entity's progress from checkpoint to checkpoint.

## *6.1.4. Scope of Traceability*

These are two types of traceability scopes.

### *6.1.4.1. van Dorp's Perspective*

Businesses and organizations typically interact with one another through supply and demand markets rather than operating as lone actors in a vast network. All of these various relationships should be taken into account when it comes to traceability since they may result in various requirements and, consequently, have different effects on the participants. An example, as seen in Figure 6.2, can be used to illustrate the business scope of traceability. A description of each perspective follows.

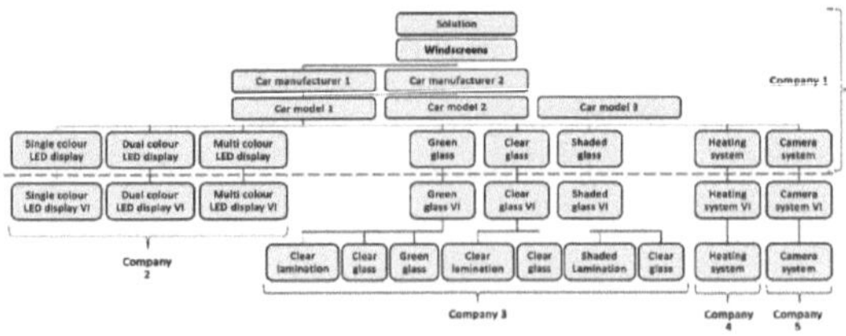


**Figure 6.2.** *Four perspectives of business scope of traceability.*

*Source: Harkonen J, Guerrero Rodriguez JM, Mustonen E. The Role of Productization in End-To-End Traceability. Eng. 2024; 5(4):2943-2965. https://doi.org/10.3390/eng5040153.*

### *Enterprise Perspective*

The first perspective is the enterprise perspective, which is based on the viewpoint of a single company and typically refers to horizontal traceability. The scope of traceability in an enterprise setting should be expanded beyond historical record-keeping, however, to achieve vertical integration, information from various management levels must be addressed. Control and planning at the strategic, tactical, and operational levels will differ in this scenario.

### *Multi-site perspective*

Due to allocation tactics, manufacturers, particularly in semi-process industries, typically have multiple plants spread across different locations, sometimes even different countries. This means that multiple material flows, information flows, and traceability information must be managed in a complex environment. As a result, greater focus is required on establishing a functional traceability system to satisfy various processing needs.

### *Supply chain perspective*

A supply chain is a network of two or more businesses that uses an integrative approach to manage the planning and control of materials from the manufacturer, supplier, and distributor to the final customer. According to this perspective, accurate management of all relevant supply chain data is a prerequisite for proper traceability.

### *External environment perspective*

To safeguard the interests of consumers, the external environment entails the official and authoritative control of products through routine inspection to ensure compliance with food laws. This viewpoint also incorporates functional product labeling to guarantee that consumers obtain the information they need about the product's composition, manufacturing process, storage techniques, and preparation, all of which contribute to the safety of the consumer. Furthermore, the literature on the manufacturer's liability for losses or damages brought on by defectiveness also discusses liability for defective products.

## *6.1.4.2. GS1 and its Perspectives*

A global non-profit organization, GS1 is committed to creating and implementing global standards and solutions that will increase the effectiveness and visibility of supply and demand chains both internationally and across industries. The most popular supply chain standards framework in the world is the GS1 system.

Integrating the physical flow of traceable goods with the information flow is a key component of supply chain traceability management. According to GS1, each component must fulfill its specific function in order for the chain as a whole to function and satisfy certain traceability requirements.

The following part will introduce each factor and describe their main characteristics are:

- **Traceable Item:** A traceable item is a tangible object for which details about its location, use, or history may need to be retrieved. As an illustration, consider shipments and logistic units, which might include pallets or containers.
- **Traceability Partner:** "All the parties involved in the supply chain, which could be enterprises, organizations, and authorities," is how one defines a traceability partner. Third-party logistics providers (3PLs), processors, manufacturers, primary producers, retailers, point-of-sale or service operators, warehouses, distribution centers, and authorities are all considered traceability partners.
- **Internal Traceability:** When a traceability partner receives one or more traceable item instances as inputs that are processed internally before one or more traceable item instances are output, this is known as internal traceability. One of the four subprocesses—transportation, transformation, storage, and destruction—must be a part of this process.
- **External Traceability:** "When a traceable item is physically handed over from one traceability partner to another," external traceability occurs. External traceability activities can be divided into two categories: actors' coordination, where traceability is used to coordinate activities between firms, and transportation, where it is used to monitor product/material monitoring during transport.
- **Physical Flow and Information Flow:** In the supply chain, traceability includes all parties involved and all activities related to the various flows. In addition to the associated information flow that travels in both upstream and downstream directions, physical flow describes the conversion of goods from raw materials to finished products and their delivery to the end user (Aung and Chang, 2014; Kamilaris et al., 2019).

### 6.1.5. Technologies in Traceability Systems

Traceability must be managed by setting up a system that can keep track of product routes and the chosen data for later use to translate the concept into reality. As a result, technologies for localization and identification become crucial instruments for traceability. When handled correctly, product identification can save a great deal of time and labor and is important for the entire product handling process, particularly when it comes to the traceability of goods. The more focused and effective the product recall can be, the less money and reputation will be lost. This is due to the traceability system's ability to record and provide more information, such as production time, batch number, and production conditions. These days, barcode labels, RFID tags, alphanumeric codes, and newly created technologies like GPS and GIS are the most popular and extensively utilized identification and localization technologies.

### 6.1.5.1. Alphanumerical Codes

Labels with alphabetical codes consist of a series of different-sized numbers and letters that are applied to products or their packaging. An alphanumeric coding system's design and underlying principle are very straightforward and cost-effective, but its main disadvantage is that it takes a lot of human resources to run and oversee for both writing and reading codes. Large labor costs are associated with this non-automated process, and performance might not be as good because a lot of manual data makes error prevention nearly impossible. Because there is no standard for alphanumeric codes and they are typically generated by the company or organization itself, sharing amongst various parties is challenging.

### 6.1.5.2. Bar Code

A barcode is a machine-readable optical representation of data that displays information about the object it is attached to. Bar codes offer a quick, precise, and affordable way to encode data that is easily readable by low-cost electronic readers. The introduction of bar codes has greatly aided traceability and increased the effectiveness of handling procedures throughout the supply chain. Simpler, more cost-effective, and more accurate traceability systems are made possible by this automatic, fast-reading, precise technology. Data can be automatically read by using bar codes, which removes the possibility of errors from manual data entry. Manufacturing, distribution, warehousing, retail, and other industries can all benefit from the use of bar codes as a practical tool for gathering, processing, transmitting, recording, and managing business data.

### 6.1.5.3. RFID

The technology known as radio frequency identification (RFID), which is frequently regarded as the next step in the evolution of barcodes, uses radio waves to automatically identify objects. An antenna in the tag is connected to a wireless microchip, which allows the microchip to send identification data to a reader. The reader then transforms the radio waves that are reflected back from the RFID tag into digital data. This allows for quick and complete automation of the reading process without the need for direct physical contact or sight positioning with the reader. RFID tags have very high data integrity and are generally hard to counterfeit. Even in harsh operating conditions like dust, snow, corrosion, and vibration, it can continue to perform effectively. At the pallet level, RFID has been adopted by major retailers like Wal-Mart and TESCO, as well as their suppliers. Unlike traditional bar codes, which require manual object scanning, RFID tags can be read automatically, saving a substantial amount of labor.

### *6.1.5.4. GIS*

A computer-based tool for managing spatial information is the geographic information system (GIS). GIS is capable of accurately providing the user with visual information and organizing, analyzing, manipulating, and managing spatial information in an intuitive manner. According to Buckley, an operational GIS consists of many crucial parts, including people, data, hardware, software, and methods that all work together to make the system function. It is possible to divide and code the cultivated areas based on the geographical information that GIS gathers because it can provide general configuration and features of the earth's surface as well as the implantation of agricultural products. Thus, the origin place of production is visualized throughout the entire supply chain management system. Each area then has a unique code serving as an ID in the database that connects with the production information, such as fertilizer management, pesticides, feeding-stuff, and ambient water quality. Additionally, since the administration department can use the data to estimate the environment's capacity and subsequently control the amount of planting, the information in the aforementioned database can also be seen as valuable data for sustainable development. When used in the context of a supply chain, GIS can provide a great deal of opportunity for agricultural practices like planning agricultural sites, managing land use, and gathering data on soil properties, climate, fertilizers, pesticides, water consumption, and production yields. These practices have the potential to improve risk assessment, food chain management, and sustainable development, making GIS a valuable tool for establishing a traceability system for sustainable agriculture products.

It is important to note that GIS offers opportunities and the potential to support traceability. Because the product information is distributed in a dynamic, real-time, and variable manner, it is insufficient to simply code a product based on its place of origin. Recording and following up on the information is also very important, but it is beyond the scope of GIS. At this point, supply chain traceability as well as distribution process traceability will be augmented by technology like RFID.

### 6.1.5.5. GPS

A constellation of 24 satellites that orbit the earth continuously forms the basis of the Global Positioning System (GPS), a satellite-based radio-positioning system. The US Department of Defense (DOD) created, funded, deployed, and ran it because of its amazing potential as a military location tool. These satellites transmit radio signals indicating the precise time and location, and they are outfitted with atomic clocks. The GPS receiver receives these radio signals and uses the known satellite positions to triangulate its exact location on Earth (longitude and latitude). A GPS receiver can determine a three-dimensional position (i.e., elevation, longitude, and latitude). The global positioning tool, GPS, has gained significant traction in civil use due to its ability to provide real-time, three-dimensional positioning and navigation throughout the year. GPS helps drivers and pilots locate themselves and prevent collisions in transportation applications. When fishing, GPS can be used as a tool of the traceability system to gather data about fishing grounds, landing times, and catching times. Such information is crucial because it is the first step in a traceability system, linked to raw material quality control, and aids in preventing overfishing for regional fishing sustainable development. By identifying each vessel's location, the fishing grounds—that is, the actual locations where the fishing took place—can be determined (Aung and Chang, 2014; Kamilaris et al., 2019).

## 6.2. INTEGRATED AGRI-SUPPLY CHAIN MANAGEMENT

Supply chains primarily deal with the movement of goods and information amongst their member companies, including the acquisition of raw materials, the conversion of raw materials into completed goods, and the delivery of those goods to final consumers. Information-driven, integrated supply chains of today are helping businesses cut expenses and inventory while increasing product value, extending resources, speeding up time to market, and retaining customers. The ability of supply chain activities to work together to increase the profitability of each link in the chain while producing value for customers is the true test of supply chain success. Supply chain management, to put it another way, is the coordinated process of creating value for the final consumer.

However, due to the inherent issues with the agriculture sector, the supply chains of various agricultural commodities in India are beset with difficulties. Numerous socioeconomic factors, such as the predominance of small and marginal farmers, disjointed supply chains, lack of scale economies, low levels of processing and value addition, inadequate marketing infrastructure, etc., influence the nation's agri-supply chain system.

Because of lower inventory and faster response times to consumer requests for goods and services, early processing-based supply chain management success included better ties between warehousing and transportation within businesses. To successfully compete in the market, supply chain management then moved into a logistics stage where other departments within businesses collaborated to integrate manufacturing, procurement, transportation, distribution, and marketing. The use of telecommunications, electronic data interfaces, and other technological advancements that improved the transparency of information transfer between companies' functional areas helped to facilitate this stage.

## 6.2.1. Food Supply Chain Networks

An organized, processing-based agri-supply chain is a component of a much more intricate network. A generic supply chain at the organizational level is shown in Figure 6.3 in the context of an entire supply-chain network. Every company is situated in a network layer and is a part of at least one supply chain, meaning that it typically has many different suppliers and customers both now and in the future.

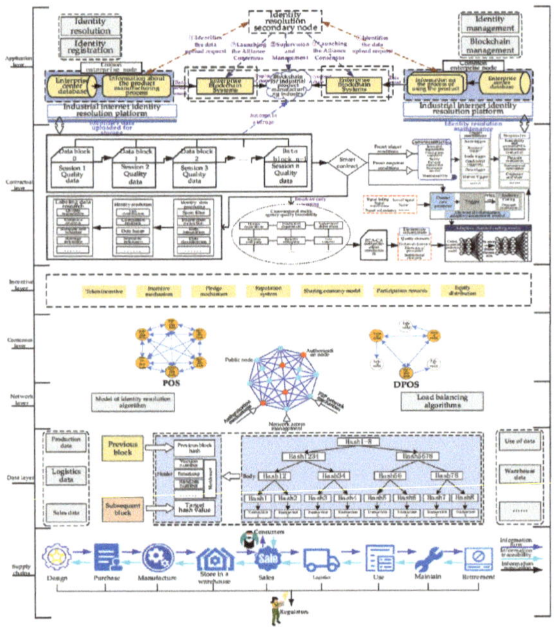


**Figure 6.3.** *Schematic diagram of a supply chain.*

*Source: Xiao L, Sun W, Chang S, Lu C, Jiang R. Research on the Construction of a Blockchain-Based Industrial Product Full Life Cycle Information Traceability System. Applied Sciences. 2024; 14(11):4569. https://doi.org/10.3390/app14114569.*

## 6.2.2. The Advantages for Supply Chain Members

Through the coordination of their value-creating activities, individual producers, marketers, and suppliers connected by a supply chain generate more value than they could in the absence of their cooperation.

Supply chains create synergies in one of three ways:

- They increase members' sales volume by expanding traditional markets beyond their initial borders;
- They reduce the delivered cost of products below the cost of competing chains and thus increase the gross margin for the working capital committed by members of the chain; and
- They target particular market segments with particular product offerings, setting themselves apart from the competition in terms of service, brand reputation, or product quality to enhance the perceived value delivered to customers. They permit chain participants to impose higher prices in this way.

Supply chains, in general, make markets more contestable at the producer and consumer ends of the chain. Chains compete mainly with one another on the consumer end through price, unique goods and services, and unique terms of sale. Supply chains compete with one another at the producer end of the chain mainly for producer affiliation and commitments from core vendors.

## 6.2.3. Components of an Agri Supply Chain

To reliably meet the needs of customers in terms of quantity, quality, and price, agribusiness supply chain management (SCM) entails managing the relationships between the businesses in charge of the efficient production and supply of products from the farm level to the consumers. In actuality, this frequently refers to managing relationships and procedures between businesses as well as both horizontal and vertical alliances. Agri-supply chains are commercial structures in which members share risks and profits. As a result, supply chains create incentives for the entire chain and enforce internal processes to guarantee that production and delivery commitments are met on time. Due to commitments regarding transaction volume, reciprocal scheduling, shared information, and product quality guarantees, they are intertwined. Agricultural products gain value from process linkages, which necessitate participant coordination in an ongoing process of improvement. The decisions made or not made at other links in the chain have a substantial impact on the costs incurred at one link. Key control processes like forecasting, purchase scheduling, production, and processing programming, sales promotion, and new market and product launches, among others, require extensive pre-planning and coordination up and down the entire chain.

Following are the components of an organized agri- supply chain:

- **Procurement or Sourcing**
- **Logistic management:**
  - Transportation;
  - Material management;
  - On the premise of supplying mostly from production not stock;
  - Warehousing; and
  - Logistics Network modeling.
- **Organizational Management:**
  - Contracting;
  - Strategic alliances and partnerships; and
  - Vertical integration:
    - Long term storage;
    - Packaging technology;
    - Cold chain management;
    - Energy efficient transport; and
    - Quality and safety.
- **Application of Efficient Consumer Response (ECR) System:**
  - Electronic scanning of price and product at the point of sale; and
  - Streamline the entire distribution chain.

### 6.2.4. Agri-Marketing and Emergence of Coordinated Supply Chains in India

The management of India's agri-supply chains is currently changing to adapt to the new marketing realities brought about by the globalization wave as well as other internal shifts like consumers' shifting dietary preferences toward higher-value items like fruits, vegetables, and animal protein and an increase in their disposable income. The country's agricultural economy has recently faced new challenges, which have prompted government agencies to pursue various legal reforms to encourage private investment in agricultural marketing infrastructure, remove various barriers to entry, and support traceability and a coordinated supply chain. The country's main agricultural marketing act, which is being implemented by India's various states, now includes enabling provisions to support private market establishment, contract farming, and direct marketing—all of which were previously prohibited. These steps will significantly help small businesses gain economies of scale by creating direct connections between farmers and processors, exporters, retailers, etc. Therefore, the measure will offer both forward and backward linkages to develop integrated supply chains for various agricultural products across the nation (Aung and Chang, 2014; Kamilaris et al., 2019).

## 6.3. TRACEABILITY AND SUSTAINABLE SUPPLY CHAIN MANAGEMENT

Sustainable supply chain management and traceability are closely related, with traceability acting as a key instrument to improve sustainability throughout the supply chain. The ability to follow goods and materials from point of origin to point of final destination while recording each stage of the process is known as traceability. By offering transparency into the environmental, social, and ethical practices involved in production and distribution, it plays a crucial role in the context of sustainability. The goal of sustainable supply chain management is to keep supply chain operations economically viable while reducing their negative effects on the environment and society. Businesses can guarantee ethical labor practices, ecologically friendly production processes, and responsibly sourced raw materials by incorporating traceability. Businesses can use traceability systems, for example, to confirm that suppliers follow organic farming methods, cut carbon emissions, or prevent deforestation. In sectors where sustainability is a major concern, such as electronics, fashion, and agriculture, this transparency is essential.

### 6.3.1. Key Aspects of Traceability

Traceability should be embedded in a fair and sustainable value chain ecosystem, encompassing the following key aspects.

#### *6.3.1.1. Individual Approach*

Procedures for traceability need to be customized for each value chain participant. From the first (farmer) to the last (consumer), traceability should be accomplished by solutions that empower each actor and offer sufficient incentives to encourage and continuously maintain each actor's participation in promoting the sustainability and equity of the value chain.

#### *6.3.1.2. Accessibility and Appropriateness of the Solution*

All supply chain participants must benefit from the solutions that are implemented. A traceability solution should be sufficiently robust and "fit for purpose," even though innovation should always be encouraged. Examples of this include data recording in remote or rural settings, issues with proper internet connectivity, etc.).

#### *6.3.1.3. Cost-effectiveness*

Cost-effectiveness is crucial in pursuing and ensuring sustainability; economies of scale can continue to operate and provide the required financial incentives for participants to fully support this common goal. Systems must be affordable and capable of managing smallholder transactions to be more accessible and inclusive.

#### *6.3.1.4. Transparency and Accountability*

The necessary accountability and transparency, as well as the incentives that go along with it, ought to encourage widespread adherence to the fundamentals of a just, long-lasting, and economical value chain. A perfect traceability system is a component of an ecosystem that guarantees information access for all value chain participants and fosters sustainability.

#### *6.3.1.5. Interoperability*

A traceability solution cannot operate independently. The successful operation of a traceability system depends on technical and content interoperability (Aung and Chang, 2014; Kamilaris et al., 2019).

### 6.3.2. Steps in Adopting Traceability

#### *6.3.2.1. Step 1: Reflecting and Understanding Motivation*

Identifying the key drivers to traceability is important so that choices can be made in terms of the objectives and expected benefits are:

- Certification requirements for tropical commodities necessitate knowledge of the chain of custody. Only participants in such a certified chain are able to manage certified volumes and sell the final product as meeting the specific standard's sustainability and other requirements.
- Traceability helps businesses manage their suppliers from an operational standpoint. Information about locations, quality, dependability, and turnover in the supply chain is important for maximizing efficiency and minimizing financial risk.
- To ensure that there are no unethical practices occurring along the value chain, consumers, investors, and/or legislators are closely monitoring product importers worldwide in the recent past. A (more) direct evaluation of the various chain actors is gradually replacing or supplementing traditional desktop-based due diligence methods. To verify that relevant sustainability principles are being followed, it is imperative that these intermediaries be identified.

- The EU Due Diligence Directive (February 2022), the German Supply Chain Due Diligence Act (LkSG, June 2021), the (draft) EU Regulation on deforestation-free products, and other laws mandate a due diligence procedure for all parties, including subcontractors, who offer goods for sale in Europe. Imports of goods that don't fit certain or minimal market requirements are prohibited. The interest in traceability solutions has significantly increased as a result of these laws and regulations.
- The ability to offer reliable and transparent information about the sustainability of agricultural commodities, the value chains that surround them, and eventually the final consumer goods is becoming more and more important. Comprehensive sustainability data is essential and frequently extends well beyond product certification.
- Additionally, systems are being developed to recognize and reward sustainability accomplishments. This entails admitting that suitable sustainability incentives are a crucial component of value chains for sustainable agriculture, necessitating improved traceability systems and reliable governance.

Traceability can thus also be viewed as a pipeline include:

- Transferring trustworthy information on sustainability and other characteristics along the value chain, from farm level to retail and consumer level; and
- While fostering and tracking the payment of a fair price and of sustainability achievement rewards as well as relevant feedback information back from the end of the value chain to the farmers and their organizations at the beginning.

#### *6.3.2.2. Step 2: Understanding and Mapping Your Processes and How Extended Traceability will Benefit Your Business*

Understanding the capacity needed on your end to facilitate a successful solution onboarding process comes next, after you have determined the essential requirements for your goal and found a viable solution.

It is very beneficial to have a more thorough understanding of your needs and expectations, as well as those of your stakeholders, before contacting traceability solution providers. Every traceability solution depends on or at least interacts with pre-existing information frameworks; none will function independently. As a result, it is prudent to start by evaluating the information management systems (databases, spreadsheets, hardcopy files, apps, servers, etc.) that are currently in use within the company and/or value chain partnership).

It is crucial to simulate (i.e., record and comprehend) the (actual and intended) flow of data between data sources and to specify who supplies what data and who may need to validate it. By doing this, you'll be able to identify areas of inefficiency and potential threats to data integrity. The analogue information flow (chain of custody paper-trail, processing reports, contracts, etc.) should also be mapped and documented. This can provide fresh perspectives on areas where further digitization of operations can lead to increased efficiency.

### *6.3.2.3. Step 3: Clarifying Your Requirements for a Traceability Solution*

To select the right traceability solution, there is first a need to define the requirements are:

- **Scope of the Solution:** Determining the solution's scope is the first step. Determine which value chain components are most important to your needs and see if a given solution meets those needs. For example, solutions with features tailored to retail and consumer countries might be more helpful if you are serving end users. On the other hand, solutions that offer features unique to production/origin countries may be more appropriate for your needs if your primary focus is on acquiring (sourcing) agricultural commodities.
- **Define Commodities of Interest:** Determining the commodities of interest is crucial because different commodities can be covered by different solutions. It may be necessary to make certain adjustments for a traceability solution to function with new commodities because it was created expressly to support them. It is imperative to recognize, nevertheless, that to account for the pertinent commodities of interest, it may be possible to broaden the range of solutions that provide technical features that are especially tailored to your requirements.
- **Chain of Custody Model:** Third, Identity Preserved, Physical Segregation, Mass Balance, and Book and Claim are the four typical chain of custody models. They vary from one another in the way that the product is managed at each stage, including the aggregation and subsequent handling of related data to make claims. Examining whether the desired traceability solution is compatible with the key chain of custody models for your present and future business is crucial. You can investigate additional criteria based on your needs in addition to these three fundamental requirements to select the traceability solution that best fits your needs.

It's time to determine whether a specific traceability solution meets your technical requirements after determining the features associated with your functional needs. Furthermore, it is always helpful to know what kind of support a solution offers and whether the terms and kind of support are sufficient for you to integrate the solution into your value chain. Finally, determine what kind of payment schedule will be needed to put the solution into practice and look for the one that best fits your budget. It should be mentioned that various solutions provide various payment plans according to the value chain type, so this is a crucial topic to talk about with the solution provider (Aung and Chang, 2014; Kamilaris et al., 2019).

#### *6.3.2.4. Step 4: Ensuring Internal Capacity and Stakeholder Involvement as Required for the Implementation*

It sounds difficult and could cost a lot of money to change traditional sourcing practices to modern supply chain traceability. If sufficient commitment is not provided by all relevant parties, initiating a comprehensive process would be a complete waste of time and energy. It is advised that you use a stakeholder approach to guarantee the necessary commitment and involvement from stakeholders and, consequently, the viability of your traceability project such as

- **Create Awareness and a Sense of Urgency:** It's critical that you comprehend your stakeholders' expectations and that they recognize the value, necessity, and advantages of efficient traceability. This entails making certain that traceability benefits your stakeholders as well.
- **Assess and Align Stakeholders' Values:** People's behavior is frequently directly correlated with their personal values. An organization or company and its stakeholders frequently reevaluate their corporate and personal values when making a strategic shift from bulk commercial operations to fully traceable operations. The personal interests of those involved in the transition may be impacted, which could make them resistant to implementing new procedures. However, this kind of change can also aid in exposing unethical business practices and ultimately reinforce those that are already in place. Adopting value chain practices that are primarily data-driven and sustainability-driven will only increase the chances of long-term financial success for all stakeholders. For instance, rewarding and incentivizing such ethical and sustainable behavior can lead to improved access to impact funding and long-lasting win-win collaborations. Supporting value chain participants during the transition process is crucial. For value chain intermediaries, this can entail creating and implementing new business models and cultures.
- **Consult on Expectations:** The implementation of these new systems presents the chance to resolve enduring inefficiencies within

the company. Stakeholders ought to take advantage of this chance to consider previous problems with their supply chain procedures. Field officers might encounter difficulties, for instance, in organizing local farmers for training. It could be difficult for suppliers to stay on top of the most active producers and their transactional history. Procurement managers might find it difficult to handle the volume of reporting in manual spreadsheets that require constant editing. It is possible that board members are unaware of how well their company complies with environmental, social, and governance (ESG) regulations. To garner more support from stakeholders, it might be thought of including such secondary features to broaden the scope of the traceability solution. The key takeaway from this is that value chain participants shouldn't have to bear any additional burdens when implementing (possibly enhanced, cutting edge, and future-proof) traceability. Integrating traceability solutions into systems will improve their productivity, sustainability, and profitability.

- **Understand Doubts and Fears:** Things that are new can be unpredictable. Employees might be concerned about the increased workload or the difficulty of utilizing digital tools. Suppliers might be concerned that if they reveal their producer base, upstream buyers will be able to avoid them. One of the most important first steps in resolving their concerns and fostering greater understanding is to listen to them. In addressing their concerns, it's critical to highlight the risks of failing to take advantage of the new business opportunities presented by cutting edge traceability.
- **Assess the Capacity of the Team:** A comprehensive understanding of the team's capabilities makes it evident which skills require improvement (and consequently investment) to increase uptake. It is necessary to have prior experience using digital devices like laptops and smartphones. Furthermore, proficiency with internal management systems is frequently a reliable indicator of success. At the most basic level, a limitation could already exist due to a failure to follow procedures and reporting requirements. To ensure successful and efficient implementation, capacity building to address this limitation from the outset would be very beneficial.
- **Select Change Agents:** There will be varying degrees of participation within an organization to lead the challenging shift to traceability. Within each stakeholder category (departments, suppliers, producers, intermediaries, transporters or service providers, etc.), it is recommended to choose change agents—individuals who are knowledgeable or have the capacity to pick things up quickly,

passionate about enacting change, and able to inspire others—during the initial stakeholder mapping process. ). These change agents are capable of taking the lead in testing, reviewing stakeholder requirements, and offering feedback.

- **Ensure Leadership Commitment:** One of the most important things for a successful implementation is making sure that the leadership is committed. The likelihood of people taking advantage of internal differences to avoid pursuing traceability is further increased by defocused leadership and a lack of commitment. When leadership itself betrays the organization's values and is seen as acting more for their own benefit than for the organization's and/or the value chain partnership's, the same risk arises. To support the change process and direct the stakeholder approach, a specialist may occasionally be required. The following actions demonstrate commitment: (i) giving core staff time to prepare, choose, and test solutions; (ii) allocating funds to hire experts and the best traceability providers and solutions (not just the cheapest ones available); and (iii) providing support to the core staff to facilitate learning and guarantee minimal disruption during the transition.
- **Offer Influence:** Throughout the implementation, establish frequent opportunities for feedback. Give stakeholders a chance to share their transitional experiences and give them the power to shape the traceability solution's actual configuration and/or design as well as its actual application at various levels.

### *6.3.2.5. Step 5: Understanding the Working Stages Required for a Successful Implementation*

It is frequently necessary to modify or even completely change operations, business concepts, and value chain relations to implement modern traceability. Therefore, it's crucial to support all parties involved in the change process and steer clear of unnecessary confusion and disruption. It could be tempting to jump right in with a full-fledged digital solution and hasten the digital conversion process. Premature implementation and roll-out, however, will probably cause confusion and disruption and have an instant detrimental impact on stakeholder commitment, trust, and morale.

A Proof of Concept is needed before rushing into implementation:

- A hardcopy paper trail and documentation of all pertinent transactions and processing steps can form the basis of a significant portion of the preparation. A thorough diagram of current procedures must be created before simulating the new process flow. To replicate the Chain of Custody steps, this data can be tracked in a spreadsheet that is updated on a regular basis by a data clerk. Running this continuously for a few days in each of the scenarios gives you a quick idea of the functionalities that the traceability solution should cover. Correcting a simulation with a spreadsheet is far simpler than making changes to a traceability system that has already been implemented.

- Nonetheless, it is frequently very dangerous to attempt to replicate all current procedures in the new traceability solution. The goal of implementing a (new or improved) traceability system shouldn't be to increase the workload associated with traceability requirements. It ought to involve a more comprehensive change in which procedures are examined and improved upon, and modern traceability is attained without adding to the value chain participants' overall administrative load. It is critical to comprehend the optimization potential that modern traceability solutions and suitable system integration make possible.
- Developing and refining the concept, then testing it in a real-world setting, is typically the next step in the learning and development process needed to obtain a Proof of Concept. It will be crucial to test in a parallel environment with actual stakeholders who are well-coached and dedicated to participating in such an exercise after simulating in the core design group.
- It is possible to group the prioritized functions of your traceability solution into clusters, as you will see. They are likely to be modules that specific users use for specific purposes within the traceability solution.

**Typical modules can be:**

- Each of these modules should be piloted as minimum viable product (MVP). MVP's offer just the core functionality without any secondary benefits, nice polishing, or extras.
- By operating this initial pilot version concurrently with the organization's current data flow, disruption can be reduced. The disadvantage of additional work is insignificant when contrasted with the demoralizing impact of a newly implemented traceability system that was implemented too soon and is therefore broken, having already supplanted earlier systems. Only after pilot versions of a particular module have been examined, tested under all conditions, and deemed flawless should the traceability solution be incorporated into the organization's main operations. From the standpoint of continuity assurance, it makes sense to keep the older (hard copy

or less developed) system operational for a little while longer. Risks frequently arise during the onboarding of new modules to the chosen solution or during later updates to the traceability solution.

- It is crucial to make sure that the new method of operation benefits all value chain participants when developing a new traceability system and once more before implementing it. All of this must be communicated to the relevant stakeholders in a clear and reliable manner. Stakeholders may be more inclined to support the implementation of a traceability solution after all the steps covered thus far have been completed.

### *6.3.2.6. Step 6: Time Required for Implementation*

With the right preparation and motivation, it becomes a lot easier to start the implementation process.

As mentioned in Step 5, each MVP will probably go through a pilot phase that includes both desktop and field reviews by a select group of end users who are part of the pilot testing team. The improvement cycle will function more effectively if end users are given the chance to provide feedback as soon as possible.

It is typical to observe that the initial draft can be evaluated and given pertinent feedback in a single day. Up to one or two weeks of field testing may be required for later iterations to examine every scenario that may arise.

Before a pilot phase works as planned, it usually takes four to seven iterations. The developer/customization team's comprehension of the user side of things and the users' awareness of their expectations beforehand will be crucial to the effectiveness (and enjoyment) of these cycles. While moving to more advanced levels of traceability or gradually adding new modules to the system can take several years, it is evident that implementation takes weeks or months when multiple modules must be customized through multiple iterations.

You must be mindful of scope creep as you proceed with implementation. Unless there was a fundamental flaw when priorities were first established, it helps to clearly prioritize developments, even though you can add specific wishes as you go. The next time the solution is implemented, new demands should be noted and followed up on. Solutions for future proof traceability that are good are adaptable and enable the activation of extra features, data fields, etc. when those are required.

The aspect of the supply chain actors who are less digitally literate's capacity for absorption needs special consideration. Spreading out the goal over several sourcing seasons would be wiser than expecting a total transformation with all modules at once. This is especially advised when collecting digital data also involves maintaining paper records, which adds to the workload (Aung and Chang, 2014; Kamilaris et al., 2019).

### 6.3.3. Traceability for Social Sustainability Dimension

Traceability for social sustainability contributes to increased social welfare.

#### *6.3.3.1. Consumer and Public Protection*

Many academic researchers based their research of traceability on these policies to improve supply chain performance and/or cope with regulatory enforcement. The regulatory framework for food safety is to control the following hazards are:

- Diseases that can be transferred from livestock to human, e.g., Bovine Spongiform Encephalopathy (BSE);
- Chemical contaminations, e.g., Medroxy Progesteron Acetate (MPA) and Dioxin; and
- Livestock epidemics, e.g., Hog cholera, foot, and mouth disease, Aujeszky's disease, Johne's disease, and Scrapie.

From a farm gate to a consumer plate, traceability tracks and traces different data to control food hazards. For instance, newborn calves in the meat industry are tagged and registered before being moved to different facilities until they reach adulthood. Every farm involved will be noted. The slaughterhouse may record the slaughter of cows and mark the product batches accordingly. However, each country has different regulations regarding livestock traceability. While some nations might only need animal origin, others might need all stages of the process.

Foods containing genetically modified organisms (GMOs) may also pose a risk to public health. Even though the effects of GMO products on human health are still unclear, consumers have the right to know or be able to contact the producers to confirm whether the food they plan to purchase is genetically modified.

Food contamination can result from mislabeled food, such as when fish is mislabeled from one type to another because some mislabeled fish originate from contaminated waterways (such as those that contain lead and mercury). Improper labeling of information may also result in allergy issues for consumers.

The extent of food-borne illness can be reduced by improving the tracking of contaminated food. By lowering illness, the public avoids losing productivity and medical costs are decreased. Companies are also compelled by traceability to make accurate information available to the public. From a liability standpoint, traceability may serve as a tool to inform consumers about the safety of food. Businesses along a supply chain are more cautious in their operations and activities since traceability reveals which party is responsible for paying for food hazard incidents. In terms of liability, traceability not only improves welfare by making companies pay fines for food hazard incidents, but it can also make them take greater responsibility for their goods by keeping a close eye on product quality to avoid issues. As a result, customers receive safer and better products. Some studies define consumer welfare in terms of other product categories besides food. Traceability throughout a supply chain allows a focal firm to keep an eye on its suppliers and regulate the quality of their manufacturing and design, as the authors previously stated in the economic dimension. By producing higher-quality, standard-compliant products, this control not only lowers supply chain costs but also improves consumer welfare. Paying for a product in accordance with its condition is another factor in consumer welfare. For instance, in the food manufacturing industry, enhanced fishing and other related practices can be used to improve product quality and safety by analyzing traceability data in cod fish. Product quality may become a deciding factor in a customer's decision. Food that is temperature-sensitive can be traced, giving customers more pricing options. Customers may be willing to pay more or less depending on how long a product lasts. Swindling is a factor that adversely affects consumers' selection, as opposed to providing them with more options. Customers would unintentionally pay more for a lower-quality product if fish was purposefully mislabeled from one type to another, such as channel catfish to southern trout. Traceability keeps manufacturers from defrauding consumers by disclosing the true information to them. RFID improves the situation with regard to counterfeit goods. They mention that it is possible to determine whether products (such as toys, electronics, etc.) are fake by following products from manufacturers to consumers. Another major factor in counterfeit products is drugs. RFID traceability is extensively used in the pharmaceutical industry due to the large number of counterfeit drug cases reported to the US Food and Drug Administration. In 2004, the market share of counterfeit medications was estimated to be between 5% and 8%. RFID tags can serve as an electronic pedigree, or e-pedigree, to verify drug movements, transactions, legal requirements, and authenticity. Since illegal issues in the drug industry are not limited to counterfeit, counterfeit drugs are kept apart from other counterfeit products. Theft, mislabeling, adulteration, illegally obtaining drugs, and illegally selling hazardous drugs are the other issues. Authentic drugs that are not acquired through legal means are referred to as illegally obtained drugs. For instance, wholesalers buy medications from hospitals or clinics at a lower cost (since patients must pay the same price to purchase them) and resell them at the going rate. Additionally, Electronic-Pedigree can determine whether medications have been speculated on previously for price incentives, such as speculating at a low price and then selling later with discounts at a higher price. To protect themselves, consumers must pay an additional fee for traceability.

#### *6.3.3.2. Local Business Protection and Competition Protection*

The domestic farm business may be impacted by global sourcing as well as the concealment of the fact that food ingredients are sourced from other nations. Traceability would aid in bringing attention to this problem and assist society in determining whether or not to protect socioeconomic assets. From the standpoint of fair competition, keeping information about inexpensive foreign suppliers secret could prevent other businesses from using those same sources. From an economic perspective, this is bad for society because it stifles competition and breeds monopolies.

#### *6.3.3.3. Animal Protection*

The social dimension encompasses moral behavior within a society in addition to human welfare. Traceability could be used to end animal abuse. Animal abuse can be prevented by closely monitoring labor activities in food processing facilities, as this can prevent both improper labor practices and improper care. Customers may also use animal treatment as a criterion to select products that are better for animals than others (quality certification schemes that require traceability can certify animal treatment). To improve animal care, food organizations all over the world have introduced many food certification programs. For instance, the Netherlands' Butlers' Choice certification program guarantees food of high quality and ethical origin. The implementation of a certification scheme can enhance animal welfare by tracing and monitoring every food processing operation to ensure food producers adhere to certification terms and regulations. Moreover, it lowers the living space density, which lowers the death rate.

### 6.3.4. Traceability for Environmental Sustainability Dimension

Traceability supports environmental welfare, which can have a beneficial impact either directly or indirectly.

#### *6.3.4.1. Controlling Genetically Modified Organism (GMO) Products*

It's still unknown how GMO products will affect the environment. To assess and stop any long-term environmental effects by removing products from the market, traceability can be used to monitor GMO products.

### *6.3.4.2. Preventing Illegal Fishing and Avoiding Fish Extinction*

Because their species are illegal to catch because of the scarcity, some fish species are mislabeled and sold as others. On this matter, traceability can be used to compel fishing companies to identify the type or origin of the fish (if it is in a prohibited area) and implement a certification program to guarantee legal operations. The International Plan of Action to Prevent, Deter, and Eliminate Illegal, Unreported, and Unregulated Fishing (IPOA-IUU) is a fishing program created by the Food and Agricultural Organization (FAO) of the United Nations. To guarantee quality, safety, and legality, this fishing plan requires that fishing operations and associated data be documented, potentially preventing the extinction of fish species.

### *6.3.4.3. Improving Environmental Resources*

One way that traceability can selectively have a positive impact on the environment is by improving marine environment resources. Jacquet et al. (2008) provide an example of how Thai shrimp raised on farms are marketed as wild-growth shrimp. Shrimp raised on farms have the potential to ruin mangrove habitats, which would harm the environment. The ability of seafood producers to offer reliable information about how their goods can enhance the marine environment is another facet of traceability. It is not against the law to raise shrimp on farms, and seafood producers are not legally obligated to preserve the marine environment, even though traceability may reveal or supply this information. Whether or not consumers wish to protect the environment is up to them. In this instance, traceability only offers visibility as a discerning feature for a better setting.

To improve environmental treatment, food organizations worldwide have launched many food certification programs. For instance, the World Wildlife Fund's Pangasius Aquaculture Dialogue guarantees both environmental and quality standards. By putting such a certification scheme into place, every food processing operation will be monitored and traced to make sure food producers abide by the certification's requirements that protect the marine environment. A study by Hamprecht et al. (2005) serves as another illustration. It demonstrates how Nestlé Company in Switzerland tracks and records dairy farmers' activities to ensure that they replenish soils with nutrients to stop deterioration. One of the main instruments for corporate social responsibility may be the use of traceability to safeguard environmental resources.

### *6.3.4.4. Preventing Pollution*

According to Shinkuma and Huong (2009), electronic waste (also known as "e-waste") from Japan does not currently have traceability to China, Vietnam, or Cambodia. Without appropriate handling, recycling or reusing e-waste would pollute the environment. Using acid to remove valuable metals from circuit boards can result in extremely toxic wastewater. Direct river water runoff will contaminate the air and soil as well. Traceability for the recycling process would lessen pollution and help enforce the correct handling of e-waste. Hazard Analysis Critical Control Point (HACCP) is a quality scheme that regulates and gives traceability-related guidelines for food manufacturing practices as a means of guaranteeing that chemicals used in the manufacturing process won't be released into the environment and cause pollution as a feature that attracts customers (Aung and Chang, 2014; Kamilaris et al., 2019).

### 6.3.5. Advantages of Sustainable Supply Chains Based on Blockchain Technology

Distributed ledger technology (DLT) includes blockchain, which stores data based on the agreement of all parties involved. An irreversible record that is impervious to manipulation by any one party is produced once the data has been saved and confirmed by all parties. A Blockchain maintains its functionality even if individual participants withdraw or file for bankruptcy, in contrast to a centralized database controlled by a single organization. Blockchain technology facilitates instantaneous transactions and allows networks to link transactions to conditions: Transaction B will be carried out automatically if Transaction A has already occurred. When the product has been delivered to the next phase of the supply chain and both parties agree that all requirements have been met, an example of a smart contract would be the instant payment of an invoice.

A digital asset that is kept on the Blockchain and has a potential real-world value is called a token. In this case, tokenization enables the connection of distinct data to products and services for a predetermined duration and can be utilized as security to obtain funding. The benefit in this case is that all payments made along the chain will be documented, and transparency is made possible by transaction traceability. Blockchain technology can provide transparency to developing and emerging markets regarding the sourcing, processing, and transportation of goods.

This is because every step in the chain of custody can be recorded in real-time and immutably onto the Blockchain. One of the more promising aspects of Blockchain technology for local producers, logistical partners, and custody is its ability to reveal the origins of consumer goods. Because of the additional cost that customers might be willing to pay and the layer of trust that a decentralized ledger creates, it might benefit those who use sustainable practices. Trade can be facilitated more quickly if all logistical information is on-chain and all parties trust the same source of information.

### *6.3.5.1. Leveraging Decentralization*

However, if transparent supply chains do help local producers, the benefits for first-mile actors must be weighed against the added value of decentralization. Larger actors in value chains that trade together without mutual trust can more easily use blockchain-powered decentralized platforms, but first-mile actors don't gain much from them. At the moment, supply chain business cases in the private sector are more likely to emphasize permissioned ledger and centralized ledger solutions to support larger businesses rather than decentralization and sustainability. Transparency, availability, the subsequent validation of all parties' data, and more environmentally friendly farming methods are all benefits of a decentralized supply chain. Decentralization of data can facilitate instantaneous payment and financial access, supporting the first-mile actor monetarily. Customers are assumed to be willing to pay more for a product sourced sustainably if production processes are transparent. To earn more money, this could encourage the first-mile actor to produce more sustainably.

### *6.3.5.2. Potential Challenges*

There isn't a solution that works for everyone. It is necessary to determine whether a supply chain can successfully use distributed ledger technology. The ability to tokenize a traded good will be crucial to the operation of supply chains powered by blockchain. The more distinctive and recognizable the product is, the more accurately its digital twin will capture its essence. By closing this representational gap between the digital token and the actual product, the person in possession of the digital and physical items can ensure that they match and haven't been altered at any stage of the supply chain. For instance, unprocessed pineapples or coconuts make simpler use cases for traceability since they are simple to mark and track both off-chain and on the Blockchain. Comparatively, it is more difficult to track fungible goods like coffee. One of the main obstacles to Blockchain use cases in agricultural supply chains is lowering the risk of such fraudulent activity along the supply chain; this is known as the "oracle problem."

Sustainable supply chains typically require third-party verification to demonstrate their ethical, social, and environmental production methods. The validation process is frequently centralized in human hands, whether they be auditors, certification issuers, or non-governmental organizations, because this third party must be trusted. This only shows that utilizing full decentralization is currently unlikely, even though it does not imply that trusted third parties need to be routinely replaced. All supply chain participants must exchange information in a standardized manner while attempting to reach a high degree of decentralization. Additionally, in this situation, a publicly accessible data model is essential to maximizing value chain collaboration. The governance is equally distributed when a decentralized model is chosen, but data credibility and quality must be guaranteed. Access to an individual actor's data that is not essential for other actors in the chain should be prohibited. Data transparency and privacy must be properly balanced (Aung and Chang, 2014; Kamilaris et al., 2019).

## REFERENCES


1. Aung, M. M., & Chang, Y. S. (2014). Traceability in a food supply chain: Safety and quality perspectives. *Food Control, 39*, 172–184.
2. Banterle, A., & Stranieri, S. (2008). The consequences of voluntary traceability system for supply chain relationships: An application of transaction cost economics. *Food Policy, 33*(6), 560–569.
3. Bechini, A., Cimino, M. G. C. A., Marcelloni, F., & Tomasi, A. (2008). Patterns and technologies for enabling supply chain traceability through collaborative e-business. *Information and Software Technology, 50*(4), 342–359.
4. Bevilacqua, M., Ciarapica, F. E., & Giacchetta, G. (2009). Business process reengineering of a supply chain and a traceability system: A case study. *Journal of Food Engineering, 93*(1), 13–22.
5. Chryssochoidis, G., Karagiannaki, A., Pramatari, K., & Kehagia, O. (2009). A cost-benefit evaluation framework of an electronic-based traceability system. *British Food Journal, 111*(6), 565–582.
6. Galimberti, A., De Mattia, F., Losa, A., Bruni, I., Federici, S., Casiraghi, M., & Labra, M. (2013). DNA barcoding as a new tool for food traceability. *Food Research International, 50*(1), 55–63.
7. Hobbs, J. E., Bailey, D., Dickinson, D. L., & Haghiri, M. (2005). Traceability in the Canadian red meat sector: Do consumers care? *Canadian Journal of Agricultural Economics, 53*(1), 47–65.
8. Kamilaris, A., Fonts, A., & Prenafeta-Bold, F. X. (2019). The rise of blockchain technology in agriculture and food supply chains. *Trends in Food Science & Technology, 91*, 640–649.
9. Kelepouris, T., Pramatari, K., & Doukidis, G. (2007). RFID-enabled traceability in the food supply chain. *Industrial Management & Data Systems, 107*(2), 183–200.
10. Lee, H., & Özer, Ö. (2007). Unlocking the value of RFID. *Production and Operations Management, 16*(1), 40–64.
11. Mehrjerdi, Y. Z. (2010). Coupling RFID with supply chain to enhance productivity. *Business Strategy Series, 11*(2), 107–123.
12. Opara, L. U. (2003). Traceability in agriculture and food supply chain: A review of basic concepts, technological implications, and future prospects. *Food, Agriculture & Environment, 1*(1), 101–106.
13. Salah, K., Nizamuddin, N., Jayaraman, R., & Omar, M. (n.d.). Blockchain-based soybean traceability in agricultural supply chain. *IEEE Access*.

14. Tian, F. (2016). An agri-food supply chain traceability system for China based on RFID & blockchain technology. In *Proceedings of the 13th International Conference on Service Systems and Service Management (ICSSSM)* (pp. 1–6).
15. Veronneau, S., & Roy, J. (2009). RFID benefits, costs, and possibilities: The economical analysis of RFID deployment in a cruise corporation global service supply chain. *International Journal of Production Economics, 122*(2), 692–702.

# CHAPTER 7

# DESIGN OF SAFETY EVALUATION AND RISK TRACEABILITY SYSTEM FOR AGRICULTURAL PRODUCT QUALITY

## Contents

# INTRODUCTION

Accurately tracing the origin of risks is challenging due to the multitude of risk factors, volume of risk information data, and circulation and disposal processes that affect the quality and safety of agricultural products. Enhancing the monitoring and supervision level of agricultural product traceability systems is necessary to achieve precise traceability of agricultural product quality and safety. This chapter sorts out the agricultural product supply chain and business process, analyzes the agricultural product risk factors influenced by heavy metals, and establishes an agricultural product quality and safety risk evaluation model. It also takes the detection data related to agricultural product safety and quality as its research object. The six-layer architecture of the equipment layer, data layer, service layer, application layer, display layer, and user layer served as the foundation for the design of the agricultural product risk traceability system. This chapter discusses a case report on China as published by Li C, Lu Y, Bian Y, Tian J, Yuan M. Design of Safety Evaluation and Risk Traceability System for Agricultural Product Quality. *Applied Sciences*. 2024; 14(7):2980. https://doi.org/10.3390/app14072980.

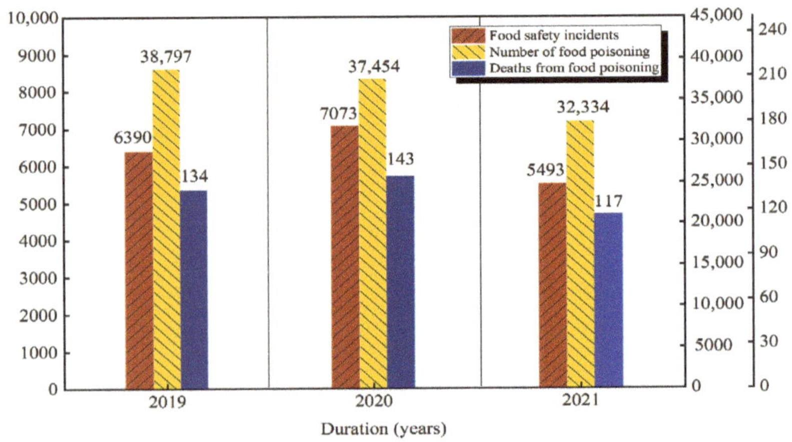


**Figure 7.1.** *Statistical chart of food safety incidents in China in the past three years.*

*Source: Li C, Lu Y, Bian Y, Tian J, Yuan M. Design of Safety Evaluation and Risk Traceability System for Agricultural Product Quality. Applied Sciences. 2024; 14(7):2980. https://doi.org/10.3390/app14072980.*

Food safety issues have become increasingly common in China in recent years, endangering people's lives and health while also posing a significant financial burden. Governments have implemented many laws and regulations to guarantee the safety and quality of agricultural products because they are very concerned about it. The traceability of agricultural products is crucial for ensuring the safety of people's lives and property as well as the peaceful and steady development of society. It makes it possible to track and trace product information in every link of the agricultural product supply chain and strengthens the supervision and management of each link. The 2019–2021 China Food Safety Annual Report contains pertinent data statistics that indicate that while the number of food poisoning cases has decreased over the last three years, it has remained relatively high. The state of food safety is still rather dire at the moment, with a high number of cases and fatalities from food poisoning, and conventional methods of agricultural product traceability being unable to satisfy the modern society's expectations regarding the safety and quality of agricultural products. Thus, it is beneficial to raise the bar for social and public safety to expedite the development of a traceability system for agricultural product quality and safety (Wang et al., 2017; Figure 7.1).

Scholars both domestically and internationally have recently investigated the use of information technology in agricultural quality and safety traceability systems from a variety of perspectives. Early in the 21st century, Chinese academics began studying traceability systems, which over time evolved into a reasonably developed system. Single information data traceability has gradually given way to systems with evaluation capabilities in traceability system design. Jihua Wang et al., discussed in detail the difficulties that Internet of Things technology faces in ensuring the traceability of the whole supply chain for agricultural products. The development significance of agricultural product quality and safety traceability systems was also illustrated, and the value of the Internet of Things technology was confirmed by examining its application in the direction of agricultural product quality and safety monitoring; Hu Jinyou et al., encountered the challenge of determining the usefulness of data in the creation of traceability systems. They were able to identify and categorize dynamic participant groups in the vegetable supply chain traceability system they had created by creating a single language model and integrating it into a system design. This served as a guide for creating the model and integrating it into the functional design of the system; Zhou Zhongwei et al., sought to raise the management level by integrating Internet of Things technology into supply chain management. They created an RFID tag-based agricultural supply chain management system by dissecting and evaluating the supply chain process. To visualize the quality of agricultural products and serve as a guide for attaining complete traceability, Song Yongjie et al., proposed a traceability model. created a framework for a traceability system based on Internet of Things and multi-agent technology to improve agricultural product quality control. The agricultural product quality and safety traceability system was integrated with agent technology, and the chain system that was created increased the effectiveness of agricultural product quality control. The first food traceability system was implemented in 1997 when the European Union began implementing a traceability system gradually in response to mad cow disease. Sven et al., to maximize the filtering capacity of data categories during automated storage, a monomer-based traceability system for agricultural chemicals in food was developed. To accomplish the task of recording and transmitting data in the traceability system, RFID tags were employed as storage devices. However, the designed traceability system finds it challenging to support high-capacity data transmission because of the requirement that the monomer architecture be implemented as a single solution with the storage capacity of RFID tags; Ganjar et al., sought to investigate the use of RFID technology in the perishable food traceability system. A traceability system was developed to efficiently track the origin of perishable food by integrating IoT sensors with an analysis of the perishable food supply chain. Nevertheless, the system's design was constrained by the electronic labels' and sensors' data storage capacities.

The application system uses microservice architecture in accordance with the business function, technical implementation, etc., and is divided into the smallest components; these components are independent of each other but collaborate with one another to complete a task, where each component or process is a microservice. Microservice architecture is a distributed system architecture. The majority of conventional agricultural quality and safety traceability systems are constructed utilizing service-oriented architecture, or SOA, or monolithic architecture. Although monolithic architectures are simpler to create, test, and implement, implementing the system program as a whole hinders the ability of the system's component modules to function separately (Velepucha and Flores, 2023).

Monolithic designs are challenging to maintain when working with larger system programs, and any updates made to any of the system's component modules require retesting and redeploying the system program as a whole. Furthermore, monolithic architectures lead to technology lock-in, where all component program development must adhere to the same technology system and framework. While all system components in a SOA architecture are independent and self-governing, microservices architecture places greater emphasis on features like load balancing, scalability, and high availability than SOA architecture. Because microservice architecture is fine-grained, it enhances development efficiency and allows system programs to operate in a highly available distributed environment with greater stability and reliability. The agricultural quality and safety traceability system of today has a vast and complex data volume, numerous user levels, and a variety of projects. Nevertheless, further research is needed to fully understand the working mechanism, and the architectural design and management guidelines also require improvement.

A significant factor influencing the safety and quality of agricultural products is the presence of heavy metals above the acceptable level. This chapter has designed a quality and safety risk evaluation system for agricultural products, constructed a quality and safety risk evaluation model for agricultural products, and realized a quantitative evaluation of the quality and safety of agricultural products to improve the level of supervision for quality and safety and the efficiency of risk traceability of agricultural products. Information technology, database technology, and a front-end framework were used in the design of the agricultural product risk traceability system for regulatory bodies, businesses, and consumers, which was based on the six-layer system architecture division. The risk traceability system and the designed agricultural product quality and safety evaluation model were integrated. The precise direction of the factors affecting the quality and safety of agricultural products was realized, and the storage and traceability of monitoring data related to agricultural products were completed, all based on the results of the risk evaluation.

## 7.1. DESIGN OF AGRICULTURAL PRODUCT QUALITY AND SAFETY EVALUATION SYSTEM

### 7.1.1. Screening of Evaluation Indicators

In terms of heavy metal enrichment, the inability of soil microorganisms to completely decompose heavy metal pollutants leads to the accumulation of a significant amount of heavy metal pollutants. Many biochemical processes become disrupted following the uptake and enrichment of heavy metals by crops. This disrupts crop growth and the safety of agricultural products, but it also leads to an accumulation of heavy metals in the human body, which can cause a variety of diseases. The issue of heavy metal pollution is particularly significant when it comes to the quality and safety of agricultural products, even though these factors are broad. When analyzing the sources of heavy metal pollution, indicators from the natural and social domains as well as qualitative and quantitative indicators are used. It is important to consider a specific level when designing each index when building the index system.

Generally speaking, the factor analysis method was used to analyze the monitoring-related indicators, such as total content of heavy metals in soil, soil pH value, soil organic matter content, soil redox potential, planted crop varieties, nutritional status, cultivation measures, heavy metal content in pesticides, fertilizer content, heavy metal content in atmosphere, and heavy metal content in irrigation water, to determine the risk of exceeding the standard of heavy metals. The available content of heavy metals in soil, the absorption coefficient of heavy metals in crops, and other sources of heavy metals were the three main components that were taken out of the total and identified as first-class indicators. The second-class indicators under each first-class indicator were further defined following a review of the literature and data analysis. Table 7.1 displays the selection of particular indicators for the risk assessment of agricultural product quality and safety.

**Table 7.1.** *Risk Assessment Indicators for Agricultural Product Quality and Safety*

| Primary Indicators | Secondary Indicators | Unit of Measurement |
|---|---|---|
| Available content of heavy metals in soil (Acs) | Total content of heavy metals in soil (a)<br>Soil pH (b)<br>Organic matter (c)<br>Redox potential (d) | mg/kg<br>/<br>g/kg<br>mV |
| Absorption coefficient of heavy metals in Crops (Ac) | Crop varieties (e)<br>Nutritional status (f)<br>Cultivation measures (g) | /<br>/<br>/ |
| Heavy metal content from other sources (Os) | Heavy metal content in pesticides (h)<br>Heavy metal content in fertilizers (i)<br>Heavy metal content in the atmosphere (j)<br>Heavy metal content in irrigation water (k) | mg/kg<br>mg/kg<br>mg/kg<br>mg/kg |

*Source: Li C, Lu Y, Bian Y, Tian J, Yuan M. Design of Safety Evaluation and Risk Traceability System for Agricultural Product Quality. Applied Sciences. 2024; 14(7):2980. https://doi.org/10.3390/app14072980.*

## 7.1.2. Design of Evaluation System

To assess the rationality of the indicator selection process, experts from agricultural product industry research institutes, pertinent universities, and industry associations were invited. The significance of each indicator was then compared using the 1–9 scale method, and a relative importance was assigned. The expert assignment led to weighted addition and equalization to determine each indicator's final weight. The researchers Li et al. (2024) made every effort to adhere to the completeness principle when building the assessment index system. Table 7.2 displays the secondary indicator results and scoring schemes for the particular agricultural product quality and safety risk evaluation model.

**Table 7.2.** *Scoring of Agricultural Product Quality and Safety Risk Assessment*

| **Risk Level Evaluation Index** | **Scoring Criteria** | **Score (Point)** |
|---|---|---|
| Total content of heavy metals in soil (a) | Exceeding the standard | 1 |
| | Not exceeding the standard | 0 |
| Risk level of soil pH (b) | pH value < 4.5, Extremely acid<br>4.5 ≤ pH value < 5.5, Strong acid<br>5.5 ≤ pH value < 6.5, acid<br>6.5 ≤ pH value < 7.5, neutral<br>7.5 ≤ pH value < 8.5, alkalinity<br>8.5 ≤ pH value < 9.5, Strong alkalinity<br>9.5 ≤ pH value, Extreme alkalinity | 6<br>5<br>4<br>3<br>2<br>1<br><br>0 |
| Risk level of soil organic matter (c) | Soil organic matter content 0%<br>Soil organic matter content < 1%<br>Soil organic matter content < 2%<br>Soil organic matter content < 3%<br>Soil organic matter content < 4%<br>Soil organic matter content < 5%<br>Soil organic matter content > 5% | 6<br>5<br>4<br>3<br>2<br>1<br>0 |
| Risk level of soil redox potential (d) | Soil redox potential ≤ −200 mv<br>−200 mv ≤ Soil redox potential < −100 mv<br>−100 mv ≤ Soil redox potential < 0 mv<br>0 mv ≤ Soil redox potential < 100 mv<br>100 mv ≤ Soil redox potential < 200 mv<br>200 mv ≤ Soil redox potential | 6<br>5<br><br>4<br><br>3<br><br>2<br><br>1 |
| Risk level of crop varieties (e) | Extremely easy to absorb heavy metals<br>Easy to absorb heavy metals<br>General to absorb heavy metals<br>Low to absorb heavy metals<br>Extremely low to absorb heavy metals | 5<br><br>4<br>3<br>2<br>1 |
| Risk level of crop nutritional status (f) | High content of nutrient elements that can resist the absorption of heavy metals<br>General content of nutrient elements that can resist the absorption of heavy metals<br>Low content of nutrient elements that can resist the absorption of heavy metals | 1<br><br>2<br><br>3 |

| | | |
|---|---|---|
| Risk level of crop Cultivation measures (g) | Application of related substances inhibiting the absorption and transport of heavy metals<br>Not application of related substances inhibiting the absorption and transport of heavy metals | 1<br><br><br>2 |
| Content of heavy metals in pesticides (h) | Detection<br>Not detected | 1<br>0 |
| Heavy metal content in fertilizer (i) | Detection<br>Not detected | 1<br>0 |
| Heavy metal content in the atmosphere (j) | Detection<br>Not detected | 1<br>0 |
| Heavy metal content in irrigation water (k) | Detection<br>Not detected | 1<br>0 |

*Source: Li C, Lu Y, Bian Y, Tian J, Yuan M. Design of Safety Evaluation and Risk Traceability System for Agricultural Product Quality. Applied Sciences. 2024; 14(7):2980. https://doi.org/10.3390/app14072980.*

### 7.1.3. Construction of the Evaluation Model

The obtained agricultural product sample testing data was statistically analyzed using the standard deviation statistical method to build the evaluation model. Table 7.3 displays the sample data and the statistical findings. The correlation between the average value and standard deviation of the obtained sample data was solved using the CRITIC method, and the fitting indicators' coefficients were then further optimized using the comprehensive expert survey method. The model is constructed using the widely used addition and multiplication mixed method because there is a close relationship between the indicators of the designed risk assessment system within the same main category, but not between the indicators of different main categories (Keshavarz-Ghorabaee et al., 2021).

**Table 7.3.** *Average Value and Standard Deviation of Agricultural Product Samples*

| **Item (Unit)** | **Sample Size** | **Average Value** | **Standard Deviation** |
|---|---|---|---|
| Soil pH | 200 | 4.46 | 0.432 |
| Soil organic matter content (g/kg) | 200 | 33.76 | 0.493 |

| | | | |
|---|---|---|---|
| Soil cation exchange capacity (cmol/kg) | 200 | 14.67 | 0.539 |
| Cadmium content (mg/kg) | 200 | 0.05 | 0.398 |
| Mercury content (mg/kg) | 200 | 0.12 | 0.425 |
| Arsenic content (mg/kg) | 200 | 4.45 | 0.512 |
| Lead content (mg/kg | 200 | 37.62 | 0.563 |

*Source: Li C, Lu Y, Bian Y, Tian J, Yuan M. Design of Safety Evaluation and Risk Traceability System for Agricultural Product Quality. Applied Sciences. 2024; 14(7):2980. https://doi.org/10.3390/app14072980.*

The primary sources of heavy metals in agricultural products are the inputs—aside from soil—that they absorb from the environment and soil. Equation 7.1 can be utilized to simplify the response model of quality and safety risk assessment through multi-factor nonlinear response analysis, provided that the coefficient is removed:

$$EM = Acs \times Ac + Os \tag{7.1}$$

Equation 7.1 contains the following terms: Os: heavy metal content from other sources; Ac: absorption coefficient of heavy metals in crops;: score value of agricultural product quality and safety risk assessment; and *Acs*: available content of heavy metals in soil.

The point allocation process is employed to maximize the model's fitting. The availability state coefficient of heavy metals in soil, which primarily included soil pH value, soil organic matter content, and soil redox potential, was a state factor affecting the existence of heavy metals in soil. The total content of heavy metals in soil was an important factor affecting the existence of heavy metals in soil. The expert scoring method was used to determine the weight coefficients of the total content of soil heavy metals, soil pH, soil organic matter, and soil redox potential after the principal component linear analysis equation and Bayesian classification. After the standardization coefficient, Equation 7.2 of the risk assessment model was obtained as follows:

$$Acs = a \times (5b + 3c + 2d) \tag{7.2}$$

The following terms are used in Equation 7.2: $a$ stands for total content of heavy metals in soil; $b$ for soil pH risk level; $c$ for soil organic matter risk level; and $d$ for soil redox potential risk level.

Under the condition that the influencing factors of the effective coefficient of heavy metals in soil have significant effects, the effective content score of heavy metals in soil will exceed the warning score of 60 points when the content of heavy metals in soil exceeds the standard.

Crop cultivating practices, irrigation techniques, and crop varieties all have a direct impact on the absorption coefficient of heavy metals by crops. The expert scoring method yielded the weight coefficients of crop varieties, crop nutrition, and crop cultivation measures following the same scoring and regression analysis. The final risk assessment model Equation 7.3 is as follows:

$$Ac = 0.6e + 0.2f + 0.2g \tag{7.3}$$

The following terms are used in Equation 7.3: $e$ stands for crop variety risk; $f$ for crop nutritional status risk; and $g$ for crop cultivation measure risk. Equation 7.3 and the scoring analysis indicate that the maximum absorption coefficient score is four and the minimum is one.

This means that, in normal circumstances, the presence of heavy metals in agricultural products can only exceed the standard when there is a risk warning of heavy metals in the external environment.

The amount of heavy metals in the environment may also be higher than expected if crop varieties, nutritional status, or incorrect farming cause an increase in the crop's absorption coefficient.

The concentration of heavy metals in pesticides, fertilizers, the atmosphere, and irrigation water were designated as four secondary indicators of heavy metals originating from other sources. Expert scoring was used to determine each secondary indicator's weight coefficient. The risk assessment model Equation 7.4 obtained after regression analysis is as follows:

$$Os = 3h + 4i + j + 2k \tag{7.4}$$

Equation 7.4 gives the following values for the concentration of heavy metals: h for pesticides; $i$ for fertilizer; $j$ for the atmosphere; and $k$ for irrigation water.

The risk of heavy metal residues in agricultural products will rise, and the risk value may increase by up to 10 points if heavy metal pollution is present in external environments in addition to soil.

Finally, Equation 7.5 of the agricultural product quality and safety risk assessment model is obtained as follows:

$$EM = a \times (5b + 3c + 2d) \times (0.6e + 0.2f + 0.2g) + (3h + 4i + j + 2k) \tag{7.5}$$

The agricultural product quality and safety risk assessment score (EM) has a risk early warning value of 60 points. It can be concluded that agricultural products with heavy metal contamination pose a high risk to quality and safety, necessitating their traceability and disposal, when the EM value is early warning, meaning the risk early warning score is greater than 60 points (Li et al., 2024).

# 7.2. CONSTRUCTION OF THE RISK TRACEABILITY SYSTEM

### *7.2.1. Systematic Traceability Methods*

Establishing a traceability system and a traceability business process that is tailored to the distribution method of agricultural products is the cornerstone of agricultural product quality and safety traceability. To meet the requirements of agricultural product safety supervision, a perfect traceability system and scientifically developed traceability methods can help to better establish an agricultural product quality and safety traceability system.

Relying on public service platforms to gather goods from production, trade, and circulation to the consumer of the entire process of the fragmented quality information is known as a traceability system. The true transmission of commodity value will be accomplished through the use of big data analysis and cloud computing, the establishment of standards and quality as the central component of the entire chain, and a closed-loop large quality management mechanism. Traceability processes can be classified as either forward or reverse, depending on the order in which the distribution of agricultural products is tracked from the upstream link to the downstream link (Shivanagowda, 2023).

Agricultural product quality and safety traceability, on the other hand, is a reverse traceability process, meaning that it starts at the consumer and ends at the planting base. Agricultural products must pass through many flow links, including processing and transportation, before reaching the hands of consumers and will simultaneously generate distinct data information.

Figure 7.2 displays the business flowchart for agricultural product quality and safety traceability. The producers, processors, wholesalers, distributors, and consumers make up the majority of the agricultural product distribution model used today. Users can query agricultural product quality and safety monitoring data that is stored in the system database, as well as the quality and safety score values that the system evaluates, by using the traceability system. Additional improvement measures can be suggested by the quantitative score analysis of the quantitative quality and safety status analysis of agricultural products, which is done using classical logic. The actual testing data of agricultural products must be entered and stored, users must query the data information of agricultural products, the system must assess the risk level of the input data, hazardous data must be labeled for early warning, and so on. These are the primary requirements of the traceability system based on this. When the system is operational, the user logs in via the interface first, then uses the AJAX engine to send HTTP requests to the server with data parameters. The server then receives the requests and uses SQL statements to get the produce traceability data. After JSON has been serialized and deserialized, the data will be returned to the client. Finally, string separation tools will parse the data into an interface and show it to the user, enabling forward tracking and reverse traceability of the agricultural product quality and safety traceability system (Surjandari, 2021).

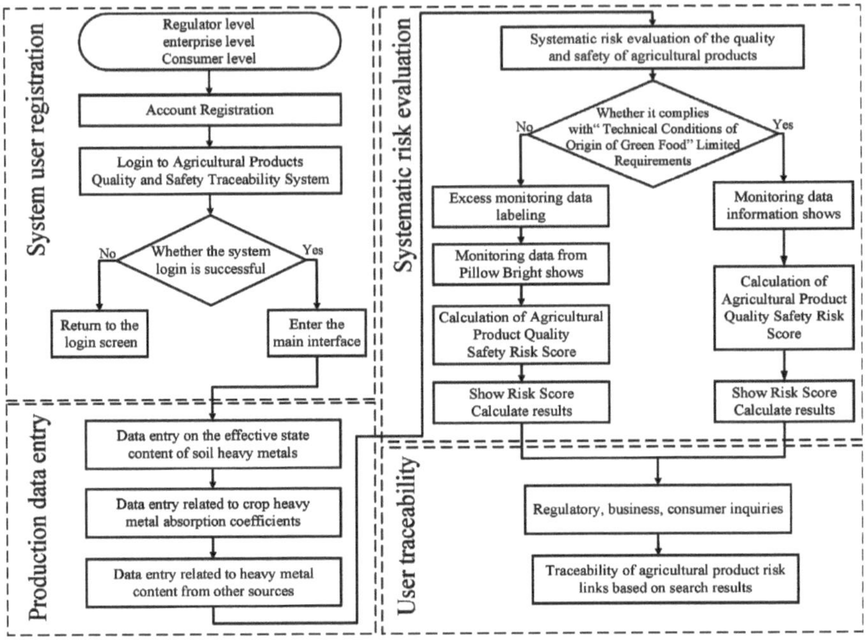


**Figure 7.2.** *Agricultural products quality and safety traceability business flow chart.*

*Source: Li C, Lu Y, Bian Y, Tian J, Yuan M. Design of Safety Evaluation and Risk Traceability System for Agricultural Product Quality. Applied Sciences. 2024; 14(7):2980. https://doi.org/10.3390/app14072980.*

## 7.2.2. Traceability System Design

### *7.2.2.1. System Architecture Design*

A six-tier development framework serves as the model for the B/S architecture used in the system architecture. B/S architecture is an advancement over C/S architecture, coinciding with the growth and development of network information technology. In system design, the six-layer architecture can more effectively realize the design principle of "high cohesion, low coupling" in contrast to the conventional three-layer architecture, which is divided into the presentation layer, business logic layer, and data access layer. This results in a more stable and compatible system. The agricultural product risk traceability system's maintainability, scalability, reusability, and architectural clarity are just a few of the benefits that the six-tier system architecture offers over the conventional three-tier system architecture. The design system's applicability and maintainability are enhanced by the six-layer system architecture. The ability to deploy individual layer modules shortens the deployment server's hardware requirements and cuts down on the development cycle. The device layer, data layer, application layer, display layer, user layer, and service layer make up the system architecture. This improves the system's structure and functionality and better fits the needs of real-world use cases. It also better adapts to the features of various monitoring access devices, such as high data flow and a variety of service objects. In Figure 7.3, the particular system architecture is displayed (Li et al., 2024).

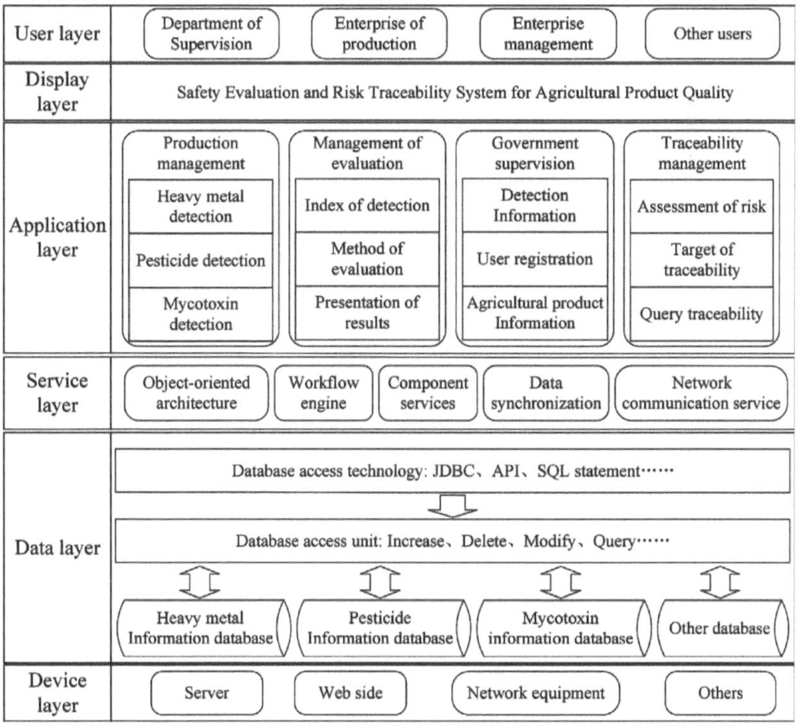


**Figure 7.3.** *System architecture diagram.*

*Source: Li C, Lu Y, Bian Y, Tian J, Yuan M. Design of Safety Evaluation and Risk Traceability System for Agricultural Product Quality. Applied Sciences. 2024; 14(7):2980. https://doi.org/10.3390/app14072980.*

B/S architecture is lighter and simpler to install on the server than C/S architecture. Since B/S architecture is essentially pure HTML technology, it integrates more effectively and shortens the development cycle significantly are:

- The foundation for system operation and construction is the equipment layer. A variety of service support devices are used in the gathering of monitoring data for agricultural product risk tracing and system maintenance. Equipment interfaces can be made clearer and system development efficiency can be increased by dividing the equipment layer apart. The server, along with any other and network devices, is at the center of the device layer. To collect data, the terminal for gathering information about agricultural products is linked to the system using real sensors. The system's primary data collection sensors include pH and gas sensors, among others.
- The actual inspection data pertaining to the safety and quality of agricultural products makes up the data layer. When agricultural products are circulated, a lot of circulation data will be produced. The design and development of an agricultural product risk traceability system also hinge on data processing. Partitioning the data layer can enhance the design of the data processing function, boost the effectiveness of data access and storage, and ultimately boost the precision and speed of traceability. The processing data is roughly divided into four directions by the data layer: detection data related to heavy metals, monitoring data related to pesticides, detection data related to mycotoxin, and other data. The pertinent regulatory bodies supply the detection data that the data layer needs to support the monitoring data of agricultural product quality and safety needed for the system's analysis and decision-making.
- The traceability system's service layer, which connects the database and application layer, is its fundamental service support layer. Data synchronization and component implementation are examples of intermediate links that are necessary for the business operations and data transmission that the risk traceability system requires. The comprehension of functions during development and the reusability of functional components in the service layer can both be enhanced by separate design and development. The data supplied by the data layer of the agricultural product risk traceability system supports the service layer. It synchronizes data with the application layer, while the back-end framework gives the front-end access to data management, user registration, risk assessment computation for quality and safety, and other business functions (Li et al., 2024).
- The business function layer of the risk traceability system is called the application layer. The traceability function division, the real-world business scenario for agricultural products, and business logic are taken into consideration when designing and developing

the application functions for the system. The development team's coordination efficiency is enhanced by the application layer. To eventually realize the system functions, developers can coordinate the deployment of various development function modules in the application layer while concentrating on the creation and upkeep of a single function. Front-end users send their request information to the application layer, which then forwards it to the back-end service layer for processing and response. For instance, when front-end users detect inputs for agricultural products, the application layer sends the data to the back end. The back end then uses the established risk evaluation model algorithm to determine the risk score by calculating the score of the received actual detection data. Production recommendations and traceability risk objectives are presented, and additional mining is carried out in accordance with the scores that were obtained.

- The user-designed interface for human–computer interaction is called the display layer. The risk traceability system's display interface achieves interactive features by utilizing JavaScript, HTML, and other development technologies. The display layer enhances the development specification of the system display interface and unifies the creation of interaction classes. The risk traceability system's operational features are made available to users through web technology. Various terminal devices can be used by different users to access the system. Web technology is more applicable, and the system's design has a greater impact on popularization.
- All users with specific needs for the agricultural product risk traceability system are referred to as being in the user layer. It explains what permissions are and how different users' operational functions differ from one another. The user layer's division makes it easier for developers to define roles and responsibilities for various user roles, such as those in the enterprise management, production, and supervision departments.

The agricultural product risk traceability system's six-layer architecture establishes each layer's duties and synchronizes their interactions with one another. The system architecture is divided so that certain layers' functions from the development phase can be reused. The six-layer architecture's division helps to simplify the effort of upcoming system updates and optimization while also making the system's structure more apparent. It is no longer required to rebuild the entire system when updating and improving it, which increases the system's practical adaptability and development efficiency. In accordance with the architecture, services are divided and clustered to improve request processing efficiency. The disjointed microservice interfaces are accessible. Mutual calls are used to form a cluster, and the registration and configuration are designed as a single, integrated unit. Figure 7.4 displays the specific microservice technology stack.

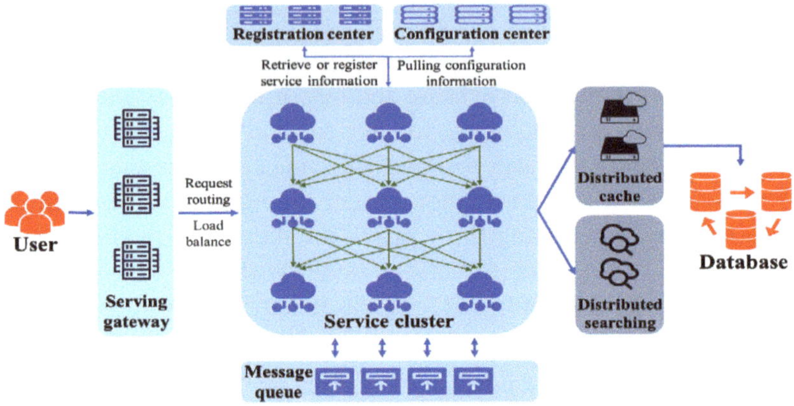


**Figure 7.4.** *System technology stack diagram.*

*Source: Li C, Lu Y, Bian Y, Tian J, Yuan M. Design of Safety Evaluation and Risk Traceability System for Agricultural Product Quality. Applied Sciences. 2024; 14(7):2980. https://doi.org/10.3390/app14072980.*

A containerized approach is selected for deployment to accomplish distributed search and storage during the actual system deployment process, taking into account the impact on system flexibility and scalability. Based on the system architecture, the design system is divided, and the independent components are deployed independently. A front-end human–computer interaction page is constructed using the Vue framework during the system's front-end design and development. Development is made easier and takes less time with the help of component-based development and a lightweight framework (Li et al., 2024).

### *7.2.2.2. System Function Module Design*

The system must be developed to fulfill the fundamental database, data collection, risk assessment, data display, user management, system configuration, and other tasks, in accordance with the system construction objectives and demand analysis.

The primary functional modules of the system are built using a modular design approach following the concept of top-level design in system engineering, as illustrated in Figure 7.5, System Functional Module Diagram. The following are the roles that the system modules play:

- **System User Module:** For three levels of users of agricultural product quality and safety traceability—supervisory department level, enterprise user level, and consumer level—the module offers registration, login, and user maintenance functions.
- **Data Acquisition Module:** This module offers the uploading of information and data from enterprise users and the supervisory department for agricultural product testing, agricultural product quality assessment, environmental testing, and other purposes. Authorities in charge of regulations and pertinent businesses assist with gathering data.

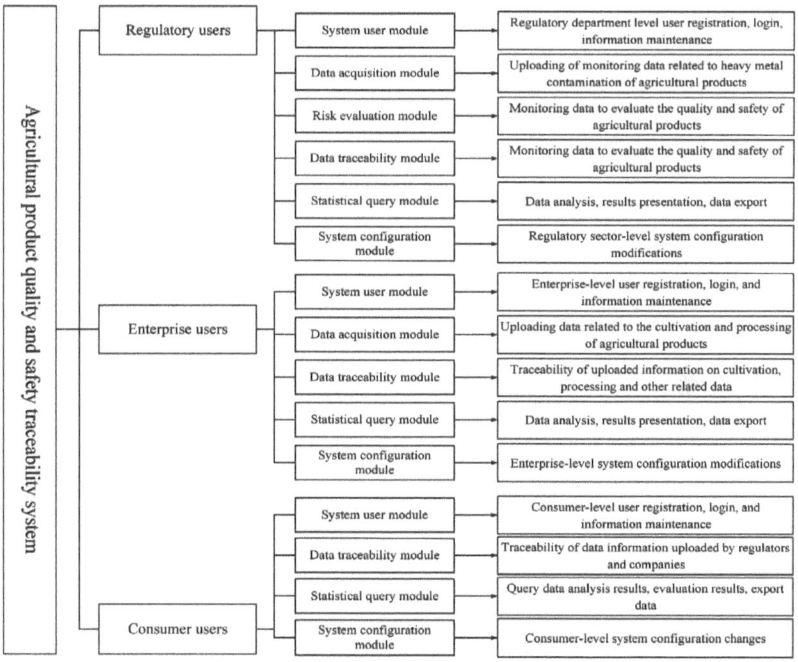


**Figure 7.5.** *Diagram of the system function module.*

*Source: Li C, Lu Y, Bian Y, Tian J, Yuan M. Design of Safety Evaluation and Risk Traceability System for Agricultural Product Quality. Applied Sciences. 2024; 14(7):2980. https://doi.org/10.3390/app14072980.*

- **Module for Risk Evaluation:** Based on the limit value standard, this module primarily addresses the risk assessment of the data obtained regarding the safety and quality of agricultural products. Its features include independent identification of risky data information and analysis of risk evaluation results.

- **Data Traceability Module:** This module offers user traceability record information at three different levels.
- **Statistical Query Module:** This module performs a scoring analysis on the quality and safety status of agricultural products using recorded agricultural product monitoring data and risk assessment techniques. Following statistical analysis, the results of the scoring process are shown.
- **Module for System Configuration:** This module primarily offers altered configuration features for system upkeep and operation.

#### *7.2.2.3. System Database Design*

The relational database MySQL was utilized to compare and analyze the system data to construct the system and realize its functional features. The system database was standardized and designed following database construction steps, based on the data flow generated by the functional business to be accomplished by the system.

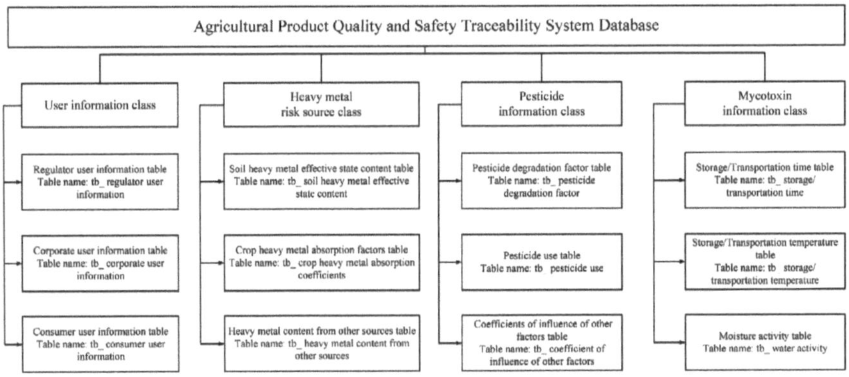


**Figure 7.6.** *System database table design diagram.*

*Source: Li C, Lu Y, Bian Y, Tian J, Yuan M. Design of Safety Evaluation and Risk Traceability System for Agricultural Product Quality. Applied Sciences. 2024; 14(7):2980. https://doi.org/10.3390/app14072980.*

Heavy metal, pesticide, and mycotoxin contamination are the three main categories into which the agricultural product hazards are divided in the traceability system of agricultural product quality and safety designed by the research of this chapter, based on the actual research and the analysis of the realized functional requirements. The possibility of surpassing the limits for each of the three hazards occurs in a methodical manner, and the degree of risk associated with agricultural products as impacted entities is established by the exposure pathways and risk sources, or impact factors. Therefore, using Navicat software, MySQL relational databases were created and the following database tables were created: user information, soil heavy metal effective state content, crop heavy metal uptake coefficients, heavy metal content from other sources, pesticide degradation coefficients, pesticide usage, other factors influence coefficients, storage, and transportation schedules, storage and transportation temperatures, and moisture activity table (Li et al., 2024). Figure 7.6 displays the specific form name design and system database table design diagram.

The aforementioned data sheet creates an electronic file for the traced agricultural products and captures the specific data needed for the assessment of agricultural products during the process from planting to marketing. The following benefits come from using information technology to trace agricultural product monitoring data: it more quickly meets business requirements for agricultural product quality and safety traceability; data can be further examined in accordance with system evaluation results; and more potent corrective actions can be proposed, resulting in unified management of agricultural product production traceability information.

## 7.3. REALIZATION OF RESULTS

### 7.3.1. System Implementation

The browser/server architecture used by the agricultural product risk traceability system is multi-layered and consists of a client, server, and database. The intended risk assessment model was incorporated into the system's data processing module during the construction phase. It was made possible for the risk traceability system to record and store user-uploaded data information and for the risk assessment model to display quantitative score results derived from the data information analysis. The regulatory department level, enterprise level, and consumer level are the three levels of simultaneous multi-user login that the system is capable of supporting. The primary business demand functions were fulfilled, including the uploading and information querying of data related to the safety and quality detection of agricultural products, the dynamic modification of user permissions, the highlighting of abnormal data related to detection and evaluation scores, and the display of results from risk assessments.

Figure 7.7 displays the login screen and system registration page. To log into the system, users must first complete the required fields for the username and password. Once completed, they can click the login button. By selecting the Register button, users who are not yet registered can register. The user information database table will contain information about the username and password (Li et al., 2024).

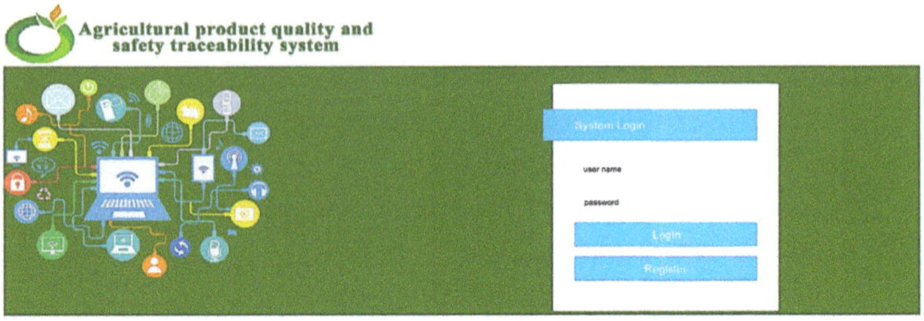


**Figure 7.7.** *System registration and login interface.*

*Source: Li C, Lu Y, Bian Y, Tian J, Yuan M. Design of Safety Evaluation and Risk Traceability System for Agricultural Product Quality. Applied Sciences. 2024; 14(7):2980. https://doi.org/10.3390/app14072980.*

In Figure 7.8, the primary system interface is displayed. Users may query testing data for relevant agricultural products using the left navigation bar. These include the following categories: pesticide information, pesticide degradation coefficients, pesticide use, and impact coefficients of other factors; mycotoxin information, storage/transportation time, storage/transportation temperature, and moisture activity; heavy metal risk sources, soil heavy metal effective state content, crop heavy metal uptake coefficients, and heavy metal content from other sources. Additionally, users have the ability to modify system configuration and administrator settings.

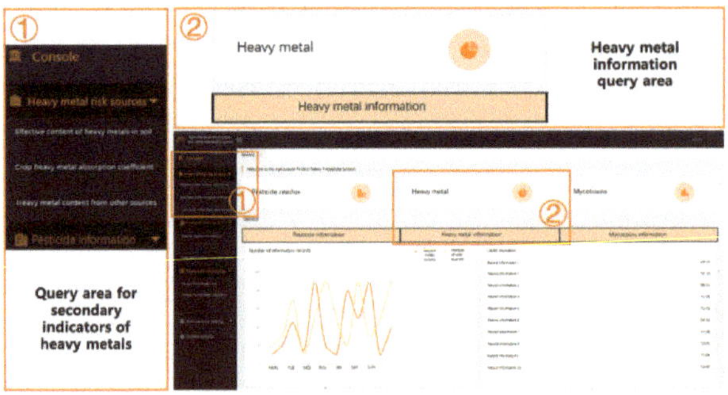


**Figure 7.8.** *Main interface of the system.*

*Source: Li C, Lu Y, Bian Y, Tian J, Yuan M. Design of Safety Evaluation and Risk Traceability System for Agricultural Product Quality. Applied Sciences. 2024; 14(7):2980. https://doi.org/10.3390/app14072980.*

The number of information records and user inquiries can be seen on the graph in the information statistics section of the main interface. Under the most recent information content and time, the most recent data can be seen; clicking on the buttons for pesticide residues, heavy metals, and fungal toxins will take you to the information entry page. Using heavy metals as an example, Figure 7.9 displays the information input page. The information input interface allows the recording of pertinent detection data for agricultural products, such as the names of the heavy metals, their concentration in the soil, pH, organic matter, redox potential, crop varieties, crop nutritional status, crop cultivation techniques, pesticide content containing heavy metals, total metal content in fertilizers, heavy metal content in the atmosphere, and irrigation content containing heavy metals. The information is kept by the system in neatly organized database tables (Li et al., 2024).

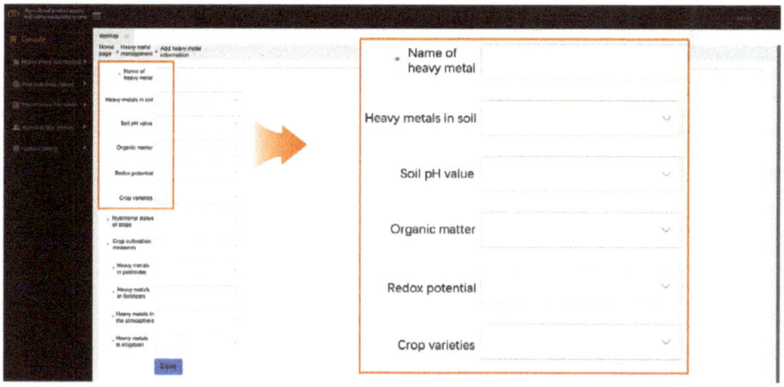


**Figure 7.9.** *Heavy metal information input interface.*

*Source: Li C, Lu Y, Bian Y, Tian J, Yuan M. Design of Safety Evaluation and Risk Traceability System for Agricultural Product Quality. Applied Sciences. 2024; 14(7):2980. https://doi.org/10.3390/app14072980.*

## 7.3.2. System Test

The sample soil underwent testing and analysis based on the experimental pilot of pertinent planting bases. The detection data of the sample soil were further applied to the system based on its implementation, and its practical significance was confirmed by utilizing the results of the risk assessment that was obtained following system operation. The soil sample that was obtained has a pH of 7.65, an organic matter content of 26.18 g/kg, and a soil oxidation-reduction potential of 158 mv. Table 7.4 lists the precise concentrations of heavy metals in soil, fertilizer, the atmosphere, and irrigation water.

**Table 7.4.** *Sample Soil Information Table*

| Category | Heavy Metal Content (mg/kg) | | | | | | | |
|---|---|---|---|---|---|---|---|---|
| | Cu | Pb | Cr | Cd | As | Hg | Ni | Zn |
| Soil | 395.1 | 97.2 | 53.5 | 2.76 | 13.5 | 0.143 | 8.18 | 49.3 |
| Fertilizer | 75.46 | 29.6 | 42.4 | 1.5 | \ | 0.551 | \ | \ |
| Atmos-phere | 48.2 | 84.4 | 82.7 | 2.25 | \ | 0.213 | 20.06 | 176.7 |
| Irrigation | \ | 0 | 0.57 | 0 | 1.46 | 0 | 0 | 0 |

*Source: Li C, Lu Y, Bian Y, Tian J, Yuan M. Design of Safety Evaluation and Risk Traceability System for Agricultural Product Quality. Applied Sciences. 2024; 14(7):2980. https://doi.org/10.3390/app14072980.*

The ability of agricultural products to enrich in heavy metals was assessed using the bioaccumulation coefficient (BCF) and transfer coefficient (BTF). The evaluation results were used to gauge how accurate the heavy metal traceability design system was.

The specific calculation formula is as follows:

$$BCF = w_{crop} / w_{soil} \tag{7.6}$$

In Equation 7.6: $wcrop$represents the mass fraction of heavy metals in agricultural objects (mg/kg);$il$represents the mass fraction of heavy metals in the rhizosphere soil of crops (mg/kg).

$$BTF = w_{above} / w_{root} \tag{7.7}$$

In Equation 7.7:$w_{above}$represents the mass fraction of heavy metals in the aboveground parts of crops (mg/kg);$_{ot}$represents the mass fraction of heavy metals in the roots of crops (mg/kg);

The soil heavy metal threshold based on crop food safety is $HMT$, and the calculation formula is:

$$HMT = E / BCF \tag{7.8}$$

The limit value standard in the corresponding health standard is represented by *E* in Equation 7.8. Using copper as an example, the sample's copper content surpasses the safety threshold, posing a serious risk of enrichment in agricultural products.

To perform a risk assessment on the detection data pertaining to Cu in the obtained sample soil, the system uses a risk assessment model for agricultural product quality and safety. It is evident from Figure 7.10 that the system's evaluation results indicate that the sample soil's Cu warning score is 63.83

points. The system will indicate it with a red warning prompt if it surpasses the predetermined warning score by 60 points. The corresponding scores for the first-level indicators are 58.30, 65.00, and 80.00. The Cu content of the sample soil will significantly affect the buildup of heavy metals in planted crops, per the system evaluation prompt. Therefore, to lessen the effect of Cu accumulation in crops, the soil environment should be improved.

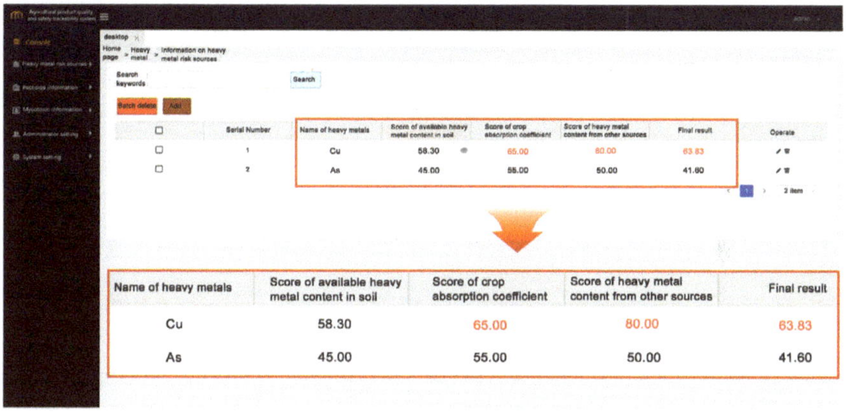


**Figure 7.10.** *Heavy metal information display interface.*

*Source: Li C, Lu Y, Bian Y, Tian J, Yuan M. Design of Safety Evaluation and Risk Traceability System for Agricultural Product Quality. Applied Sciences. 2024; 14(7):2980. https://doi.org/10.3390/app14072980.*

With the evaluation indicator with the highest risk indicated in red and the evaluation indicator with the highest risk highlighted in blue, the interface can show monitoring data for each indicator.

Several indicators are highlighted when examining the heavy metal content of other sources, and the findings indicate that Cu was found in soil, fertilizer, and the atmosphere. To ensure the quality and safety of agricultural products, appropriate measures to reduce the Cu content in the planting environment are urgently needed. This is because the system's red warning score of 80 points indicates that the amount of Cu metal in the planting environment of agricultural products is too high. Using the search box on the interface's upper side, heavy metals can be fuzzily looked up during the traceability process. The high-risk connections between quality and safety problems for the problematic products can be identified using the information that has been highlighted, and production or processing techniques can be enhanced to lower the likelihood of quality and safety problems in agricultural products. The system's evaluation prompts allow for a precise determination of the risk of heavy metal enrichment in agricultural products. A quantitative assessment of the risk of heavy metal enrichment can be successfully accomplished using the developed agricultural product quality and safety risk assessment model and the established agricultural product quality and safety risk assessment system. The ecological environment will be indirectly protected and the growth of green agriculture will be encouraged by developing a system for assessing the safety and quality of agricultural products and a risk tracing system to keep track of the environmental factors influencing crop growth. Multi-stage supervision helps prevent food quality and safety accidents, which lowers production costs and improves response to potential risks. It also improves factory management by identifying risk sources of quality and safety issues in agricultural products in a timely manner (Li et al., 2024).

# REFERENCES


1. Alfian, G., Syafrudin, M., Farooq, U., Ma'Arif, M. R., Syaekhoni, M. A., Fitriyani, N. L., Lee, J., & Rhee, J. (2020). Improving efficiency of RFID-based traceability system for perishable food by utilizing IoT sensors and machine learning model. *Food Control, 110*, 107016.
2. Cocco, L., Mannaro, K., Tonelli, R., Mariani, L., Lodi, M. B., Melis, A., Simone, M., & Fanti, A. (2021). A blockchain-based traceability system in agri-food SME: Case study of a traditional bakery. *IEEE Access, 9*, 62899–62915.
3. Demestichas, K., Peppes, N., Alexakis, T., Alexakis, T., & Adamopoulou, E. (2020). Blockchain in agriculture traceability systems: A review. *Applied Sciences, 10*, 4113.
4. Dong, X. W., Qi, Q., & Fang, S. S. (2014). The optimization and countermeasures research of agricultural products quality and safety supervision mode. *Advanced Materials Research, 1061–1062*, 1271–1274.
5. Govindan, K., & Kannan, D. (2018). Sustainable consumption and production in the food supply chain: A conceptual framework. *International Journal of Production Economics, 195*, 419–431.
6. Hu, J., Zhang, X., Moga, L., & Neculita, M. (2013). Modeling and implementation of the vegetable supply chain traceability system. *Food Control, 30*, 341–353.
7. Jaa, B., Rmt, C., Mdk, D., Aeh, A., & Atn, A. (2020). Evaluating food safety management systems in Singapore: A controlled interrupted time-series analysis of foodborne disease outbreak reports. *Food Control, 117*, 107324.
8. Keshavarz-Ghorabaee, M., Amiri, M., Zavadskas, E. K., Turskis, Z., & Antucheviciene, J. (2021). Determination of objective weights using a new method based on the removal effects of criteria (MEREC). *Symmetry, 13*, 525.
9. Krishnan, A. R., Kasim, M. M., Hamid, R., & Ghazali, M. F. (2021). A modified CRITIC method to estimate the objective weights of decision criteria. *Symmetry, 13*, 973.
10. Li, C., Lu, Y., Bian, Y., Tian, J., & Yuan, M. (2024). Design of safety evaluation and risk traceability system for agricultural product quality. *Applied Sciences, 14*(7), 2980.
11. Namiot, D., & Sneps-Sneppe, M. (2014). On micro-services architecture. *International Journal of Open Information Technologies, 2*, 24–27.
12. Peets, S., Gasparin, C. P., Blackburn, D., & Godwin, R. J. (2009). RFID tags for identifying and verifying agrochemicals in food traceability systems. *Precision Agriculture, 10*, 382–394.

13. Ping, H., Wang, J., Ma, Z., & Du, Y. (2018). Mini-review of application of IoT technology in monitoring agricultural products quality and safety. *International Journal of Agricultural and Biological Engineering, 11*, 35–45.
14. Raj, V., & Sadam, R. (2021). Patterns for migration of SOA based applications to microservices architecture. *Journal of Web Engineering, 20*, 1229–1245.
15. Schwägele, F. (2005). Traceability from a European perspective. *Meat Science, 71*, 164–173.
16. Shivanagowda, G. P., Chitimalla, R., Karabasanavar, N., & Sen, A. R. (2023). A database for buffalo meat traceability in India. *Buffalo Bulletin, 42*, 437–447.
17. Song, Y., Lv, C., & Liu, J. (2018). Quality and safety traceability system of agricultural products based on multi-agent. *Journal of Intelligent & Fuzzy Systems, 35*, 2731–2740.
18. Söylemez, M., Tekinerdogan, B., & Tarhan, A. K. (2022). Challenges and solution directions of microservice architectures: A systematic literature review. *Applied Sciences, 12*, 5507.
19. Surjandari, I., Yusuf, H., Laoh, E., & Maulida, R. (2021). Designing a permissioned blockchain network for the halal industry using Hyperledger Fabric with multiple channels and the raft consensus mechanism. *Journal of Big Data, 8*, 10.
20. Velepucha, V., & Flores, P. (2023). A survey on microservices architecture: Principles, patterns, and migration challenges. *IEEE Access, 11*, 88339–88358.
21. Wales, C., Harvey, M., & Warde, A. (2006). Recuperating from BSE: The shifting UK institutional basis for trust in food. *Appetite, 47*, 187–195.
22. Wang, J., Yue, H., & Zhou, Z. (2017). An improved traceability system for food quality assurance and evaluation based on fuzzy classification and neural network. *Food Control, 79*, 363–370.
23. Wong, D. R., Bhattacharya, S., & Butte, A. J. (2019). Prototype of running clinical trials in an untrustworthy environment using blockchain. *Nature Communications, 10*, 10–18.
24. Yang, L., Zhu, J., Huang, Z., & Geng, J. (2019). A systematic review of China's food safety management since reform and opening up. *Emerging Markets Finance & Trade, 55*, 3472–3489.
25. Zhang, X., Guo, Q., Shen, X., Yu, S., & Qiu, G. (2015). Water quality, agriculture and food safety in China: Current situation, trends, interdependencies, and management. *Journal of Integrative Agriculture, 11*, 2365–2379.

26. Zhang, X., Zhang, J., Liu, F., Fu, Z., & Mu, W. (2010). Strengths and limitations on the operating mechanisms of traceability systems in agro-food, China. *Food Control, 21*, 825–829.

27. Zhou, Z., & Zhou, Z. (2012). Application of internet of things in agricultural products supply chain management. In *Proceedings of the 2012 International Conference on Control Engineering and Communication Technology* (pp. 266–269). Shenyang, China.

# CHAPTER 8

# TRACEABILITY AND CONSUMER TRUST

## Contents

# INTRODUCTION

Customers are demanding more transparency about the origins and production methods of the food they buy as they grow more conscious of the ethical and safety concerns surrounding food production. Customer confidence is critical to the agriculture industry. According to survey, only 12% of consumers completely trust the safety of their food, even though 91% of consumers think it's important to know where it comes from. This disparity draws attention to a big problem facing manufacturers: in addition to offering top-notch goods, they also need to make sure that customers are comfortable with their safety and ethical sourcing. Farmers are urged to put in place comprehensive traceability programs that record every step of their production processes to foster this trust. These programs may contain thorough documentation of farming methods, acquired certifications, and product labels that include vital details like lot numbers and best-before dates. Producers can reassure consumers about the integrity of their products by providing this information easily.

## 8.1. UNDERSTANDING CONSUMER PERCEPTION OF AGRICULTURAL PRODUCTS

Growing interest in sustainable agriculture products is a result of the trend toward greener and sustainable products, especially in the agricultural sector. For producers, marketers, and legislators to satisfy consumer demand, consumer perception and preference are essential. Climate change, food safety, health issues, and environmental concerns are some of the factors influencing this interest. Digital platforms and greater transparency have also made it easier to make well-informed decisions. However, different perspectives on sustainability, cultural norms, socioeconomic status, and individual experiences make it difficult to meet consumer expectations. Preferences may also be influenced by trade-offs between sustainability and other product features.

For many reasons, it is critical to investigate consumer perceptions of sustainable agriculture products such as:

- **Environmental Conservation:** Understanding consumer perceptions of products derived from sustainable agriculture can help determine market demand, which in turn motivates manufacturers to employ environmentally friendly practices. The goal of this study is to increase consumer awareness and enjoyment of sustainability, which will help to mitigate climate change, reduce pollution, and conserve natural resources.
- **Market Demand:** Customer preferences play a major role in shaping the market's dynamics. Analyzing consumer perceptions of goods originating

from sustainable agriculture aids in determining market demand, which aids in the development of sustainable supply chains and the introduction of new goods that align with consumer values. This could encourage innovation and market growth in the field of sustainable agriculture.

- **Economic Sustainability:** Sustainable agricultural methods are often linked to higher initial production costs. However, growing consumer demand for environmentally friendly products might also encourage farmers to adopt sustainable farming practices and open up new business opportunities. Finding effective ways to convey the value proposition of sustainable agriculture to consumers requires an understanding of consumer attitudes, which is crucial for promoting the financial sustainability of sustainable farming practices.
- **Social Responsibility:** Fair labor practices, community involvement, and equitable resource access are all part of sustainable agriculture's tenets of social responsibility. Researchers can determine how much consumers prioritize these social aspects of sustainability in their purchasing decisions by looking at their perceptions. These observations can help guide initiatives to support social responsibility and ethical sourcing in the agriculture sector.
- **Policy Implications:** Decisions about labeling specifications, sustainability programs, and agricultural regulations are influenced by consumer preferences. Policymakers can create policies that support sustainable practices in the agricultural industry and are in line with public values by using evidence-based recommendations from research on consumer perceptions of sustainable agricultural products (Bredahl et al., 2001).

## 8.2. IMPACT OF TRACEABILITY ON CONSUMER PERCEPTION

Traceability of agricultural products has a profound and complex effect on consumer perception. The ability to follow food products' production, processing, and distribution through different supply chain phases is known as traceability. As consumer awareness of food safety, quality, and environmental sustainability has grown, so too has the importance of this capability in contemporary agriculture.

### 8.2.1. Enhancing Trust Through Transparency

The ability of agricultural product traceability to increase consumer trust is one of its main effects. Customers can confirm the origins and handling procedures of their food when they have access to traceability information. They feel more secure about the quality and safety of the goods they buy because of this transparency. Research indicates that when consumers believe there is sufficient information available about a product's journey from farm to table, they are more likely to develop "quick trust.."

### 8.2.2. Influence on Purchasing Behavior

Purchase behavior is greatly influenced by the availability of traceability information. Customers are more likely to make purchases that reflect their values when they are aware of the production methods used to produce their food, such as whether it follows sustainable or organic standards. According to the study, traceability directly correlates with more purchases made by customers who place a high priority on environmental sustainability in addition to fostering a favorable perception.

### 8.2.3. Perceived Environmental Information Transparency

Consumer perceptions of agricultural products are greatly influenced by perceived environmental information transparency, which is a result of traceability. Customers are better able to make decisions that reflect their preferences for environmentally friendly options when they are aware of the environmental impact of the foods they choose, such as carbon footprints or pesticide use.

### *8.2.4. Mediating Effects on Consumer Behavior*

Perceived environmental information transparency acts as a mediator in the relationship between consumer behavior and traceability information. Customers' willingness to buy products is influenced by their increased trust in brands as they learn more about how those products impact the environment. Therefore, clear communication about traceability can result in both positive perceptions and noticeable adjustments to purchasing patterns.

To put it briefly, traceability of agricultural products has a big impact on consumer perception because it increases trust through transparency, influences consumer behavior based on environmental factors, and mediates these effects through perceived transparency of environmental information.

## 8.3. COMMUNICATING TRACEABILITY INFORMATION TO END-USERS

Supply chain traceability is now a crucial requirement to safeguard the integrity of textile companies due to the challenges faced by the fashion industry. Fashion companies face two challenges as a result of the complexity of supply chains: first, they must comprehend the product's journey to identify and address risks; second, they have a responsibility to accurately and transparently convey this journey to end users.

## 8.3.1. Understanding Supply Chain Traceability

The internationally recognized definition of supply chain traceability, as provided by the International Organization for Standardization (ISO), will be modified and used in this section along with the crucial addition of a sustainability focus:

### *8.3.1.1. Definition of Supply Chain Traceability*

The capacity to track the origins, uses, and locations of the subject matter to verify the veracity of assertions made concerning labor (including health and safety), the environment, and human rights. This can apply to products and include information about the product's distribution and location after delivery, the processing history, and the place of origin of materials and parts. Within the realm of sustainability, supply chain traceability serves as an instrument to confirm and validate sustainability assertions linked to products and materials, guaranteeing optimal practices for both human welfare and the environment. The process of tracking the origin of goods and their inputs from the start of the supply chain through manufacturing and end-use is known as supply chain traceability. In the end, traceability of materials, parts, and products guarantees the accuracy of the data, which includes:

- Origin;
- Composition;
- Processing history;
- Quality;
- Safety; and
- Labeling.

Businesses can use that data to improve customer service, spot wasteful resource use, innovate quickly, react to demand changes more quickly, and certify sustainable products and processes.

### *8.3.1.2. Traceability for Sustainability*

Companies are adopting traceability as part of their sustainability initiatives, and it is becoming more widespread. In addition to offering a tool to track goods and materials as they move through the supply chain to make sure that ethical social and environmental practices are applied at every stage, it is crucial in confirming sustainability claims associated with their products. Although a sustainable product system's traceability serves as the basis for confirming social and environmental claims (e.g., carbon neutral, organic, certified), it is unable to provide this confirmation by itself. Verifying such information requires the use of both new technologies and certification programs.

Verification is crucial. Even in supply chains that use blockchain technology, we still want our system to be independently audited. And this is how we win the confidence and trust of every link in our value chain.

To facilitate the shift to a low-carbon economy, governments are also looking to supply chain traceability. The urgent adoption of an EU-wide regulation requiring businesses to identify, address, and mitigate the environmental, social, and governance (ESG) risks in their supply chains was one of the major developments of 2021. Beyond procurement, governments can encourage certification or use sustainability standards to supplement regulation through laws like the EU Renewable Energy Directive and the US Lacey Act (Bredahl et al., 2001).

## 8.3.2. Supply Chain Transparency

Transparency is the process of sharing information about a product's origin or material composition, whereas traceability is the process of creating such data. Companies must be aware of what is going on upstream in the supply chain and inform downstream stakeholders of this information to maintain supply chain transparency. A modified version of the Business for Social Responsibility (BSR) nonprofit organization's definition of supply chain transparency will be used in this section. Transparency is defined by BSR as the information you will disclose, to whom, and when.

- **Who:** Who needs to have access to supply chain information?
- **What:** What information is being shared, in what format, and for what purposes?
- **When:** When (in which conditions) should the information be shared?

From sharing a code of conduct to revealing traceability from the supply chain's raw materials stage, the degree of disclosure can vary. Your company determines how transparent you want to be. Greater supply chain transparency over time can assist businesses in establishing and preserving worthwhile alliances, enhancing their reputation as a brand, and boosting customer confidence. Customers are willing to pay 2–10% more for products when the supply chain is more transparent.

### 8.3.3. Without Traceability There is No Transparency

Due to the relocation of current supply chains, goods are transported around the globe, going through numerous hands, factories, and processes. The industry's fragmentation poses significant risks to the environment and society, which also erodes consumer confidence. We are standing on the cusp of a significant shift in which the industry has prioritized supply chain knowledge and transparency.

### 8.3.4. The Relevance of Communicating the Product Journey

However, as a fashion company, you must not only fully understand the traceability of your products but also be able to explain this journey to your customers to guarantee brand authenticity and promote transparency. It's critical to bolster your communications with facts to back up your statements. By offering information about every phase of your supply chain, you can increase the precision and openness of your communications, ensuring the validity of your assertions and preventing greenwashing. The objective is to share truthful tales that uplift your final customer, enhance the value of your offering, and increase public confidence in your company (Bredahl et al., 2001).

### 8.3.5. Closing the Misinformation Gap Through Technology

Fortunately, with the correct tools and technology, sharing these insightful observations is made easier. With the help of the e-commerce integration, you can automatically provide your customers with pertinent information that helps tell the story of each of your products, such as where the various supply chain stages are located:

- **Production:** Where all of the materials that are processed into each of your products are assembled.
- **Completing:** The nation in which the material is processed to enhance its functionality, handling, and appearance.
- **Processing:** The place where your raw materials are converted into particular materials.

### 8.3.6. Legislative Pressure to Guarantee Transparency

Beyond customer demand, this degree of openness is now necessary to comply with laws like AGEC and is no longer merely a nice-to-have. This law, which applies to all companies doing business in France, aims to provide accurate, comprehensible, and transparent information about the features of waste-producing products. It includes requirements like disclosing the location of the manufacturing, finishing, and processing stages. AGEC is not the only law, though, requiring the fashion industry to disclose details about its goods and production methods. Similar laws have been put into place in many nations and areas, with particular focus on the European Union, where tighter laws are already planned to advance sustainability in the near and medium future.

An end-to-end solution enables you to take charge of your supply chain, properly convey the path taken by your goods, and adhere to legal requirements. As a fashion business, sharing data in an automated and tailored manner helps you to differentiate yourself from competitors, win back customer trust, and improve transparency.

## 8.3.7. Tips for Traceability and Transparency Implementation

Although supply chain transparency and traceability have been hailed as being resolved by blockchain, this issue cannot be resolved by a single technology. For any solution to be successful, the proper combination of people, knowledge, and technology must be used. For instance, both internal and external stakeholders should be involved, and the technology used should support appropriate decision-making and include solutions that generate, collect, translate, and communicate useful data across supply chains.

### *8.3.7.1. Suppliers*

Knowing what information is too sensitive to divulge and what can be shared publicly is important for suppliers. You must also evaluate the information's accessibility within your company. Where can I find it? Is it in your ERP system or in Excel files? Does the requested information even exist? To turn traceability and transparency into a significant competitive advantage, it is critical to begin this process early and collaborate closely with your clients and technology providers.

### *8.3.7.2. Brands and OEMs*

It is your responsibility as a brand owner to explain to suppliers the advantages of traceability and transparency for them. Along with educating them about the traceability and transparency schemes and their responsibilities, you also need to determine whether they can meet the requirements. Listening to them is crucial. You can position your business and its supply chain for success by establishing early and consistent connections and communication.

## 8.3.8. The Future

Global supply chains today need new levels of visibility as businesses develop new goods, enlist new partners, enter new markets, and attempt to satisfy constantly changing legal and consumer requirements. Transparency and traceability are becoming essential for successful business operations. Supply chain visibility helps businesses better understand opportunities and respond to risks in their intricate, multi-tiered, highly regulated supply chains. Benefits include increased security and improved communication of data pertaining to sustainability and ethical sourcing. Blockchain technology has the potential to significantly increase global supply chain visibility in the future, but it cannot achieve this goal on its own. Improved collaboration between manufacturers, suppliers, brands, regulators, and customers themselves will bring about the shift. Companies will have a competitive advantage in satisfying the ever-increasing demands of customers and regulatory requirements if they invest in enhanced transparency, traceability, and data measurements along supply chains (Bredahl et al., 2001).

## 8.4. CONSUMER-DRIVEN TRACEABILITY DEMANDS

Consumers' concerns regarding the safety and quality of their food have grown in recent years. Increased knowledge of food recalls, allergies, and contamination problems is a major factor in this change in consumer behavior. Traceability from farm to fork has become more important as a result of the growing demand for transparency in the agricultural supply chain.

### 8.4.1. Food Safety and Food Quality

Customers now make more discriminating decisions about what they eat. Concerns about food safety and quality have gained prominence in industry strategies, consumer concerns, and occasionally government policy. Diverse initiatives in the private sector and public policy domain have surfaced regarding traceability, partially aimed at mitigating the information asymmetry that consumers possess concerning food safety and quality characteristics. The function of traceability systems in the food industry is discussed in this section, along with the differences between ex post traceback and ex ante quality verification systems. The trade ramifications of required traceability and labeling are covered, along with examples of voluntary private sector livestock traceability systems and public sector traceability initiatives.

A series of high profile food safety scares has heightened public awareness and concerns over food safety, for example the Bovine Spongiform Encephalopathy (BSE) crisis in the UK beef industry; in the US, *E. coli* O157:H7 outbreaks attributed to ground beef (Jack-in-the-Box; Tyson meats); in Scotland, the deaths of 22 senior citizens linked to *E. coli* O157:H7 contaminated meat purchased from a local butchers shop; in Belgium, the scare created by the contamination of animal feed with potentially cancerous dioxins in 1999; in Australia, the

'Garibaldi' incident in 1995 in which *E. coli* O157:H7 from a contaminated meat sausage resulted in several illnesses and the death of a child.

The quality of food has gained importance, especially in light of its expanding definition. A portion of this is supply-driven due to advancements in technology and more opportunities for product differentiation. Demand is a contributing factor, as consumers are becoming more interested in a greater range of intrinsic and extrinsic food qualities. Palatability, nutrition, food's functional (healthy) qualities, and other aspects are examples of intrinsic quality attributes. Organic food, environmental stewardship, farm animal welfare, and other aspects of the production environment are examples of extrinsic quality attributes. Many of these are credence attributes that consumers are unable to recognize in the absence of a quality signal, like a label.

Food safety is frequently the same. Customers are unable to use their senses to identify food safety risks before making a purchase unless there has been significant product deterioration. Food safety may occasionally be considered an experience attribute in the event that a customer contracts a foodborne illness right after consuming a particular food item. Food safety typically possesses attributes of credibility. This was undoubtedly the situation with BSE in the UK beef market; consumers could not immediately determine if they had eaten beef contaminated with BSE.

To ascertain whether an experience or credence attribute may be present, consumers must pay information costs. If the market unfavorably chooses lower-quality (or unsafe) food in the absence of information signals to consumers, market failure may result from information asymmetry. If quality disclosure is free of cost, then Grossman (1981) contends that the market has a self-correcting mechanism. High-quality product vendors have an obligation to disclose their standards. Non-disclosure therefore suggests poor quality. This self-correcting process depends on the ability to (economistically) confirm ex post product quality disclosures. McCluskey (2000) demonstrates how manufacturers who prioritize profit can benefit from misleading customers with exaggerated claims of quality. Markets for high-quality credence goods require third-party monitoring and repeat purchase relationships to function effectively. Credible product quality signals are necessary for efficient markets in credibility goods. Systems of traceability make it easier to provide consumers with quality signals. However, many of the newly developed livestock traceability systems seem poorly built to give consumers reliable ex ante quality signals.

## 8.4.2. Private Sector Traceability Programs

Credence attributes can be identified through voluntary labeling by businesses, which is occasionally complemented by third-party certification. Businesses that produce goods with higher levels of food safety have an incentive to include this feature on a label if there is a market premium for "safer" food. In the US, irradiated meat products are a prime example. To lower the possibility of cheating by mislabeling, a reliable monitoring and enforcement system is required. If companies that produce "high quality" (or clearly safer) food can punish companies that take advantage of the certification program by using false or misleading labeling, then a self-policing industry quality assurance or safety labeling program may work. For products with poor quality or safety attributes, the market failure issue still exists in the absence of an efficient self-policing mechanism. A company won't willingly reveal poor quality.

Both industry-wide programs and individual supply chain initiatives are examples of private sector traceability initiatives in the livestock sector. The UK beef industry has seen the emergence of supply chain partnerships that provide traceability, partly due to the decline in consumer confidence that followed the BSE crisis. One such is Tracesafe, a tiny, farmer-owned business that has established a network of finishers and breeders of cattle that raise animals in accordance with particular production standards. The production protocols include a comprehensive system of on-farm record keeping and stipulate that feed must be purchased from a contracted set of feed mills. The ability of Tracesafe's beef to track individual meat cuts back to the original animal, along with an implied guarantee of safety, sets it apart from competitors. Under the Tracesafe brand, the beef is offered for sale in restaurants and specialty retail stores (Fearne, 1998).

The meat processing industry is also aware of how traceability can be used as a strategy for product differentiation and to increase consumer trust in food safety. Michael McCain is Maple Leaf Foods Inc.'s president and chief executive officer. (a significant Canadian processor of pork and poultry) recently called traceability the "holy grail of the food supply chain.." To make it easier to trace meat back to its original farm, Maple Leaf is currently investing in the development of DNA identification technology (Powell, 2002). This development seems to be largely driven by pressure from export markets, especially the Japanese market.

In other instances, pressure from downstream food retailers has led to private sector traceability initiatives. The drivers are a desire to lower transaction costs associated with monitoring product quality or downstream production methods, or a desire to lower risk exposure. This does not imply that consumers are provided with traceability information on retail packaging. Even though the meat might not be able to be tracked back to the exact farm of origin, UK supermarkets require their beef suppliers to be part of approved quality assurance programs. Sobeys, a Canadian retailer, demands proof from its suppliers that certain handling, transportation, processing, and production procedures have been followed. Although it may not be a specific requirement, traceability to the farm may be a prerequisite for disclosing details about production and processing practices (Hobbs, 1996; Fearne, 1998; Bredahl et al., 2001).

Industry associations and producer groups have also introduced industry-wide private sector traceability programs. CCIA, or the Canadian Cattle Identification Agency, was founded by the Canadian Cattlemen's Association. To make it easier to trace back cattle in the event of a food safety issue, the CCIA implemented a nationwide cattle identification system in July 2002. As a risk-reduction tactic, the industry launched CCIA. The identification and tracing of animals in the event of a major crisis such as the BSE outbreak would have been nearly impossible without the implementation of the traceback system.

A distinct ID number is given to cattle departing from their home herd and is shown on a CCIA tag with a barcode. Authorized service centers that keep track of which ID numbers belong to which producers distribute tags. Up until the point of carcass inspection in the packing plant, the distinct ID number is kept current. Producers who fail to comply may face financial penalties. The Canadian Food Inspection Agency (CFIA), a federal government organization, starts a traceback process when there is a food safety issue. The CFIA determines the animal's last location and the herd of origin using data from the CCIA database (CCIA, 2002). Cattle movements are tracked using this data, both forward and backward in the supply chain. It is not necessary for producers to keep records. The Canadian system differs significantly from the UK's "cattle passport" system in this regard. According to the UK system, farmers must notify the national identification agency of every cattle movement that occurs on or off a farm. The distinct animal ID number in the UK system ought to make it possible to quickly identify every farm where the cattle have been found. The Canadian system only permits the identification of the cattle's final location and herd of origin; any traceback beyond those two points is dependent upon the producers' ability to furnish this data.

A voluntary quality assurance program, comprising a national identification program with DNA sampling for traceback, is in place in the Australian beef industry. The Meat and Livestock Agency (MLA), a government agency, is in charge of the Australian system, which is an example of a partnership between industry and government. Under the general heading of "Cattle Care," many quality management guidelines pertaining to production, handling, and processing were created. When selling cattle, a producer must sign a National Vendor Declaration form with their name and basic production details (e.g., details regarding the feeding regimen, whether the cattle received growth-promoting hormone treatment, etc.). In the event of an issue, this data along with a cattle tag allows cattle to be tracked back (Lawrence, 2002).

Meat Standards Australia (MSA), a voluntary grading system, predicts the eating quality of meat using many pre- and post-slaughter metrics. Every carcass that meets the requirements for the MSA program has blood samples drawn from it while it can still be linked to a seller. A consumer's complaint regarding a poor meat-grade MSA experience can be resolved by comparing a meat sample's DNA with a carcass blood sample. Meat cuts can be tracked back to the farm of origin and throughout the supply chain in this manner. The MSA system's traceback is more concerned with quality than food safety. It makes it possible to directly connect eating quality with production and processing techniques. According to Lawrence (2002), it can help identify areas that might benefit from improvement or sellers who routinely misrepresent cattle on their National Vendor Declaration form. The quality assurance and identification systems in Australia are optional. They set up the informational framework that separate supply chains can use to add their own premium-branded beef initiatives. The MSA system is employed by many Australian branded beef initiatives as a component of their strategy for product differentiation (Lawrence, 2002).

### 8.4.3. The Growth of Organic and Sustainable Markets

The market for organic food and drink has experienced tremendous expansion; it is expected to reach a valuation of over $230 billion in 2023 and increase by 13.9% by 2030. This increase is indicative of the demand from customers for goods that are sourced ethically in addition to being safe. Customers who place a premium on organic, fair-trade, and non-GMO products are especially demanding traceability. These customers want proof that the goods they buy reflect their beliefs about sustainability in the environment and human health.

### 8.4.4. The Role of Consumers in Demand for Traceability

The need for traceability in the food industry is largely driven by consumers. Customers are asking for more information about the products they use because they are more concerned about food safety and transparency. For many consumers who want to know where their food comes from, how it's made, and what ingredients are used, traceability has become essential. This section will examine the role that consumers play in driving traceability demands and how they affect the food industry.

#### *8.4.4.1. Consumer Awareness and Education*

Education and consumer awareness have played a major role in increasing demand for traceability. Growing consumer awareness of the food industry has led to increased demands for transparency from food producers. Education campaigns, like those conducted by nonprofits, have contributed to a greater understanding of the significance of traceability in the food industry. Nowadays, consumers are more inclined to select transparent products, and they are even prepared to pay a premium for those that offer traceability data.

#### *8.4.4.2. Consumer Demand **for** Food Safety*

Consumers are very concerned about food safety, and traceability can help to guarantee the safety of food products. Products with extensive testing and quality control have a higher chance of being bought by consumers. Customers can follow a product's path from the farm to the fork thanks to traceability, which guarantees that it was made in an environmentally friendly and safe manner. A product's supply chain is more likely to be transparent and traceable, which increases consumer trust.

#### ***8.4.4.3. Impact of Traceability on Consumer Behavior***

Consumer behavior is significantly impacted by traceability. Product selection is more likely for those with an open and traceable supply chain. Additionally, they are more inclined to spend more for goods that include traceability data. 75% of consumers stated they would move to a brand that offers more detailed product information in a Food Marketing Institute study. Consumer purchase decisions now heavily consider traceability.

#### ***8.4.4.4. The Role of Technology in Meeting Consumer Demand***

The need for traceability among consumers has been met in large part by technology. Technology breakthroughs have made it feasible to track goods from the farm to the fork in the supply chain. Technologies such as blockchain, RFID, and barcoding have made it easier to track products and provide consumers with traceability information. Businesses that make technology investments to offer traceability data have a higher chance of satisfying customer demand and gaining a competitive edge.

#### ***8.4.4.5. The Challenges of Implementing Traceability***

For food producers, implementing traceability can be difficult. Implementing traceability can be expensive and necessitates large infrastructural and technological investments. It is imperative for food producers to adhere to regulations and standards regarding traceability. But there are much more advantages to traceability implementation than disadvantages. Businesses that make traceability investments can outperform their competitors by satisfying customer demands for safety and transparency. The need for traceability in the food industry is largely driven by consumers. The need for food safety, consumer education and awareness, and the influence of traceability on consumer behavior have all increased demand for traceability. The need for traceability has been met in large part by technology, and businesses that use technology to offer traceability information stand to benefit financially. Although putting traceability into practice can be difficult, the advantages are far greater than the drawbacks.

## 8.4.5. Regulatory Traceability & Labeling Programs

In certain nations, mandatory labeling and traceability programs have been implemented. By 2003, all Member States must have mandatory beef labeling and traceability systems in place, according to a beef labeling regulation from the European Union (EU). The regulation consists of three parts. First, a national cattle identification and registration system will be in place in each Member State. Second, a traceability number that identifies the origin of the meat—including the birth, rearing, slaughter, and processing locations of the animals—will be attached to the label of beef products. Third, the rule establishes guidelines for optional labeling of extra information (for example, production information, animal welfare information, etc.).

Retail-level country of origin labeling for beef, lamb, pork, fish, fresh and frozen fruits and vegetables, and peanuts was implemented by the US Farm Bill of 2002. For now, this is optional, but by September 30, 2004, a law requiring it will be in place. Cattle have to be born, raised, and killed in the US to be designated as US country of origin. The traceability, record keeping, and information systems that will be necessary to support the "born, reared, and slaughtered" claim for any meat product will be significantly impacted by mandatory country of origin labeling.

Numerous instances of both public and private sector initiatives that provide varying levels of traceability and quality assurances with regard to credence attributes have been demonstrated by these examples. Therefore, it is reasonable to ask whether traceability and origin labeling should be the responsibility of the public or private sector, depending on whether there is a market failure and, if so, whether requiring traceability and/or labeling benefits consumers more than it costs. It is helpful to think about the purposes and possible financial advantages of a traceability system to evaluate these issues.

## 8.4.6. Trade Implications

A greater emphasis on traceability and quality assurances, such as retail labeling of product origin and/or production methods, has been prompted by consumer interest in food safety and quality. These are frequently the private sector's reactions to chances for product differentiation. Alternatively, they might be private sector tactics to lower risk exposure and preserve customer trust. International trade is hampered by regulatory actions that require labeling of production methods, origin, or traceability. The "right to know" of consumers has been used as justification for mandatory labeling of credibility attributes, such as country of origin and genetically modified foods.

The World Trade Organization (WTO) system is particularly challenged by the labeling of production and processing methods (PPM). Three ideas form the foundation of the non-discrimination principle. Since goods are exchanged rather than PPMs, the WTO regulations mandate that "like" goods be treated equally. It mandates that all foreign-like goods produced in member nations receive equal market access and that they be given the same "national treatment" as domestic goods. The difficulty comes when domestic concerns ask for PPM regulations that go against one of the three non-discrimination bases. For instance, if PPMs vary between nations, a domestic regulation might disadvantage foreign producers or go against the principle of the most favored nation (Isaac and Kerr, 2002). Regulations requiring country of origin labeling or mandatory traceability and labeling may be challenged at the WTO on the grounds that they impede trade and violate WTO guidelines for permissible labels.

In response to the controversy surrounding PPM-related trade barriers, the WTO has allowed "legitimate" violations of the non-discrimination principle. PPMs were separated into product-related and non-product-related PPMs as a result of a negotiated compromise during the Tokyo round of GATT negotiations. The Agreement on Sanitary and Phyto-Sanitary Measures (SPS) addresses trade barriers pertaining to safety. The Agreement on Technical Barriers to Trade (TBT) addresses trade barriers pertaining to PPM products. Under the TBT agreement, a PPM-based trade barrier might be allowed if the PPM produces a unique product that eliminates the need for the "like" product designation. The Most-Favored Nation and National Treatment principles are still relevant, though. PPMs unrelated to products are not covered by the TBT agreement. Trade restrictions based on nonproduct-related PPMs, such as the use of intensive agricultural production techniques instead of an organic system to produce the cotton used to make shirts, cannot be justified by TBT measures (Isaac and Kerr, 2002).

If a non-product-related attribute is being labeled and the labeling policy violates one of the three pillars of non-discrimination, mandatory labeling policies based on consumers' right to know violate WTO regulations. For instance, it would be against WTO regulations to require GM food to be labeled according to consumers' right to know about a technology or processing method (instead of a claim that the product was novel or a safety objection) (Isaac and Kerr, 2002). One could also argue that, if the Non-Discrimination principle is broken, requiring the labeling of traceability information and/or product origin based on consumers' right to know would also be against WTO regulations.

In the US, proponents of mandatory country-of-origin labeling contend that it legally protects consumers' right to know where products come from. A mandatory regulation, according to its detractors, serves the domestic protectionist interests of producers. It costs the food industry needlessly and will have a negative impact on (e.g., livestock) exports from other nations, especially since US processors will have to include a proportionate amount of information about the country of origin on food labels. To let consumers know if there is a high demand for food produced domestically, the voluntary labeling of "US origin" would be adequate. A mandatory country of origin labeling requirement violates the Non-Discrimination principle because it aims to differentiate similar products through labeling, even though the country of origin has no effect on the product (beef is beef).

The WTO was not established to address consumer demands for import regulation or protection, which is at the core of this problem. According to the straightforward neoclassical theory of a world with perfect information, trade liberalization always benefits consumers by lowering costs and increasing their options. After accounting for information asymmetry concerning credence attributes (country of origin, on-farm production methods, etc.) and concerns about food safety, some customers might ask for stronger labeling laws or protection from imports (e.g., genetically modified food). Renegotiating WTO regulations to permit trade restrictions based on consumer preferences is something the EU has expressed a wish to do. The difficulty is in telling the difference between policies implemented to address valid consumer concerns and trade restrictions motivated by vested producer interests masquerading as "consumer interests."

Gaining a better understanding of consumer attitudes regarding food safety and quality issues is a first step in making this distinction. Understanding how customers react to various traceability, labeling, and quality assurance initiatives is part of this. The degree to which these efforts are appropriate for the private sector as creative methods of product differentiation or the degree to which a strong case for requiring traceability and labeling on the grounds of lessening consumer information asymmetry can be made for market failure are of immediate interest.

Despite potentially serving as a helpful ex post cost reduction mechanism in the event of a food safety issue, it seems that the EU's mandatory beef labeling and traceability regulation does little to lessen consumer information asymmetry ex ante. Customers will not be able to learn about other credence attributes outside of the country of origin under the new US regulations requiring the mandatory provision of country of origin information at the retail level. When mandatory labeling laws negatively impact exports, they raise the possibility of tensions in global trade.

However, industry-driven traceability systems have surfaced, frequently with government support and encouragement. In these situations, the private sector is reacting to the need to lower the possible expenses of a food safety incident, lower the costs of supply chain relationship monitoring and enforcement, preserve consumer trust, or have identified a chance for product differentiation. The traceability system frequently serves as a platform for offering customers further quality assurances. Without extending traceability labeling to the retail counter, traceability, and ex ante quality verification across the supply chain may provide enough financial advantages.

In the absence of clear evidence of direct consumer benefits, mandatory retail labeling of traceability and product origin information is likely to have a substantial negative economic impact on industry and cause tensions in international trade. This does not negate the significance of traceability. Conversely, it can be crucial in lowering transaction costs associated with supply chain relationship management, lowering risk, and bolstering tort liability incentives for food safety. Therefore, it is critical to differentiate between the different functions of a traceability/quality verification system and to assess the degree to which it contributes to supply chain management or consumer education.

## 8.4.7. Data Traceability Mechanisms

Agricultural producers are implementing data traceability mechanisms, such as unique identifiers like GS1's Global Trade Item Numbers (GTINs) and Global Location Numbers (GLNs), to satisfy consumer demands for traceability. Stakeholders are able to precisely follow products through different production and distribution phases thanks to these identifiers. Producers can substantiate their claims about sustainability by gathering vital data inputs related to farming practices, such as crop water usage or animal grazing locations.

## 8.4.8. Technological Innovations Supporting Traceability

Improvements in technology are essential for improving agricultural supply chain traceability. Electronic Product Code Information Systems (EPCIS) are one tool that makes it possible for stakeholders to share data easily. As products move through the supply chain, real-time tracking is made possible by barcodes, RFID tags, and Internet of Things sensors. This degree of transparency about product origins not only helps producers comply with regulations, but it also fosters consumer trust.

### 8.4.9. Consumer Empowerment Through Transparency

Customers are expecting brands to be more transparent about their sourcing methods as they grow more aware of the origins of their food. Producers are encouraged by this trend to implement sustainable practices that are verifiable by traceable data systems. For example, traceability data linked to unique product identifiers can be relied upon by consumers seeking to determine whether a product is truly organic or non-GMO.

The push for consumer-driven agricultural product traceability signifies a big change in the direction of increased accountability in the food supply chain. Agricultural producers will need to invest in strong traceability systems that not only comply with regulations but also build customer trust as consumers prioritize sustainability and quality in their purchases (Fearne, 1998).

## REFERENCES


1. Bredahl, M. E., Northen, J. R., Boecker, A., & Normile, M. A. (2001). Consumer demand sparks the growth of quality assurance schemes in the European food sector. In A. Regmi (Ed.), *Changing Structure of Global Food Consumption and Trade* (pp. 90–102). U.S. Department of Agriculture, Market and Trade Economics Division, Economic Research Service.
2. Canadian Cattle Identification Agency (CCIA). (2002). *http://www.canadaid.com*.
3. Dickinson, D. L., & Bailey, D. (2002). Meat traceability: Are U.S. consumers willing to pay for it? *Journal of Agricultural and Resource Economics, 27*(2).
4. Fearne, A. (1998). The evolution of partnerships in the meat supply chain: Insights from the British beef industry. *Supply Chain Management: An International Journal, 3*(4), 214–231.
5. Grossman, S. J. (1981). The informational role of warranties and private disclosure about product quality. *Journal of Law and Economics, 28*(3), 461–489.
6. Hayes, D. J., Shogren, J. F., Shin, S. Y., & Kliebenstein, J. B. (1995). Valuing food safety in experimental auction markets. *American Journal of Agricultural Economics, 77*(February), 40–53.
7. Hobbs, J. E. (1996). A transaction cost analysis of quality, traceability and animal welfare issues in UK beef retailing. *British Food Journal, 98*(6), 16–26.
8. Isaac, G., & Kerr, W. A. (2002). Genetically modified organisms and trade rules: Identifying important challenges for the WTO. *The World Economy*.
9. Lawrence, J. D. (2002). Quality assurance "down under": Market access and product differentiation. *MATRIC Briefing Paper 02-MBP 1*, Midwest Agribusiness Trade Research and Information Center, Iowa State University. http://www.matric.iastate.edu.
10. McCluskey, J. J. (2000). A game theoretic approach to organic foods: An analysis of asymmetric information and policy. *Agricultural and Resource Economics Review, 29*(1), 1–9.
11. Powell, G. (2002). Leverage the Canadian brand by delivering a gold standard in food safety assurance. *Food in Canada, September*. http://www.foodincanada.com (accessed on 14 December 2024).
12. Shogren, J. F., Shin, S. Y., Hayes, D. J., & Kliebenstein, J. B. (1994). Resolving differences in willingness to pay and willingness to accept. *American Journal of Agricultural Economics, 84*(March), 255–270.

# CHAPTER 9

# ECONOMIC IMPLICATIONS OF TRACEABILITY

## Contents

# INTRODUCTION

The ability to trace food products from farm to table, including their location, handling, and history, is known as traceability in agriculture. As consumers want more details about the provenance, safety, and quality of their food, this system is becoming more and more important. Traceability implementation in agriculture has many economic ramifications that affect consumers, producers, and the supply chain as a whole. These ramifications can be both positive and negative. Enhancing food safety is one of the primary economic benefits of traceability. Agricultural products can be tracked to make it simpler to find the sources of contamination or outbreaks, resulting in faster and more effective recalls. As a result, there are fewer potential financial losses for companies as a result of widespread recalls and reputational harm to their brands. Traceability systems, for example, can identify the precise farm or processing facility that is in charge of a batch of produce that turns out to be contaminated, thereby isolating the issue and minimizing the impact on other producers. Increased consumer confidence and ultimately higher sales can result from this capacity to more effectively manage food safety risks.

Traceability also makes it easier to follow international laws and standards. Traceability systems assist producers in proving compliance with the safety and quality standards that are increasingly mandated by many nations for agricultural products that are imported. Exporters of agricultural products will have improved access to global markets, which could result in higher profits. For example, strict traceability requirements mandated by EU regulations can lead to profitable opportunities for exporters from developing nations. Traceability can therefore be viewed as a tool for market differentiation, providing producers with a competitive edge who can demonstrate that their products are safe, sourced ethically, or organic. Lower supply chain transaction costs are another financial benefit of traceability. It is simpler to track shipments, manage inventories, and confirm the legitimacy of goods when a system is traceable. This is especially advantageous for high-value agricultural products where customers are prepared to pay more for products that have been verified to be of superior quality, like organic produce or specialty coffee. Product tracing can help businesses run more efficiently, improve logistics, and cut down on waste, all of which lower costs and increase profitability.

But there can be serious financial obstacles to overcome when putting traceability systems into place, especially for small-scale farmers. Installing these systems frequently necessitates expensive investments in infrastructure, training, and technology. Adopting advanced tracking systems may be more expensive for smaller farms than it is beneficial, particularly if their profit margins are already narrow. Smaller producers may also be excluded from some markets as a result of these costs because they lack the means to meet traceability regulations. As a result, there may be less competition and higher prices for consumers as the market consolidates and only larger, better-capitalized producers are able to compete. The potential rise in production costs as a result of the requirement for precise labeling, data management, and record-keeping is another cause for concern. Continuous recording of the production process is necessary to maintain traceability; this includes recording the kind of seeds used, the chemicals used, and the growing conditions of the crops. Although gathering this data can yield insightful information about enhancing sustainability and productivity, it also adds another level of labor and administrative expenses. These higher expenses may be transferred to customers further down the supply chain, possibly raising their prices.

In addition, traceability may have financial ramifications for data security and intellectual property. There may be worries about the misuse of the comprehensive information that agricultural producers are required to share about their supply chains and farming methods. Large agribusinesses, for example, might learn about a farmer's operations and take advantage of this knowledge, which could lead to market imbalances. To maintain fair competition, it is necessary to address the issues of data ownership and protection that are brought up by traceability, even though it provides transparency. Traceability in agriculture has the potential to advance ethical and sustainable practices, which may yield long-term financial gains, notwithstanding the obstacles. Demand for sustainably sourced products can be increased by traceability, which informs consumers about the source of their food. Producers are encouraged to embrace eco-friendly practices by this change in consumer preferences, which may raise costs at first but result in long-term savings through increased resource efficiency. To further ensure that resources are used responsibly and lessen farming's environmental impact, governments and non-governmental organizations can use traceability data to monitor and enforce sustainable agricultural practices.

## 9.1. COSTS, BENEFITS, AND BUSINESS VALUE OF TRACEABILITY

A relatively recent process called traceability enables business partners to follow the progress of finished goods through the supply chain and track backwards through the procurement and processing of input. It is not inherently novel to be able to trace and monitor the flow of products. When it comes to the degree that commercial activities have involved the ability to determine a product's provenance, traceability between buyers and sellers has almost always existed. An information flow linking process from the start to the finish of a supply chain is the new feature that the traceability concept brings to business practices.

Because it deals primarily with perishable goods, traceability in the food industry is especially special. Producing, processing, distributing, and retailing food all have serious safety concerns about perishable goods. Food companies operate with thin margins by selling large volumes of perishable product, which makes the food industry a unique sector of the consumer packaging goods industry.

Compared to other industrial sectors, the food industry has unique characteristics that make traceability a more demanding and complicated process to execute. The ability to track a food, feed, food-producing animal, or substance that is intended to be, or expected to be, incorporated into a food or feed, through all stages of production, processing, and distribution is known as traceability, according to the Food Standards Agency (2004). This definition, which connects all business partners from the start of a supply chain to its conclusion, demonstrates the intricacy of a traceability system. Traceability procedures are increasingly being used in business operations within the food industry. Various factors are driving the adoption of traceability systems that are given below:

- **Food Safety Issues:** Major illnesses like BSE in the US and Canada and foot and mouth disease in the UK have led to a demand for higher food safety certification. Businesses in the food sector may request a higher standard of safety through the use of traceability systems at various points in the supply chain. Retailers might ask for guarantees from wholesalers or processors, and processors might ask for these guarantees from farmers or other processors. Thus, it becomes crucial to track goods and animals throughout the beef supply chain.
- **Regulatory Compliance:** Governments and regulatory bodies are also calling for the food industry to adhere to stricter safety standards. Hobbs (1996a) proposed traceability to the UK Ministry of Agriculture, Fisheries, and Food in 1996 as "part of the Government's strategy to restore consumer confidence in British beef and to improve animal identification and traceability" in response to the issue of regulatory compliance. All food and food ingredients should be traceable, and food that poses a serious risk to consumer health should have appropriate recall procedures in place, according to a 2000 White Paper on Food Safety from the European Commission.
- **Recall and Reverse Logistics:** One of the most important components of food safety in the food industry is the ability to recall products quickly and efficiently. For this reason, it's critical that the reverse logistics process—which involves recalling or withdrawing products—be executed effectively and that the origin and destination

of product units be identified. Recalls are expensive and can have a detrimental effect on business. By limiting the amount of a production lot that can be recalled, effective reverse logistics or recall capability can lower the likelihood of a food safety issue and recall costs.

- **Consumer Demand for Transparency:** The modern consumer is better informed than the consumer of the past and knows about features of products like provenance and production methods. The development of traceability systems is partially being driven by consumers' growing demands for food safety and transparency. Take the significance of consumer demands for openness, for instance, when speaking with different companies involved in the food supply chain. Consumer demand to know the origins and production methods of their food was the main driver behind study participants' adoption of traceability practices. Due to increased knowledge about production procedures, food safety regulations, livestock housing, and medication use, among other topics, greater transparency is anticipated to boost consumer confidence in food safety. By increasing transparency, traceability can have an impact on the choices made by consumers when making purchases.
- **Supply Chain Management:** Business partners now need to reconsider how they approach their relationships with suppliers and customers due to the industry's rapid changes and high level of dynamic activity. Businesses work together to improve operational efficiency and strategic positioning through supply chain management, also known as value chain or demand chain management. As opposed to minimizing the costs of each business partner independently, supply chain management places a strong emphasis on the significance of minimizing the costs of at least two business partners—or the entire chain.
- **Certification of Specific Attributes:** Nowadays, a lot of focus is placed on catering to particular market segments by food companies. Promoting a product's unique features to a group of customers who share a similar profile is a common strategy used in marketing campaigns. This highlights how crucial it is for food manufacturers, distributors, and processors to be able to certify both their goods and their ingredients. For process credibility attributes or content attributes that are expensive or difficult to measure, traceability is an essential system in any market. Maintaining documentation that certifies these characteristics' creation and preservation is the only method to confirm their existence. "The set of measures taken to preserve and communicate the exact identity and source of food and

food ingredients to the end user" is the definition of traceability, the system that enables products to be certified for identity preservation. The aforementioned aspects underscore the significance of the food industry adopting traceability to monitor the production, processing, and distribution of its products. Few studies have thoroughly examined the factors influencing the benefits of traceability systems, despite the fact that numerous studies have highlighted the advantages of these systems for the food industry.

### 9.1.1. Challenges faced by Traceability

For many food processors in Canada, putting in place a traceability system is a significant challenge. Traceability, according to Verdenius (2006), is frequently viewed as an imposed requirement that demands investment but does not enhance competitiveness or profitability. This explains in part why the food industry has been slow to adopt traceability. More so than food supply chains' desire to monitor product flow for business logistics, the process's growing popularity is a result of tighter food safety laws. Traceability systems are not always an activity that adds value for the end user, and they can be expensive to set up and run. Traceability is generally seen by the Canadian food industry as raising operating costs without increasing sales or profits. Many food manufacturers only think of traceability as a recall mechanism that lowers the possibility of a problem with food safety. The fact that food companies are running on thin margins and that traceability could drastically lower profitability is the basis for this opinion. According to a survey, a large number of industry participants—including larger corporations—invest mainly in response to legal requirements. It is possible that many of these unfavorable opinions regarding traceability are unfounded. The uncertainty surrounding traceability's costs and benefits has never been empirically examined at the industry level, which is the primary economic issue. Assessing the elements influencing traceability's business value could help researchers and industry managers in the following ways: The findings of an empirical study of traceability practices could help managers in the industry improve the implementation and operation of traceability systems. Furthermore, it would be advantageous for researchers who are looking to create more empirical studies that examine the performance of traceability systems.

### 9.1.2. The Process of Traceability

A system called traceability is used to follow and trace goods and their constituent parts along the supply chain. This indicates that the development of traceability is an end-to-end procedure that is comprehensible in terms of process guidelines. The notions of track and trace, which are described as follows, are two directions of traceability established by ECR Europe are:

1. **Tracking:** Positioning a product at any stage of the supply chain according to predetermined standards is known as tracking. The ability for businesses to identify and locate their products is essential for any traceability system, as it allows them to withdraw or recall them as needed, in accordance with the one-step forward legal principle.
2. **Tracing:** The ability to pinpoint a product's origin and features using standards established at every stage of the supply chain is known as tracing. This is also a crucial component of a traceability system since, according to the one-step backward legal principle, businesses need to be able to quickly and accurately ascertain the identity and source of products they receive.

The position of an intermediary in the supply chain determines the product's location at any given time. Determining the intermediaries and where they are in the chain is crucial. The idea of upstream and downstream traceability can divide the supply chain into two distinct groups, making the placement of intermediaries easier. According to a supply chain partner's perspective, there are two types of traceability: downstream and upstream. Typically, this partner is a business or a processing location.

Upstream traceability pertains to the protocols and instruments employed to identify a past event that transpired before the partner in question assuming legal or physical liability for the products.

The process and instruments used to identify an event that took place following the physical transfer of goods from a partner to a third party or following the transfer of property are referred to as downstream traceability.

Product location and application within the operations of a particular intermediary are two additional reasons why traceability is developed: to trace and track products throughout the entire chain. The internal area refers to the traceability component found inside an organization's operations. Here is how it can be described:

Internal traceability refers to the traceability that is used during the partner's processing or transformation. Internal traceability occurs apart from business associates. The accuracy of the system each middleman uses determines how effective a traceability system is. The effectiveness of the chain as a whole will ultimately be determined by how well each operator tracks and traces products. Actually, the goals of every intermediary might vary when it comes to what motivates a business to put such a system in place. Furthermore, the range of goals that propel system adoption will also influence the amount of data that is gathered. As a consequence, the efficiency of traceability will vary depending on the various food industry sectors and supply chain levels. Each traceability system's level of accuracy is determined by its goals. Disparities in the scope, complexity, and accuracy of traceability systems are partly due to differing objectives. These three variables can be defined as follows which explain the traceability accuracy:

- Breadth describes the amount of information the traceability system records;
- Depth of a traceability system refers to how far backward or forward the system tracks in the supply chain; and,
- Precision reflects the degree of assurance with which the tracing system can pinpoint a particular food product's movement or characteristics.

It is crucial to note that these factors will ultimately define the precision of a given supply chain. They also represent the system that each intermediary has designed and put into place. It is necessary to come to an understanding regarding the fundamental data that buyers and suppliers must exchange and retain. Linking data and product flow is necessary for efficient tracking and tracing. The ability to track a product from its point of origin to the retail or foodservice level seems to depend on this linkage being necessary for all participants in the supply chain. In contrast, the product's ability to be tracked back through the supply chain is also guaranteed by this connection.

### 9.1.3. Ways Traceability Systems Benefit the Agribusiness

Preventing risks to food safety and excessive food waste is a major challenge for agribusinesses nowadays. The supply chain system's lack of technological solutions and transparency is to blame for this. Agri-food companies have developed traceability systems, a technological solution that tackles the security and quality issues in the food supply chain, to address these challenges. Blockchain-based agricultural traceability is becoming more and more popular as a reliable means of acting quickly and precisely in the event of a food recall. Gaining the trust of consumers and addressing serious food safety issues can both benefit greatly from this. Similar to this, traceability in agriculture can help optimize supply chain operations. It also gives agribusinesses the credibility they need by enabling them to track their entire lifecycle from farm to fork with just a QR code scan. All parties involved in the agricultural process stand to gain from traceability. Continue reading to find out more about the top 5 advantages of traceability for your agricultural enterprises.

#### *9.1.3.1. Quick Response to Food Recall Incidents*

You must find the source of any risk or anomaly in a food product before you can stop it from being sold to customers. In agriculture, products that are contaminated or damaged can be easily identified by using a batch or serial number thanks to traceability. Furthermore, by following the food supply chain from beginning to end, an efficient traceability system can locate the issue's origin. Planning a food recall scenario can be facilitated by traceability in agriculture, which can give you a clear picture of the supply chain during emergency food recalls. Furthermore, rapidly determining the batch of the affected products streamlines the entire process. Guaranteeing that the products in the stores are safe and of high quality contributes to the development of consumer trust. You can identify the underlying cause of the issue and stop the harm from occurring again with the help of a traceability system.

### *9.1.3.2. Food Safety and Compliance with Quality Standards*

Ensuring safety and quality compliance is one of the main advantages of traceability in agriculture. Food safety, however, is dependent on two things: a well-defined production process for high-quality yields, and an efficient transportation system to guarantee that perishable agricultural products arrive at the appropriate warehouses or retail locations on schedule. If this is not done carefully, it can lead to serious food-borne illnesses as well as a decline in the quality of agriproducts. In agriculture, traceability can serve as a useful remedy for this dilemma. The stakeholders benefit from the visibility traceability systems give them in identifying and stopping any unsanitary practices. Additionally, it helps identify inefficiencies in any of the supply chain's processes, empowering agribusinesses to take precise preventive measures. In this way, agricultural traceability contributes to the guarantee that the end products fulfill the required standards of quality.

### *9.1.3.3. Streamlined Supply Chain Operations*

Increasing the productivity of supply chain operations is another clear benefit of digitalizing supply chains and implementing traceability in agriculture. Whatever its nature and sector, every company strives to continuously improve the quality of its output through operational optimization.

Agro-producers' ability to accomplish this can be greatly aided by traceability in agriculture.

Using digital tools to track and analyze supply chain activities in the agriculture industry can help identify patterns and improve processes for better results. In this sense, traceability systems give stakeholders more visibility and control over the activities within the supply chain. This in turn assists them in addressing multiple areas for quality and food waste improvement as well as opportunities to increase process efficiency.

#### *9.1.3.4. Higher Customer Trust and Confidence*

Nowadays' consumers are very aware of and worried about the food items they eat. Initially, they want to confirm that the agriproducts they plan to purchase adhere to the appropriate standards for quality and safety. Additionally, they are curious to learn more about the effects those agriproducts' transportation and cultivation have on the environment. In agriculture, traceability can be very important in giving customers these vital details. Customers can now track an agriproduct's entire journey, from seed to finished product available in stores, with the aid of a QR code. You can easily obtain information about the farm, the farmer, the product's quality, and many other details by scanning this QR code. This contributes to the consumer's increased confidence in your agricultural product.

#### *9.1.3.5. Much Better Market Prices for Agriproducts*

Producers typically struggle to persuade consumers that their agriproducts meet the highest quality standards to obtain the best market price. In agriculture, traceability aids producers in verifying the precision and caliber of their agriproduct production. It allows them to demonstrate to clients that they have complied with all quality and safety requirements set forth by certifying bodies. These days, consumers will even pay more for the food products they eat if they are provided with comprehensive information regarding the product's manufacturing process. Everything from seed to production, harvest, post-production, and shipping is included in this. In this sense, agricultural traceability can enable farmers to bargain for better market rates for the superior quality of their agriproducts. Producers, agribusinesses, and other supply chain participants need to strive to satisfy consumer demands as consumers around the world become more informed about food safety and quality hazards. This is where quality and safety throughout the digitally managed supply chain are preserved utilizing a seed-to-shelf traceability system.

### 9.1.4. Traceability within the Agriculture and Food Sectors

The need for food safety and transparency among consumers is driving the development of a traceability system, as is the requirement of food regulations. These two factors alone might justify the installation of such a system. It's crucial to comprehend the additional incentives pushing the food industry to implement traceability, though. This consequently makes it possible to more accurately evaluate the variables influencing the costs and advantages of traceability. Food manufacturers are also being encouraged to adopt traceability systems by many factors, including the need for particular certification, streamlining the recall procedure, and optimizing logistics. Many techniques are employed in the social sciences, business administration, economics, biology, food sciences, medicine, and engineering, among other fields, to analyze traceability. However, business administration and economics are the fields most suited to its goals when evaluating the business value of traceability implementation and practices. More specifically, since traceability is assumed to be a managerial task, a cost-benefit analysis will be conducted. Under the headings of fields like willingness to pay, transaction costs economics, logistics, and supply chain management, and so on, pertinent research findings are presented.

### *9.1.4.1. Business Logistics and Supply Chain Management*

Logistics, supply chain management, and inventory management can all benefit from traceability procedures. From a logistics standpoint, traceability is the capacity to follow a raw material from the point of production to the final product, from the supplier to the end user along the supply chain. It is also the capacity to locate the enclosure in which the item is located (e.g., pallet, outer, carton, pack, case, etc. ) as well as to trace and monitor the logistics' constituents, including personnel, airplanes, ships, cars, and forklifts. Tracing is a transportation management process used in supply chain logistics management, specifically in freight control, to find missing or delayed shipments. The ability to audit—either in real time or as a historical audit trail—is necessary for traceability.

The development of traceability has also resulted from initiatives to increase supply chain profitability through the use of new technologies. For example, the use of information technologies like bar codes, online freight information systems, satellite, and internet-based communications has greatly reduced traditional problems associated with tracing and expediting. Since the early 1990s, traceability has been seen from this angle as a system that can improve supply chain management. Technologies for traceability are a useful tool for enhancing product flows. Systems for high-tech traceability can also be used to calculate cost centers and profit margins. Traceability was already seen as a foundation for creating and evaluating supply chain member partnerships in the early 1990s. Acknowledging that supply chain processes should be reengineered rather than just automated to cut costs is equally crucial. The following are the primary ways that traceability can improve business operations.

Enabling the creation of logistics partnership through the creation of seamless supply chains;

- Introducing paperless and reduced paper systems with improvements in information accuracy and reductions in waste;
- Providing the opportunity for throughput accounting techniques that assist in accurately measuring where costs are incurred as well as where profits are made;

- Offering the opportunity to reduce costs by automating processes and reducing manual effort.

These chances demonstrate how, in the early 1990s, supply chain management was thought to be enhanced by traceability, a dynamic tool. During this time, traceability systems were first implemented in the commercial and industrial domains. Even though these management-wise may seem ambitious, they offer intriguing ways to combine supply chain management and traceability. The addition of traceability to a particular function can improve many aspects of the business process.

To help implement paperless systems, for instance, tracking systems can be used administratively. This will enhance information accuracy and cut down on waste. For instance, they could be the foundation for automatically paying haulers. Data collection can also offer significant and pertinent input into a management information system, which can be used to verify whether the process quality is still acceptable and to identify cost and profit centers.

Practices for traceability are frequently incorporated into particular business operations. Traceability has been shown to enhance certain supply chain functions in many ways. When compared to trucks without this technology, trucks equipped with an on-board computer (OBC) have a utilization capacity that is 13.5% higher.

The primary reason for this increase seems to be the advanced capabilities that enable dispatchers to track trucks in real time and facilitate communication between dispatchers and drivers while the drivers are in their trucks. Here's an example from the transportation industry that shows how a tracking system can increase productivity by keeping trucks with loads on the road for a greater percentage of the time.

The implementation of Continuous Product Replenishment (CPR) in the supermarket sector serves as another illustration. Improving supply chain inventory management is the primary goal of CPR. CPR enables a retailer to send its inventory information to the supplier electronically each day via electronic data interchange (EDI).

The retailer receives the decision about how much inventory to ship from the supplier. The effect of operating costs and inventory level on a CPR system. When CPR is implemented, a retailer's distribution center's inventory can be cut by up to two-thirds overall. The retailer's cost of goods sold may be reduced by just over 1% as a result of this. In a sector where net profit margins are typically less than 3% of sales, this represents a significant opportunity for cost savings.

### 9.1.5. Adoption of Precision Agriculture Technologies: Cost-Benefit Analysis for Producers

Utilizing cutting-edge instruments and data-driven methods to optimize farming practices, precision agriculture makes use of tools like drones, GPS-guided machinery, soil sensors, and data analytics. These technologies help farmers make more accurate decisions by giving them comprehensive information about input requirements, crop health, and field conditions. Precision agriculture can have significant costs associated with it, despite its many advantages. As such, a comprehensive cost-benefit analysis is necessary to determine whether or not precision agriculture is economically viable for farmers. The initial outlay for infrastructure, software, and equipment is the main expense associated with implementing precision agriculture technologies. For example, investing a significant amount of money can be necessary to buy GPS-guided tractors, drones, or soil sensors. The integration of these instruments with current farming systems could result in extra expenses, as it might require modernizing outdated equipment to be in line with emerging technologies. It is also necessary to include in the cost analysis recurring costs like maintenance, software subscriptions, and data storage fees. It might also be necessary for farmers to spend money on training so that they or their employees can use and understand the data from these cutting-edge instruments. It is crucial to determine whether the upfront costs are justified in light of the long-term advantages, as small-scale producers with tight budgets may find the upfront costs to be a major deterrent to adoption.

The possibility of greater operational complexity is a crucial factor to take into account in the cost analysis. Technical proficiency in data management and analysis is frequently needed for precision agriculture. There can be a steep learning curve for producers who aren't used to using digital tools, which would mean hiring experts or providing extra training—both of which would raise overall expenses. Additionally, merging data from several sources across various technologies (e.g., soil sensors, satellite imagery, and meteorological data) into a coherent system can be difficult and may need additional funding for software platforms that can handle a variety of data streams. Particularly in the early phases of adoption, these additional complications may cause brief interruptions to routine farming operations. Precision farming has potential for significant cost savings, increased yields, and environmental sustainability; these advantages outweigh the costs and complexity of the practice. Optimizing the use of input is one of the main benefits. By giving precise information on where and how much to use, precision tools can help farmers apply pesticides, fertilizers, and water more effectively. Soil sensors, for instance, can identify the moisture content of the soil. This information allows irrigation systems to water only the areas that require it, saving water. Comparably, variable-rate technology (VRT) can guarantee that fertilizers are applied precisely, cutting input costs and waste. By minimizing the excessive use of chemicals and water, which can cause soil erosion and water pollution, these efficiencies not only reduce production costs but also have a positive environmental impact.

The potential for higher crop yields is a key advantage of precision agriculture. Farmers can identify early indicators of disease, pest infestations, or nutrient deficiencies by closely monitoring crop health using drone surveillance or satellite imagery. Prompt resolution of these problems can reduce yield losses and raise the overall standard of the produce. Furthermore, data-driven decisions about planting, crop rotation, and harvesting can be made by farmers with the help of precision agriculture, which can improve crop management techniques and ultimately increase productivity. To increase outputs per hectare, data analytics, for example, can assist in determining the best planting density, timing, and variety selection based on historical yield data and predictive models. Precision farming can improve resource management and sustainability as well, which will benefit the economy in the long run. For instance, producers can save operating costs and promote environmental conservation at the same time by using less water, fertilizer, and pesticides. Sustainable farming methods can also lead to new market opportunities for farmers, particularly as consumers grow more aware of the environmental impact of the food they eat. By labeling their products as sustainably grown, producers can increase profitability by charging premium prices for their products thanks to precision farming techniques. Furthermore, improved resource management can result in long-term enhancements to soil health, guaranteeing the long-term viability of farming activities.

Certain situations may allow farmers to access new sources of income through precision agriculture. There are several ways to monetize the insights provided by the data gathered from precision instruments. Farmer collaboration can take the form of farmers sharing anonymized data with agribusinesses, agricultural researchers, or tech companies to develop new farming technologies or improved crop models. Precision farming also helps with sustainability certifications and standards compliance, which can be necessary to gain access to some foreign markets that demand evidence of sustainable farming methods. This broadens the pool of potential customers and puts farmers in a position to gain from government grants and incentives that support sustainable farming. But depending on how big the farming operation is, precision agriculture may or may not be financially feasible. Large farms can accomplish economies of scale that increase the return on investment and frequently have the funds to purchase cutting-edge machinery. Nonetheless, small and medium-sized farms might find it difficult to justify the expenses unless they have access to reasonable loans, grants, or cooperative agreements that let them pool resources and knowledge.

Governments and agricultural associations can be very helpful in promoting the use of precision agriculture by offering smallholders financial support, educational opportunities, and technical support. This will help to ensure that the advantages of these technologies are shared more widely throughout the agricultural industry.

## 9.1.6. Cost-Benefit Analysis of Organic Certification for Agricultural Producers and Suppliers

With the growing demand for organic products from consumers, organic certification has emerged as a significant factor for agricultural producers and suppliers. This accreditation is evidence that a farm or product conforms to particular organic farming guidelines, which generally forbid the use of artificial chemicals, genetically modified organisms (GMOs), and specific agricultural techniques. Although there are many financial advantages to organic certification, there are also high expenses involved. To find out if getting organic certification is a wise and profitable investment, producers, and suppliers must perform a cost-benefit analysis.

### *9.1.6.1. Cost Analysis*

For producers switching from conventional farming practices, organic certification can come with significant costs. Application, inspection, and annual renewal fees are all included in the certification fee, which is one of the biggest expenses. The certifying organization, farm size, and farming operation complexity can all affect these expenses. Small-scale farmers may find the cost of certification prohibitive, particularly if their profit margins are already narrow. A transition period, typically lasting three years, is also necessary for producers to follow organic farming practices without marketing their products as organic to receive certification as organic. This could result in short-term financial strain as they bear the higher expenses of organic farming without the instant advantages of premium prices.

The necessity of altering farming methods is an additional substantial expense. In general, organic farming involves more labor-intensive methods than conventional farming. Crop rotation, organic pest control, and manual weeding require time and resources from producers, which can raise labor costs. In addition, they might have to spend more on organic inputs like compost and seeds than their conventional counterparts. Additionally, to prove that they are adhering to organic standards, producers must keep thorough records. This requirement raises administrative burdens and may call for new record-keeping software or staff hiring, both of which can raise operating expenses.

### 9.1.6.2. Benefit Analysis

Notwithstanding the expenses, organic certification can have a significant positive impact on producers' and suppliers' bottom lines, market accessibility, and environmental sustainability. Being able to charge premium prices for products is one of the biggest advantages. Owing to consumer perceptions of increased safety, health, and environmental friendliness, organic products typically fetch higher prices than conventional ones. In addition to partially offsetting the higher production costs connected with organic farming, this price premium can result in higher revenue for producers. Producers are strongly encouraged to obtain certification because organic products can command prices in some markets that are 20–50% higher than those of their non-organic counterparts.

An additional advantage of organic certification is easier access to markets. To satisfy consumer expectations, a growing number of retailers and distributors choose to collaborate with certified suppliers as the demand for organic products rises globally. Obtaining certification can lead to access to new markets, particularly export markets where there is a strong demand for organic products. For instance, organic certification is frequently necessary in areas like North America and Europe to gain access to major retail chains. Producers and suppliers can establish themselves as reliable partners in the organic supply chain and possibly grow their market share and sales volume by obtaining certification. Sustainable farming methods are supported by organic certification as well, and these can have long-term positive economic effects. Ecological balance, biodiversity, and soil health are prioritized in organic farming. Crop rotation, the use of organic fertilizers, and avoiding synthetic chemicals are some of the practices that help maintain the soil's fertility over time and eventually lower the need for expensive inputs. Better yields and more resilient crops can result from healthier soil, which can be especially significant in areas where environmental degradation and climate change are prevalent. Furthermore, organic farming lowers the possibility of water contamination and pesticide runoff, which can save expenses related to environmental harm and regulatory compliance. Although switching to organic farming methods may be expensive at first, in the long run, these practices can result in more steady and predictable production costs.

The reputation of a farm's brand can also be improved by organic certification. Customers are searching for brands that share their values more and more as they become aware of the effects their food choices have on the environment and their health. As a marketing tool, organic certification lets customers know that a product is devoid of dangerous additives, GMOs, and artificial chemicals. This has the potential to increase client loyalty and draw in more customers. Strong brand recognition can benefit manufacturers and suppliers by boosting sales, encouraging repeat business, and creating chances to expand their product offerings to include things like organic cosmetics and processed foods.

### *9.1.6.3. Challenges and Considerations*

Despite the obvious advantages of organic certification, choosing to pursue certification necessitates carefully weighing many factors. The fluctuation of consumer demand is one of the primary obstacles. The demand for organic products has been rising steadily, but it is still susceptible to changes in the economy. Demand may fluctuate as a result of consumers prioritizing cost savings over organic options during recessions. If market conditions change, producers need to determine whether their target markets can afford the price premiums associated with organic products and whether they can continue to be profitable. Maintaining certification is complicated, which is another factor to take into account. The strict requirements of organic standards necessitate ongoing adherence, which includes frequent audits and inspections. The advantages of the investment would be lost if non-compliance resulted in the loss of certification. To handle compliance and record-keeping, producers must make sure they have strong systems in place, which may call for continual assistance and training. Further raising costs are regional variations in certification procedures, which may require producers who want to export their goods to obtain multiple certifications to satisfy various national requirements (Buhr, 2002; Bolzani et al., 2010).

## 9.2. SUPPLY CHAIN TRACEABILITY IN SUSTAINABLE ECONOMIC GROWTH

With environmental awareness rising and sustainable practices becoming more and more important, the world economy is at a crossroads. Originally seen as an added benefit, the idea of sustainability is now a basic prerequisite for any business model that looks forward. Supply chain traceability is a key component of this change and not only a technical necessity but also an essential facilitator of long-term, sustainable economic growth. A key element in the goal of sustainable economic growth is supply chain traceability. When we talk about supply chain transparency, we mean the capacity to trace and keep an eye on the movement of goods and resources from the point of origin to the point of final destination. Due to rising consumer awareness, tighter regulations, and the growing pressure on businesses to ensure ethical sourcing and production methods, the idea of traceability has gained popularity in recent years. Its promotion of moral business conduct, enhancement of resource efficiency, and support of social and environmental sustainability all contribute significantly to the possibility of sustainable economic growth.

Fundamentally, supply chain traceability makes sure that every step of the manufacturing process—from the extraction of raw materials to the final product—is recorded and verifiable. This degree of openness is especially crucial in sectors like food, fashion, electronics, and agriculture where sustainability concerns are common. Consumers can learn where their food comes from, how it was made, and whether it complies with safety and ethical standards, for instance, thanks to traceability in the food industry. This promotes customer loyalty and trust, which can fuel long-term company growth. The ability of supply chain traceability to encourage ethical sourcing is one of its main advantages. Businesses can confirm that raw materials are sourced ethically and following labor and environmental standards by monitoring their origins. For example, traceability can assist fashion brands in avoiding the use of materials resulting from environmental degradation or exploitative labor practices. Businesses can gain a competitive edge in the market by demonstrating their dedication to ethical sourcing, which can improve their reputation. Additionally, consumers are becoming more prepared to pay more for goods that are certified to be produced sustainably and ethically, which gives businesses an incentive to implement traceability procedures.

Enhancing resource efficiency is another important function of supply chain traceability. Businesses can find waste, bottlenecks, and inefficiencies in the supply chain by having a thorough understanding of it. This enables them to streamline operations and cut expenses. Traceability, for instance, can assist in tracking waste generation, water use, and energy consumption across the supply chain, allowing companies to adopt more environmentally friendly procedures. This lowers the overall carbon footprint and aids in resource conservation, both of which are critical for reaching long-term sustainability objectives. By reducing environmental impact and guaranteeing responsible resource use, effective resource management promotes sustainable economic growth. Traceability of the supply chain can also help with regulatory compliance. Stricter laws are being put in place by governments worldwide to address social and environmental concerns, and businesses must abide by these laws to operate lawfully. By offering thorough documentation of their supply chain operations, traceability systems assist businesses in proving adherence to these standards. For example, laws such as the Conflict Minerals Rule in the US and the Due Diligence Act in the EU require businesses to confirm and disclose the source of specific raw materials. Businesses benefit from traceability by avoiding fines and harm to their reputation, which could otherwise interfere with operations and impede economic expansion. Better risk management can also be facilitated by supply chain traceability. Supply chains are frequently intricate and involve numerous stakeholders from various geographical locations in an increasingly globalized economy. Vulnerabilities like product recalls, supply interruptions, and reputational hazards may result from this complexity. Businesses can promptly locate the cause of issues, like contamination or faulty materials, and take prompt corrective action by putting traceability systems in place. Being able to react quickly can shield the brand's reputation and avert large financial losses. Additionally, traceability can offer information about supplier performance, assisting businesses in choosing or switching suppliers with knowledge and lowering the chance of supply chain interruptions.

The potential of supply chain traceability has been further increased by technological integration. The use of cutting-edge technologies like blockchain, artificial intelligence (AI), and the Internet of Things (IoT) to enhance traceability systems is growing.

For instance, blockchain provides a decentralized, unchangeable ledger that documents each supply chain transaction, guaranteeing security and transparency. Real-time product location, temperature, and condition tracking by IoT devices provides useful information for supply chain operations monitoring and enhancement. AI can examine vast amounts of data to find trends and anticipate possible problems before they materialize, allowing businesses to take preventative action. The efficiency, dependability, and scalability of traceability systems are increased by these technological developments, which also promote sustainable economic growth.

Supply chain traceability serves both businesses and larger societal objectives in the framework of sustainable economic growth. Traceability supports local communities, encourages fair labor practices, and safeguards the environment by guaranteeing that products are sourced and produced responsibly. For example, by confirming that timber products are harvested responsibly and lawfully, traceability can help reduce deforestation. By guaranteeing that they are fairly compensated for their products, it can also help small-scale farmers and producers maintain their standard of living. The goals of sustainable development, which aim to strike a balance between economic growth and the welfare of people and the environment, are in line with these social and environmental advantages (Buhr, 2002; Bolzani et al., 2010).

### 9.2.1. Understanding Traceability in Sustainability

The ability to follow materials and products through their whole lifecycle—from raw materials to completed goods and, eventually, to the final customer—is known as traceability.

Businesses can identify, manage, and minimize environmental, social, and economic risks and impacts throughout their supply chains thanks to traceability in the context of sustainability. Traceability can help achieve several Sustainable Development Goals (SDGs), including encouraging ethical production and consumption (SDG 12) and fair employment and economic growth (SDG 8). Traceability can assist organizations in addressing sustainability challenges and supporting the global effort to achieve the SDGs by improving transparency and accountability in supply chains.

#### *9.2.1.1. Benefits of Traceability in Supply Chain Sustainability*

Implementing traceability systems offers numerous benefits for organizations, including:

- **Risk Mitigation:** Organizations can minimize adverse effects and enhance overall sustainability performance by using traceability to detect and control supply chain risks related to environmental, social, and economic concerns.
- **Regulatory Compliance:** By providing greater visibility into supply chain practices, traceability can help organizations comply with regulatory requirements and industry standards related to sustainability.
- **Improved Reputation:** An organization's reputation among customers, investors, and other stakeholders can be improved by showcasing a dedication to sustainability and transparency; this could result in a rise in market share and brand value.
- **Enhanced Stakeholder Engagement:** Systems for traceability can help suppliers, customers, and other stakeholders communicate and work together more effectively, which will strengthen relationships and advance common sustainability goals.
- **Increased Efficiency and Cost Savings:** By providing insights into supply chain operations, traceability can help organizations identify opportunities for efficiency improvements, cost reductions, and waste minimization.

#### *9.2.1.2. Key Concepts in Traceability Sustainability*

To effectively implement traceability systems in the context of sustainability, organizations should be familiar with several key concepts, including:

- **Chain of Custody:** The sequence of entities that take custody of a product or material throughout its lifecycle, from raw materials to finished goods.
- **Product Integrity:** The assurance that a product's sustainability attributes, such as environmental or social certifications, are maintained and verified throughout the supply chain.

- **Mass Balance:** A method used to track the flow of sustainable materials within a supply chain, ensuring that the quantity of certified sustainable input material matches the output of certified sustainable products.

### *9.2.1.3. Best Practices for Implementing Traceability Systems*

To advance traceability sustainability, organizations should consider the following best practices:

- **Define Clear Goals and Objectives:** Establish specific sustainability goals and objectives that can be achieved through enhanced traceability, such as reducing environmental impacts or improving labor conditions.
- **Engage Suppliers and Stakeholders:** Collaborate with suppliers, customers, and other stakeholders to develop and implement traceability systems that meet shared sustainability objectives.
- **Adopt Appropriate Technology:** Utilize technology solutions, such as barcodes, radio frequency identification (RFID), or blockchain, to enable more efficient and accurate tracking of products and materials throughout the supply chain.
- **Establish and Enforce Standards:** Develop and implement clear standards and protocols for traceability, ensuring consistency and accuracy in data collection, management, and reporting.
- **Monitor and Verify Data:** Regularly monitor and verify traceability data to ensure its accuracy, reliability, and completeness, identifying and addressing any discrepancies or inconsistencies.
- **Train and Educate Employees:** Provide training and education to employees at all levels of the organization on the importance of traceability and their role in supporting sustainability objectives.
- **Communicate with Transparency:** Clearly communicate traceability efforts and sustainability progress to stakeholders, including customers, investors, and regulators, demonstrating a commitment to transparency and accountability.
- **Continuously Improve:** Regularly assess and evaluate traceability systems to identify opportunities for improvement, incorporating lessons learned and best practices to enhance overall sustainability performance.

#### 9.2.1.4. Challenges and Potential Solutions in Traceability Sustainability

Implementing traceability systems in the context of sustainability can present several challenges, including:

- **Complexity of Global Supply Chains:** The increasing complexity of global supply chains can make it difficult to achieve full traceability, particularly when dealing with multiple suppliers, intermediaries, and jurisdictions.
- **Potential Solution:** Collaborate with industry peers, trade associations, and multi-stakeholder initiatives to develop standardized traceability requirements and practices that can be applied across complex supply chains.
- **Data Management and Interoperability:** Collecting, managing, and sharing traceability data can be challenging, particularly when different organizations and systems are involved.
- **Potential Solution:** Utilize technology solutions, such as cloud-based platforms or blockchain, to improve data management and enable seamless data sharing and collaboration among supply chain partners.
- **Cost and Resource Constraints:** Implementing traceability systems can be resource-intensive, requiring significant financial and human resources.
- **Potential Solution:** Identify opportunities to leverage existing systems, tools, and partnerships to reduce costs and increase the efficiency of traceability efforts.
- **Resistance to Change:** Some suppliers or stakeholders may be resistant to adopting new traceability practices or sharing sensitive information.
- **Potential Solution:** Communicate the benefits of traceability for all supply chain partners, emphasizing the potential for risk mitigation, cost savings, and enhanced sustainability performance.

Traceability sustainability plays a critical role in advancing responsible business practices and promoting environmental, social, and economic well-being in global supply chains. By understanding key concepts, benefits, and best practices, organizations can implement effective traceability systems that support the achievement of the Sustainable Development Goals and contribute to a more sustainable future. By leveraging the insights provided in this guide, SDG Resources can help organizations navigate the complexities of traceability, fostering more transparent, accountable, and sustainable supply chains (Buhr, 2002; Bolzani et al., 2010).

## 9.2.2. The Imperative of Traceability in a Globalized Economy

Supply chains have developed into intricate networks spanning several nations and industries in today's globally interconnected world. There are advantages and disadvantages to this complexity. On the one hand, it gives companies access to a worldwide market of clients and resources. However, it makes it more challenging to guarantee that every link in the supply chain follows moral and environmentally friendly procedures. Supply chain traceability becomes essential at this point. Businesses can follow the origin, transit, and transformation of products across the supply chain thanks to traceability. It guarantees the accountability and transparency of each link in the chain. This transparency is essential in the context of sustainability to ensure that labor practices are moral, environmental impacts are kept to a minimum, and raw materials are sourced responsibly. Without traceability, companies run the risk of participating in unsustainable practices, which can harm their brand, result in fines, and erode customer confidence.

## 9.2.3. Wealth Distribution and Economic Equity

The ability of supply chain traceability to encourage fair wealth distribution is among its most important advantages. Value in traditional supply chains is frequently concentrated at the top, with big businesses benefiting financially while smaller producers and suppliers fight for survival. This disparity hinders economic growth and innovation in addition to making global inequality worse. By guaranteeing that each supply chain participant receives just compensation for their contributions, traceability can aid in redressing this imbalance. For instance, companies can guarantee that small-scale farmers or producers receive a fair price for their goods by tracking the origin of raw materials. In addition to raising these people's standard of living, this boosts regional economies and promotes environmentally friendly behavior. Traceability can help support rural communities and advance economic equity in the biofuels industry by guaranteeing that farmers who provide feedstocks, like spent cooking oil, receive a fair portion of the profits. Additionally, traceability promotes openness, which aids in the prevention of unfair labor practices and exploitation. Businesses support a fairer and just global economy when they can confirm that child labor, forced labor, and other unethical practices are not present in their supply chains. As a result, more people are choosing to support brands that share their values, which increases consumer loyalty and trust.

## 9.2.4. Valuing Used Products and the Circular Economy

The ability of supply chain traceability to advance the circular economy—a system in which resources are used as long as possible, products are reused, and waste is reduced—is another crucial component of this technology. In this situation, traceability makes sure that used goods are valued and repurposed into the supply chain as opposed to being thrown away as waste. Using spent cooking oil as an example, this fuel is essential to the manufacturing of sustainable aviation fuel. Used cooking oil is less likely to be used in high-value applications without traceability since it would be challenging to confirm the source and quality of the oil. But with strong traceability protocols in place, companies can safely integrate spent cooking oil into their operations, cutting waste, saving money, and lessening their environmental effect. This strategy encourages the creation of new technologies and business models that have the potential to turn waste into useful resources in addition to optimizing the value of used goods. For example, by making certain that used cooking oil can be tracked (Buhr, 2002; Bolzani et al., 2010).

### 9.2.5. Innovation and the Creation of Alternative Input Materials

Another strong force behind innovation is supply chain traceability. Businesses are searching more and more for environmentally friendly, ethical, and renewable alternative input materials as they work to meet sustainability goals. Through traceability, businesses can confirm that these materials are sustainable, meet all requirements, and can be incorporated into current supply chains. For instance, the pursuit of alternative feedstock in the biofuels sector has prompted research into substances such as carbon dioxide, algae, and agricultural waste. However, the ability to track and confirm the source and sustainability of these inputs is essential to the success of these innovations. Without traceability, there is a serious risk of "green washing," in which businesses make exaggerated claims to be ecologically friendly, which damages the legitimacy of sustainable innovations. Traceability also promotes cooperation between sectors and industries. Businesses can collaborate to create new technologies and procedures that increase the sustainability of their supply chains by exchanging traceability data. In addition to speeding up innovation, this cooperative strategy opens up fresh avenues for expansion and value generation in the global economy.

### 9.2.6. Maximizing Value within the Economy

Fundamentally, supply chain traceability aims to maximize economic value, including monetary gains as well as social and environmental advantages. Traceability guarantees that resources are used effectively, waste is reduced, and all supply chain participants receive just compensation by fostering accountability and transparency. The economy as a whole is included in this value maximization, not just specific businesses. Supply chains support the general well-being and robustness of the economy when they are open and sustainable. For instance, the industry can support the shift to a low-carbon economy, lower greenhouse gas emissions, and increase energy security by making sure that biofuels are made from sustainable and traceable feedstock. Furthermore, traceability can support risk mitigation and resilience building in the face of global issues like resource scarcity and climate change. Businesses can spot possible weaknesses and take preventative action to fix them by having visibility into the complete supply chain. This helps to maintain the stability and sustainability of the global economy in addition to shielding the company from interruptions.

## 9.2.7. Driving Consumer Confidence and Demand

Customers are more knowledgeable and aware than ever in the current market. Consumers are placing an increasing emphasis on high-quality goods that are sourced responsibly and ethically. Because supply chain traceability offers the transparency that customers desire, it is essential to satisfying this demand. Businesses gain the trust of customers when they can prove that their products are traceable. When a product's sustainability is confirmed, this trust can even result in premium pricing for that product and foster brand loyalty. For instance, traceability in the biofuels industry can reassure airlines and other customers that the fuel they are using does not contribute to deforestation, violations of human rights, or other adverse effects. Traceability is essentially a potent instrument for coordinating corporate operations with customer values. Businesses can stand out from the competition, draw in more clients, and eventually increase demand for sustainable products by embracing traceability. Consequently, this upholds the overarching objective of shifting towards a more equitable and sustainable global economy.

## 9.2.8. Importance

It is impossible to exaggerate the significance of supply chain traceability. It is an essential part of any plan meant to achieve long-term economic expansion. Traceability guarantees that supply chains are not only effective but also morally and ecologically conscious by encouraging openness, responsibility, and equity. Traceability must be a top priority for companies if the world economy is to continue expanding in a sustainable manner. The long-term viability and prosperity of economies and businesses alike will depend on this strategic imperative, which goes beyond simple risk management or regulatory compliance. It is our duty as pioneers in the biofuels sector and beyond to promote traceability and incorporate it into all facets of our supply chains. By doing this, we can promote economic equity, open up new avenues for innovation, and build a more sustainable future for everybody. Now is the moment to take action. It is essential to the future of both our planet and the world economy.

### 9.2.9. Benefits Resulting from the Implementation of Traceability

A company may be motivated to implement traceability by a variety of factors, which largely influence the technology selection for traceability: i.e. they either use a simple traceability system or a sophisticated one. Notwithstanding these technologies' distinct advantages, the traceability system's full benefits can only be realized by efficiently combining and integrating the traceability resources. Businesses must be sufficiently motivated by the potential benefits of implementing traceability systems to outweigh the obstacles and difficulties they will face along the way (Buhr, 2002; Bolzani et al., 2010).

#### *9.2.9.1. Impact on Crisis Management*

The speed and effectiveness that traceability systems offer SC stakeholders in the case of a crisis or disruption falls under this category of traceability benefits. As a result, it encompasses the set of advantages associated with crisis prevention, crisis "containment," and rapid crisis recovery. Improved visibility and monitoring are the main advantages, as listed in Table 9.1, since traceability systems make it easier to keep an eye on the physical attributes (such as weight, temperature, and texture) of a product while it is being transported or stored. The visibility of "long" SCs, which provide information on "who made it," "what was made," "when it was made," and its current location—all crucial for crisis management—requires traceability in particular. Effective decision-making will be hampered, in particular, by a lack of visibility and control procedures since there is no comprehensive knowledge of what is occurring at other SC locations.

**Table 9.1.** *Elements of Industrial Product Quality Traceability System*

| Quality Information | | | Traceability Stage | Departments | Work Requirements for Session Participants |
|---|---|---|---|---|---|
| Earlier stage | Design | Material selection, process, labeling, and packaging design, including the product's basic function, performance indicators, and use characteristics, approved time, competent leadership | Traceability of decisions | Research and development department | Dress code, safety training, industry norms |

| **Quality Information** | | | **Traceability Stage** | **Departments** | **Work Requirements for Session Participants** |
|---|---|---|---|---|---|
| | Procurement | Vendor information, purchase contracts and orders, product completeness, selling time, selling address, product quantity, price, and payment information | Supply traceability | Supply chain management and procurement | Dress code, skill requirements, industry norms |
| Medium-term | Manufacturing | Processes, manufacturing equipment, engineering, quality control, safety, productivity, cost control, standards, and specifications | Production process and early warning traceability | Manufacturing and production operations | Dress code, skill requirements, basic knowledge requirements, safety training |
| | Warehouse | Product batches, inventory management system (WMS), classification, and identification of goods, storage methods, inventory scheduling and optimization, incoming inspection reports | Reserve traceability | Logistics and supply chain management or warehouse management department | Safety training, skill requirements, industry norms |
| Late stage | Sales | Product quality standards, packaging material sources, packaging material qualification certificates, product batch feedback improvement mechanism, continuous improvement and innovation | Packaging traceability | Sales and marketing | Skill requirements, knowledge requirements, industry norms |
| | Logistics | Product batches, temperature, and humidity monitoring reports, sources of packaging materials, certificates of conformity of packaging materials, product quality information, entry into and exit from storage | Transportation traceability | Logistics and supply chain management sector | Dress code, skill requirements, industry norms |
| | Use | Product procurement records, qualified inspection reports, safety performance records, records on the use of the environment and conditions, the type of manufacturer, product reserves | Condition monitoring traceability | Production operations | Dress code, skill requirements, basic knowledge requirements, safety training |

| Quality Information | | | Traceability Stage | Departments | Work Requirements for Session Participants |
|---|---|---|---|---|---|
| | Maintenance | Maintenance records and plans, operating practices, maintenance records and procedures, part replacement records, maintenance cost records, personnel safety training records, feedback information | Inspection and maintenance traceability | Equipment operations and maintenance department | Professional knowledge requirements, skill requirements, dress code requirements, safety training |
| | Retirement | Records of reasons for end of life, decommissioning procedures and plans, records of environmentally friendly disposal, records of departmental approvals, records of product data, risk assessment and management | Recycling and remanufacturing traceability | Equipment and asset management | Industry standards, specialized knowledge requirements, skill requirements |

*Source: Xiao L, Sun W, Chang S, Lu C, Jiang R. Research on the Construction of a Blockchain-Based Industrial Product Full Life Cycle Information Traceability System. Applied Sciences. 2024; 14(11):4569. https://doi.org/10.3390/app14114569*

Note: The first, middle, and last stages of the traceability phase represent the preparatory monitoring, production practice, and life cycle management phases, respectively.

For manufacturers and their SCs, product recalls are a serious crisis situation that can negatively impact shareholder value, a product's brand, business profits, and customer satisfaction, particularly when there is widespread media coverage. Therefore, to reduce the impact, it is necessary to be ready to quickly remove flaws from the market. Product recall management works best when it effectively integrates problem identification, risk mitigation, and recall learning. Researchers have claimed that to identify an issue, SC partners must work together to share timely information on possible malfunction issues. It cannot be overstated that a traceability system helps control the consequences by facilitating quick recalls of dangerous products, even though it is said that it does not lessen the likelihood of a SC crisis. For instance, traceability systems' logistics data (batch quantity, origin, destination, and dispatch date) guarantees that the most recent location of a product is known, allowing for a prompt removal from the market. Businesses can use traceability systems as a platform for quality verification to reassure all parties involved that the product has been manufactured according to the correct standards and is therefore safe for use. By protecting the integrity of the SC through efficient monitoring and making sure that SC actors have access to the appropriate information at the appropriate time to confirm the product's authenticity, it can also prevent fraud and counterfeiting. The food, beverage, and pharmaceutical industries—where the health and safety of consumers are of utmost importance—benefit greatly from traceability.

### *9.2.9.2. Impact on Firm and SC Performance*

Both the immediate firm's and its larger SC's operations are thought to benefit from this category. These advantages fall into two categories: internal to the SC and internal to a firm. This category offers advantages such as lower operating expenses due to fewer internal inefficiencies (e.g., shortages, stock mistakes, thefts, and shrinkages) and enhanced product recall control by means of real-time logistics data sharing. The results of traceability, such as current inventory data and the capacity of SC partners (including manufacturers, distributors, suppliers, and retailers), also enhance decision-making at the firm and SC levels. At any level, the increased visibility provided by this track-and-trace data makes it easier to predict possible disruptions in production, transportation, or warehousing capacity. In the worst situation, this aids SC firms in using SCRes to swiftly recover from disruptions and in developing strategies in advance of them. Systems for traceability improve asset tracking in real time, which makes it easier for SC partners to share and utilize resources and assets for their mutual advantage. In the medical field, for instance, traceability has been helpful in effectively tracking costly, long-lasting, and mobile devices like wheelchairs, infusion pumps, blood supplies, and other items like medications, surgical trays, and supplies. It thus helps to ensure that healthcare is delivered at a high level of service.

There is a direct correlation between the degree of trust and confidence in a SC and the quality of its information. To establish long-term cooperative relationships between SC actors and consumers, traceability establishes an effective communication protocol that fosters confidence and trust. Accurate point-of-sale data is another benefit of traceability technology, which can then be used to reconcile inventory records with actual inventories inside the company. Through the quick transfer of inventory data throughout the SC, this degree of traceability improves security and reliability by reducing inventory errors and ultimately reducing the bullwhip effect in the SC. It offers trustworthy transaction history for financial audits as well. Because of the stability and immutability of advanced traceability systems, especially the blockchain, information about previous transactions can be readily retrieved, guaranteeing justice and trust and lowering corruption. Systems for traceability are also a dependable way to boost brand competitiveness. A crucial step that can either save or ruin the company's reputation is alerting the relevant parties—government organizations, customers, and SC partners—when problems occur in the SC operations. Businesses can reassure stakeholders that they have complete control of the situation by using traceability as an effective communication tool. The brand's reputation is enhanced and consumer trust in the SC's integrity is increased when counterfeits are eradicated.

### 9.2.9.3. Impact on Consumers and Society

The third category of benefits addresses customer expectations and has a positive social impact in addition to firm and SC improvements. For instance, traceability fosters consumer trust and may increase their willingness to pay more for safer goods, particularly those related to food and medicine. The ability of traceability systems to provide such evidence keeps current customers and draws in new ones because consumers are becoming more interested in knowing the source and composition of the food they eat due to concerns about ethics, health, and safety, and religious beliefs. For instance, by identifying counterfeiters early on, ongoing drug distribution and sales monitoring contributes to patient safety. Additionally, traceability aids in handling returns of recyclable artifacts and unsuitable products. The effectiveness of these reverse logistics operations greatly contributes to boosting customer confidence and lowering the possibility of lost sales. Since traceability information provides verifiable details about the product's location and safe handling practices, it also aids in lowering the direct and indirect costs related to reverse logistics.

### 9.2.9.4. Attaining the Full Potential of a Traceability System

Even though traceability systems have many advantages, there is a chance that these advantages will be accompanied in the near run by a sharp rise in operating costs and strain in the relationships between SC actors. Furthermore, the advantages may outweigh the costs associated with adopting a new, more sophisticated system in favor of making relatively inexpensive complementary improvements to an already-existing traceability system. Furthermore, it is crucial to remember that the mere collection of tools and systems does not ensure the intended outcomes unless these resources are skillfully combined and incorporated. In other words, while technologies vary in what they can do, they only serve to facilitate the traceability system; therefore, to fully realize the advantages of traceability, other components like resource organization and user skills must be integrated with focus. Although research suggests that traceability fosters inter-firm trust and confidence, it's crucial to remember that firms are more likely to implement traceability systems to reduce internal risks than external ones in the SC, which suggests the presence of opportunistic behavior in the implementation process. As a result, bigger brands might put more pressure on smaller businesses to comply with strict traceability requirements to coordinate their vertical transactions. As a result, the SC may not always reap all of its potential advantages (Buhr, 2002; Bolzani et al., 2010).

## 9.3. CHALLENGES AND OPPORTUNITIES IN AGRICULTURAL VALUE CHAINS

In many developing nations, low agricultural productivity persists despite the combined efforts of numerous governments and international organizations. Particularly in sub-Saharan Africa, smallholder farmers' actual yields and potential yields differ significantly. Yields in Africa increased by only 30% between 1980 and 2018, whereas farmers in South Asia increased their yields by nearly 133%, presumably due to the "Green Revolution." This is probably due to the fact that Africa was mainly left out of the Green Revolution. Additionally, because various stakeholders throughout the value chain collect rents, farmers frequently do not receive the full value of the products they produce. High transaction costs are typical in many developing nations, despite the fact that the limitations facing agricultural value chains vary by context. Liquidity limitations and difficulties in transporting crops are linked to these transaction costs. Lack of knowledge about input access and restricted access to capital frequently present difficulties for farmers when it comes to purchasing inputs. Other issues along the crop value chain include low-quality storage facilities that result in significant losses after harvest, weak market connections, inadequate safety regulations, and low consumer confidence in the caliber of agricultural products. A lack of bargaining power, small plot sizes, low capital investments, and inadequate access to global markets are additional disadvantages faced by smallholder farmers.

Many smallholder farmers in LMICs rely heavily on livestock for their livelihoods, and the outcome of food and nutrition for these populations is closely related to the livestock value chain. Because of shifting dietary patterns and growing incomes, the demand for livestock products is expected to more than fivefold globally between 2030 and 2050. Livestock value chains still face significant obstacles, such as the inability to meet quality and sanitation standards, the lack of animal identification and traceability systems, the frequency of animal disease outbreaks, and the extreme vulnerability to climate shocks, even though they have begun to become more competitive in many countries. Lastly, the distribution of resources, roles, and responsibilities both on and off the farm, decision-making, and control over farm income returns are all examples of gendered behavior patterns that frequently define agricultural value chains. Women are frequently restricted to specific value chain nodes that call for unskilled labor by current gender norms and labor market structures, which hinders their ability to advance in their careers. Integrating environmental and equity considerations will ultimately be necessary for the sustainability of productive agricultural value chains to guarantee that these value chains support resilient livelihoods and food security. Based on the results of three value chain assessment case studies carried out in Uganda, Mozambique, and Georgia by the American Institutes for Research (AIR), this brief offers evidence-based policy recommendations to enhance value chain programming in LMICs.

## 9.3.1. What Has Worked Well Across Agricultural Value Chains?

### *9.3.1.1. Quality Standards and Certification*

Establishing a certification procedure is a crucial first step in enhancing the production of safe, superior livestock and agricultural products in the majority of animal value chains. Farmers can raise the prices they charge for their goods when the quality of their output improves, which leads to more investments and business advancements.

In most situations, it has been successful to concentrate on the establishment of training programs, standards, and certification, as well as the establishment of auxiliary services, particularly advisory service providers and certification bodies. Furthermore, there is now a window of opportunity for improved quality standards and certification in value chains due to the growing consumer demand for high-quality products. By incorporating certification procedures into value chains, farmers have been encouraged to improve their products through healthy competition in the agricultural sector, which has the potential to boost profits and consumer demand.

### *9.3.1.2. Market Access and Linkages*

The majority of value chain projects funded by donors have emphasized strengthening connections throughout the agricultural value chain. Every actor in the value chain has benefited from expanding market access by creating connections to other players, such as new consumers for farmers and cooperatives and retailers for processors to purchase their goods.

For example, putting farmers in touch with veterinary clinics and input suppliers has raised demand for supplements and raised farmers' knowledge of preventive measures to safeguard animal health. In other cases, connecting farmers with cooperatives has resulted in higher sales, better prices, and higher-quality products. Marketing opportunities and connections to value chain actors have been successfully provided by linkage-building strategies, such as enabling farmer and business participation in conferences, exhibitions, trade fairs, and meetings.

### *9.3.1.3. Trainings and Access to Information*

Trainings that incorporate content that is pertinent to the local context and suited to the needs of value chain stakeholders have proven to be extremely effective. Farmers have also benefited from instances where trainings were videotaped and shared on social media to encourage additional uptake and dissemination. Farmers generally responded favorably to trainings on vaccinations, artificial insemination, animal care, and appropriate feeding techniques. Other successful trainings that farmers have reported to have increased sales include those on produce storage, the use of contemporary technologies in processing, and food safety standards and quality management systems, which include recommendations for improved sanitation and hygiene practices. Well-tailored and context-specific trainings have the potential to improve farming and business practices such as altering the housing conditions of cattle, enhancing the ventilation systems in handling facilities to prevent bacterial growth and multiplication, initiating vaccinations for cattle, switching the feed from silos to green fodder, upgrading the procedures for collecting and storing milk, and altering the processes for making cheese.

## 9.3.2. Common Structural Challenges

### *9.3.2.1. Financial Constraints*

For smallholder farmers, access to capital to buy essential inputs and technology is typically a challenge. Farmers' capacity to invest in agriculture is hampered by credit restrictions. Some farmers are also forced by a lack of funding to sell their goods directly from the farm rather than at a nearby market, which drives down the prices they can charge. In addition to needing money to start planting crops or raising cattle, funding is necessary at every point in the value chain due to the expenses incurred by various actors. Donor-funded initiatives help to lessen this difficulty to some degree by giving farmers grants for co-financing, which lessens their financial burden. However, small farmers, who make up the majority of the sector, frequently find these co-financing obligations—which are typically set at 50% of the grants—to be excessively burdensome. Lack of financing is a significant barrier to the development of these value chains because of the ongoing difficulties farmers have obtaining capital, which include a dearth of appropriate financial products, loans, and low-interest loans.

### *9.3.2.2. Women Continue to Face Challenges in Value Chains*

The production and sales volumes and values of farms owned by women are generally higher than average. With rare exceptions, though, farms owned by women are generally smaller in terms of land or herd sizes and have a harder time getting credit than farms owned by men to grow. Women may also be less available for farm activities since they are frequently expected to take care of the home and children. Subsequent secondary research indicates that women are often left out of the financial aspects of marketing their livestock or crops.

#### *9.3.2.3. Challenging Regulatory Environment*

A recurring theme in agricultural value chains is the encounter of major regulatory obstacles. For instance, in Georgia, where small, unregistered farms dominate the market, the ability to monitor and control certification procedures was a major obstacle affecting the certification process's sustainability. Getting laws pertaining to dairy, such as the National Dairy Strategy and the Livestock Health Legislation, approved in Mozambique was difficult. Along the value chains, government representatives also observed that, even after legislation was approved, its implementation remained difficult due to a lack of political will, unclear operational guidelines, a lack of coordination amongst the various stakeholders involved, and a lack of financial and human resources.

### 9.3.3. Policy Recommendations

Here are three key policy recommendations.

#### *9.3.2.1. Addressing Financial Barriers*

With less stringent requirements, regional and local governments can take over co-financing grants after a donor-funded project concludes. This strategy necessitates sufficient financial resources at all governmental levels in addition to the political will to give smallholder farmer investments top priority. Many farmers find that high co-financing levels prevent them from participating, which is especially problematic given the financial hardships many of them experienced during the COVID-19 pandemic. Stakeholders also proposed expanding access to funding by promoting the involvement of Village Savings and Loan Associations (VSLAs) to alleviate some of the obstacles and enable farmers to obtain loans with low interest rates. Addressing financial barriers also requires an understanding of the obstacles to VSLA participation and making sure that all farmers can take advantage of VSLA loans.

#### *9.3.2.2. Making Value Chains Work for Women*

Gender norms for labor force participation in particular will determine how women's challenges in agricultural value chains are addressed, and this will vary by country and context. For women farmers to succeed in value chains, there must be a clear focus on them. This can involve providing women farmers with specialized training and resources to help them get past their obstacles and become more commercial by growing their businesses and implementing better business practices. The difficulties faced by female farmers may also be lessened by public policies that permit the provision of extra resources to obtain childcare, if it is available, or assistance in setting up childcare. Furthermore, depending on the value chain they are involved in, government initiatives and future donor-funded interventions could specifically target women to assist them in growing their herd sizes and/or acreage.

#### *9.3.2.3. Reducing Bureaucracy in Regulatory Processes*

The majority of agricultural regulatory laws must be approved by several government ministries and stakeholders before they can be passed. Acquiring sufficient support from government agencies regarding the necessity and significance of such legislation is another challenge. Government ministries and stakeholders would greatly benefit from a clear focus on cutting down on the amount of time and bureaucracy required at every stage of the regulatory process. Moreover, although approval is a crucial first step, it is only the beginning; a crucial next step would be for government stakeholders to actively participate in identifying and addressing any potential implementation obstacles.

### 9.3.4. Digital Transformation in Agricultural Value Chains

The term "digital transformation" in agricultural value chains describes how digital technologies are incorporated into different phases of agricultural production, processing, marketing, and distribution. The way agricultural value chains function is being radically altered by this change, becoming more resilient, transparent, and efficient. Digital tools like blockchain, data analytics, mobile apps, AI, and IoT have emerged as key players in recent years in tackling long-standing issues in agriculture, from increasing productivity to improving traceability and cutting waste.

With the growing demand for food around the world, digital transformation presents promising ways to empower farmers, support sustainability, and optimize agricultural value chains. The potential to increase productivity and efficiency is among the biggest advantages of digital transformation in agriculture. Farmers can make well-informed decisions regarding planting, irrigation, fertilization, and pest control by utilizing data-driven technologies in precision agriculture. For instance, farmers can apply the proper amount of water and fertilizer only where necessary by using sensors and drones to monitor crop health, soil moisture, and nutrient levels. This helps boost crop yields and lessens resource waste. AI-powered tools can also evaluate data from multiple sources, including field sensors, satellite photos, and weather forecasts, to offer insights that help farmers improve their operations and make better plans for upcoming seasons. A more resilient and effective agricultural value chain can result from these abilities to increase farming's predictability and profitability.

Enhancing the transparency and traceability of the supply chain is a critical component of digital transformation. A growing number of transparent and safe systems for tracking goods from farm to customer are being developed using blockchain technology. Blockchain guarantees that all parties involved have access to precise information about the provenance, caliber, and flow of agricultural goods by logging each transaction on an unchangeable ledger. This is crucial for industries where consumers want guarantees that products are sourced ethically and sustainably, like the organic food industry. As it enables prompt identification and isolation of contaminated products in the event of a recall, improved traceability also aids in combating food fraud and guaranteeing food safety.

The sustainability of agricultural value chains depends on the trust that is fostered by this degree of transparency between producers, suppliers, and consumers. Farmers' access to markets and financial services is also being revolutionized by digital platforms. Historically, small-scale farmers have encountered difficulties in establishing direct connections with purchasers, frequently depending on middlemen who demand a sizeable portion of the earnings. Farmers can sell their produce at fair prices by connecting them directly to markets through digital platforms, which can fill this gap. For instance, farmers can increase the size of their market by connecting with consumers outside of their immediate area through e-commerce platforms made especially for agricultural goods. Farmers can now obtain credit, savings accounts, and insurance products more easily thanks to digital financial services like digital wallets and mobile banking. Farmers now have safe and effective ways to make payments thanks to mobile money solutions, which have been especially revolutionary in areas with weak traditional banking infrastructure. To further improve productivity and economic stability within agricultural value chains, farmers are empowered by this increased financial inclusion to invest in better inputs, technology, and infrastructure.

Better risk management is also made possible by digital transformation, which makes early warning systems and more accurate forecasting possible. Due to climate change, farmers now face greater risk of droughts, floods, and other extreme weather events, making agriculture more unpredictable. By giving farmers access to real-time data on weather patterns, pest outbreaks, and other possible threats, digital technologies can help reduce these risks. Farmers can modify their planting schedules by using AI and machine learning algorithms, which can analyze weather patterns to predict rainfall and droughts. Likewise, digital platforms have the ability to notify farmers about pest infestations so they can take preventative action before their crops suffer damage. To preserve the integrity of value chains, farmers must be able to minimize losses, safeguard their investments, and guarantee a steady supply of agricultural products. These innovations help farmers achieve these goals. Digital transformation in agriculture not only improves productivity and risk management, but also benefits sustainability. Utilizing data-driven technologies can minimize the environmental impact of farming by optimizing the use of resources like water, energy, and fertilizers. For instance, precision irrigation systems make sure that water is used effectively, reducing waste and protecting this essential resource. Additionally, by giving farmers access to knowledge and training, digital solutions can encourage the adoption of sustainable practices. To help farmers adopt more sustainable farming practices, mobile apps, for example, can provide educational content on integrated pest management, organic farming, and soil health management. These eco-friendly methods help the ecosystem and also appeal to customers who are buying more and more sustainably produced goods.

Since smallholder farmers are the backbone of agriculture in many developing nations, the digital transformation of agricultural value chains also has the potential to empower them. Digital tools can help level the playing field by giving small-scale farmers access to information, markets, and financial services, enabling them to compete with larger agribusinesses. For instance, farmers can make better decisions by using mobile-based advisory services to get professional advice on pest control, crop management, and market trends. By combining their resources and exchanging information, farmers can enhance their negotiating power through digital cooperatives, which will enable them to get better prices for their produce. Digital transformation can support more inclusive and equitable economic growth by mitigating the disparities in the agriculture sector. To be successful, digital transformation in agricultural value chains must, however, overcome many obstacles. The digital divide—the difference in access to digital technologies between those with and without it—is one of the primary challenges. The internet is still not widely available in many rural areas, particularly in developing nations, and farmers may not be digitally literate enough to use sophisticated tools. Investments in digital infrastructure, educational initiatives, and reasonably priced access to technology are necessary to address this problem. Furthermore, to create and execute digital solutions that are suited to farmers' needs and contextually specific, partnerships between public, private, and non-governmental organizations are required (Buhr, 2002; Bolzani et al., 2010).

## REFERENCES

1. Bolzani, D., de Villard, S., & de Pryck, J. D. (2010). Agricultural value chain development: Threat or opportunity for women's employment? [Gender and Rural Employment Policy Brief #4]. International Labour Organization.
2. Buhr, B. (2002). Understanding the retail sector: Towards traceability in the meat production chain. Department of Applied Economics, University of Minnesota. *London Swine Conference–Conquering the Challenges*, 11, 12 April 2002.
3. Carter, M., Laajaj, R., & Yang, D. (2021). Subsidies and the African green revolution: Direct effects and social network spillovers of randomized input subsidies in Mozambique. *American Economic Journal: Applied Economics, 13*(2), 206–229.
4. de Brauw, A., & Bulte, E. (2021). African farmers, value chains and agricultural development: An economic and institutional perspective. *Palgrave Studies in Agricultural Economics and Food Policy*. Palgrave Macmillan.
5. De Janvry, A., & Sadoulet, E. (2005). Achieving success in rural development: Toward implementation of an integral approach. *Agricultural Economics, 32*, 75–89.
6. Dickinson, D. L., & Bailey, D. (2005). Willingness to pay for information: Experimental evidence on product traceability in the U.S.A, Canada, the UK, and Japan. *Journal of Agricultural and Applied Economics, 37*(3), 537–548.
7. Ehui, S., & Pender, J. (2005). Resource degradation, low agricultural productivity, and poverty in sub Saharan Africa: Pathways out of the spiral. *Agricultural Economics, 32*(s1), 225–242.
8. Fan, S., Hazell, P., & Thorat, S. (2000). Government spending, growth, and poverty in rural India. *American Journal of Agricultural Economics, 82*(4), 1038–1051.
9. Fearne, A., & García Martinez, M. (1999). The assured combinable crop scheme in England and Wales: Carrot or stick? *Farm Management, 10*(5), 243–261. Cited in Smyth, S., & Phillips, P.W.B. (2002). Product differentiation alternatives: Identity preservation, segregation, and traceability. *Journal of Agrobiotechnology Management and Economics, 5*(2), 30–42. *AgBioForum*. Available at: http://www.agbioforum.org (accessed on 14 December 2024).
10. Gledhill, J. (2002). Tracing the line: Using information technology to reduce costs while meeting industry requirements. *Food Processing*

(March Issue), 48–57. In Smith, G. C., Tatum, J. D., Belk, K. E., Scanga, J. A., Grandin, T., & Sofos, J. N. (2005). Traceability from a US perspective. *Meat Science, 71*, 174–193.

11. Hardgrave, B. C., Waller, M., & Miller, R. (2005). Does RFID reduce out of stocks? A preliminary analysis. Information Technology Research Institute, Sam M. Warton College of Business, University of Arkansas, Arkansas.
12. Häsler, B., Dominguez-Salas, P., Fornace, K., Garza, M., Grace, D., & Rushton, J. (2017). Where food safety meets nutrition outcomes in livestock and fish value chains: A conceptual approach. *Food Security, 9*(5), 1001–1017.
13. Hobbs, J. E. (1996). A transaction cost analysis of quality, traceability and animal welfare issues in UK beef retailing. *British Food Journal, 98*(6), 16–26.
14. Liddell, S., & Bailey, D. V. (2001). Market opportunities and threats to the US pork industry posed by traceability systems. *International Food and Agribusiness Management Review, 4*, 287–302.
15. Verdenius, F. (2006). Using traceability systems to optimise business performance. In I. Smith & T. Furness (Eds.), *Improving Traceability in Food Processing and Distribution* (Chapter 2). Woodhead Publishing.

# CHAPTER 10

# INNOVATIONS IN TRACEABILITY FOR SUSTAINABILITY

## Contents

# INTRODUCTION

A key innovation in advancing sustainability in many industries, especially supply chain management, food production, and agriculture, is traceability. Traceability systems offer a clear picture of the path products take from point of origin to customer, which is helpful as consumers grow more conscious of the ethical and environmental effects of the decisions they make. Businesses can efficiently monitor and confirm the sustainability of their products by leveraging technologies like blockchain, the Internet of Things (IoT), and artificial intelligence. Every transaction in the supply chain is recorded in a decentralized, immutable ledger made possible by blockchain technology, guaranteeing that everyone involved has access to the same data. Customers gain trust as a result of this transparency, which empowers them to base their decisions on businesses' sustainability policies. Customers can, for example, track the origin of ingredients in their food products to make sure they originate from ethical farms that place a high value on environmentally friendly methods. In addition to encouraging ethical consumption, this degree of transparency motivates producers to use more sustainable practices because they are aware that their actions will be closely monitored. IoT devices, which provide information on variables like temperature, humidity, and location, are essential to the real-time monitoring of supply chains. Businesses can reduce waste and the transportation sector's carbon footprint by using this information to manage resources more effectively. IoT sensors, for instance, can monitor the freshness of fish in the seafood sector from the ocean to the plate, guaranteeing that only products sourced ethically and sustainably make it to consumers.

Artificial intelligence improves supply chain traceability by spotting trends and anticipating possible hazards through the analysis of massive volumes of data. With the help of this predictive capability, businesses can proactively handle problems, such as supply chain bottlenecks or disruptions brought on by external factors. AI can also improve logistics, guaranteeing that goods are delivered as efficiently as possible and assisting in the pursuit of sustainability objectives.

Initiatives aimed at promoting traceability can also foster industry-wide cooperation by encouraging businesses to exchange best practices and develop guidelines for environmentally friendly procurement. By working together, we can increase the impact of individual efforts and bring about more significant systemic changes. Through increasing transparency, boosting stakeholder collaboration, and optimizing resource management, traceability innovations play a critical role in advancing sustainability. These technologies have the power to transform industries and pave the way for a more sustainable future that benefits both producers and consumers as they develop further.

## 10.1. SUSTAINABLE AGRICULTURE AND TRACEABILITY LINKAGES

The goal of sustainable agriculture is to meet society's current food and textile needs while preserving the capacity of present or future generations to meet their own needs. It might be predicated on an awareness of ecosystem services. There are numerous ways to improve agriculture's sustainability. The development of adaptable farming techniques and business procedures is crucial for the advancement of agriculture within sustainable food systems. Agriculture has a massive environmental impact, contributing significantly to climate change (one-third of anthropogenic greenhouse gas emissions are caused by food systems), water scarcity, water pollution, land degradation, deforestation, and other processes.

It also both causes and is affected by environmental changes. Sustainable agriculture refers to farming practices that are kind to the environment and enable the production of crops or livestock without endangering human or natural systems. It entails avoiding negative effects on the land, water, biodiversity, and downstream or surrounding resources, in addition to people who live or work nearby or on the farm. Crop rotation, multiple cropping, mixed farming, agroforestry, and permaculture are examples of sustainable agricultural practices.

Traceability and sustainable agriculture are inextricably linked, promoting a more accountable and transparent food system. By enabling customers to follow the path of their food from farm to table, traceability makes sure that farming methods adhere to sustainability standards.

Farmers can show their commitment to ecologically friendly methods, like organic farming and conscientious land management, by putting traceability systems into place. In addition to fostering consumer trust, this openness pushes manufacturers to use sustainable practices to satisfy the expanding demands of the market. Additionally, traceability reduces food waste by revealing inefficiencies in the supply chain, which facilitates improved resource management and allocation.

By enabling brands to keep an eye on working conditions across the supply chain and guarantee adherence to human rights standards, it also promotes ethical labor practices. By offering safe and easily accessible information on product origins and production processes, cutting-edge technologies like blockchain and IoT improve traceability efforts. In the end, incorporating traceability into sustainable agriculture encourages responsibility and gives consumers the power to make knowledgeable decisions, propelling the transition to a more environmentally and socially responsible food system.

### 10.1.1. Traceability Linkages for Sustainability

Traceability has become an essential tool in today's global supply chains, helping to advance sustainability in a range of industries, including manufacturing and agriculture. Traceability is the capacity to follow a product's path through all phases of its lifecycle, from the procurement of raw materials to ultimate consumption. Because of this accountability-fostering transparency, stakeholders are better equipped to make decisions that support sustainable practices.

Moreover, traceability is essential to cutting waste. In the fashion and electronics industries, for example, understanding the origin and path of materials can greatly reduce waste. By putting traceability systems in place, brands can find inefficiencies in their supply chains and minimize excess production and resource usage. This is especially important when considering the concepts of the circular economy, which emphasize material reuse and waste reduction to support more sustainable production methods.

Furthermore, through promoting moral work practices and defending workers' rights, traceability links support social sustainability. Businesses can monitor working conditions throughout their supply chains and make sure that labor practices adhere to human rights standards when they have strong traceability systems in place. This attention to detail not only safeguards employees who are at risk but also cultivates a sense of loyalty and trust among customers who value ethical consumption. In markets where ethical considerations are becoming more and more important, brands that are able to track the labor conditions of their products tend to have an advantage over their competitors.

Traceability also aids in risk management and regulatory compliance. Traceability systems allow businesses to effectively demonstrate compliance with increasingly stringent sustainability regulations that are being implemented by governments and organizations. By taking a proactive stance, businesses not only reduce legal risks but also improve their reputation as industry leaders in sustainability. Additionally, traceability enables businesses to respond quickly in the event of a product recall, reducing possible harm to customers and the environment.

Technology integration transforms traceability in a revolutionary way. The way data is gathered, stored, and shared throughout supply chains has completely changed as a result of innovations like blockchain, IoT, and AI. For instance, blockchain ensures that all parties involved have access to verified information about a product's journey by providing an immutable ledger of transactions. This degree of openness fosters more sustainable consumption habits and increases consumer confidence. To advance sustainability in contemporary supply chains, traceability links are essential. Through the promotion of ethical labor practices, waste reduction, responsible sourcing, and regulatory compliance, traceability enables stakeholders to make well-informed decisions that have a positive impact on society and the environment. Businesses hoping to attain sustainable business practices will need to implement strong traceability systems as consumers demand more accountability and transparency from brands. Adopting this paradigm promotes a more sustainable future for future generations while also improving operational efficiency.

## 10.2. USING TRACEABILITY FOR CARBON FOOTPRINT TRACKING

Organizations are facing mounting pressure to precisely measure and lower their carbon footprints in an era increasingly characterized by environmental sustainability and climate change. The application of traceability systems is one of the best ways to accomplish this. The ability to follow a material or product from its raw materials to the point of end-user consumption is known as traceability. Through the utilization of traceability, corporations can acquire a more comprehensive comprehension of the carbon emissions linked to every phase of their product lifecycle, which in turn facilitates better decision-making and focused mitigation tactics.

The first step in the process is gathering data from all points along the supply chain. This contains details on distribution, transportation, manufacturing procedures, and raw materials. Cutting-edge technologies like blockchain, data analytics, and the Internet of Things (IoT) are essential for gathering real-time data that can be utilized to precisely calculate emissions. IoT sensors, for example, can track factory energy usage, and blockchain technology can guarantee data security and immutability, fostering transparency and confidence in the information gathered.

Organizations can assess the environmental impact of their products through lifecycle assessments (LCAs) once data has been gathered. Because LCAs account for emissions at every stage of production, businesses are able to pinpoint the hotspots where emissions are most concentrated. With this knowledge, they can decide which interventions to prioritize, such as changing to more environmentally friendly materials, streamlining production procedures, or enhancing logistics to cut down on emissions from transportation. Additionally, having traceability in place enables businesses to report on progress transparently and set believable sustainability goals, improving their standing and fostering stakeholder trust. Furthermore, traceability can improve supply chain cooperation. By disclosing carbon footprint information to vendors and clients, businesses can encourage a group mindset toward sustainability. For example, if suppliers are aware of how their emissions affect the final product's overall carbon footprint, they might be encouraged to adopt greener practices. When businesses cooperate to create sustainable alternatives and more effective procedures, innovation may result.

Furthermore, traceability is becoming more and more necessary for compliance as regulatory frameworks require emissions tracking and reporting, making it more than just a best practice. Strong traceability systems will put businesses in a better position to comply with these rules and stay out of trouble. These systems can also help identify opportunities and risks within the framework of carbon markets, allowing companies to take advantage of new developments in the carbon offset and credit space.

Traceability is becoming more and more important as consumers demand sustainability. The environmental impact of consumer purchases is a growing concern for modern consumers, who are more inclined to support brands that can demonstrate responsible sourcing and production practices. Businesses can meet this expanding market demand and gain a competitive edge by using traceability to provide clear information about the carbon footprint of their products.

Tracking carbon footprints with traceability is a highly effective approach with many advantages. It satisfies customer demand for sustainability, guarantees regulatory compliance, encourages collaboration across supply chains, and gives organizations the data they need to make educated decisions. Adopting traceability will help businesses not only lessen their environmental impact but also position them as leaders in the shift to a more sustainable future as the world struggles with climate change.

### 10.2.1. The Importance of Carbon Traceability

Although they are the backbone of the world's food systems, agribusinesses also significantly increase carbon emissions. Every step of the agricultural supply chain produces carbon emissions, from energy-intensive processes like food processing and transportation to land-use changes like deforestation for crop production. With the world coming to terms with climate change realities, there is a growing emphasis on comprehending and reducing these emissions. Therefore, feeding the world and doing so in a way that minimizes their impact on the environment are the dual responsibilities of agribusinesses. This entails precisely monitoring their carbon footprint, locating emission hotspots, and devising mitigation strategies. In the global push towards sustainability, agribusinesses run the risk of lagging behind if they don't have a clear understanding of their carbon footprint. Accurate emission tracking and reporting is essential to the credibility of carbon project developers and goes beyond legal requirements. These developers work on initiatives like reforestation programs, renewable energy projects, or sustainable farming methods that aim to lower or offset carbon emissions. They must offer precise information on the amount of carbon being saved or sequestered to demonstrate the effectiveness of their efforts. The project's integrity is compromised by inaccurate or ambiguous reporting, which also erodes the confidence of the public, regulators, and investors. Carbon credits need to be supported by solid data to be relevant and commercially viable. Meaningful climate action is made possible by accurate carbon monitoring, which guarantees that every ton of carbon offset is actual, quantifiable, and verifiable. ..

#### *10.2.1.1. Risks of Inaccurate Data*

When it comes to carbon traceability, the stakes are very high. For both agribusinesses and the developers of carbon projects, inaccurate or insufficient data can result in many issues. Inadequate traceability may lead to environmental regulations being broken, which may result in steep fines and legal issues. However, the risk to a company's reputation may be even more detrimental. Nowadays, stakeholders, investors, and customers are paying more and more attention to how businesses affect the environment. A company may be swiftly accused of "greenwashing" if it is discovered that it is deceiving the public about its carbon footprint, either on purpose or as a result of poor tracking. This phrase describes the practice of making inflated or inaccurate statements regarding a business's environmental initiatives. Greenwashing can have serious negative effects, including a damaged company reputation, a decline in brand value, and a loss of consumer trust. On the other hand, companies that place a high priority on precise carbon traceability can steer clear of these problems and establish themselves as industry leaders in sustainability, giving them a competitive advantage.

### 10.2.2. A Novel Credible Carbon Footprint Traceability System for Low Carbon Economy Using Blockchain Technology

Greenhouse gas emissions are rising in tandem with economic expansion, and environmental problems are having an impact on every aspect of every nation on the planet. Emissions of greenhouse gases have a detrimental effect on the environment and intensify climate change. Controlling human activity is therefore crucial to lessening the environmental impact of greenhouse gas emissions. Here, a case study by Ju C, Shen Z, Bao F, Weng P, Xu Y, Xu C. A Novel Credible Carbon Footprint Traceability System for Low Carbon Economy Using Blockchain Technology. *International Journal of Environmental*

*Research and Public Health*. 2022; 19(16):10316. https://doi.org/10.3390/ijerph191610316 is presented as an example.

Among these suggestions is the implementation of an emission trading system (ETS). In its capacity as an authorized center, the ETS sets a reduction goal for carbon emissions to oversee the trading of carbon emission permits (CEPs). Demonstrated the effectiveness of carbon emission trading policies in lowering carbon emissions in pilot cities and the necessity of a carbon emission trading system to meet the carbon emission reduction target. Nevertheless, there are numerous obstacles to overcome before carbon emission trading can be fully implemented, such as the following: (1) the market for carbon emission trading is corrupt and opaque; (2) the process of allocating CEP is complicated; (3) the computation of carbon budgets is vulnerable to manipulation and abuse; and (4) there are no tracking measurements for carbon emission trading.

Carbon footprints can be created on the distributed super ledger by employing blockchain technology, creating a system for tracking carbon footprints. This tracking system is essential to eradicating corruption and guaranteeing the integrity of the records to avoid manipulation. Therefore, a technical architecture that can be used for carbon resource management and traceability is needed to meet the aforementioned system requirements. Blockchain technology is a promising solution that can ensure complete supply chain transparency and establish the credibility of carbon emission and carbon footprint data. Enterprises can achieve a real carbon neutral supply chain and negative carbon supply chain by implementing effective carbon disclosure and developing corresponding carbon compensation schemes, all made possible by transparent and reliable data. All measurable carbon emission data can be safely shared, and information can be verified while remaining invisible thanks to blockchain and privacy computing technologies. Enterprise data security concerns can be removed, allowing carbon emission data to be shared and circulated at the industrial level and its value to be fully utilized. This will accelerate the process of reaching carbon neutrality. The prevalence of falsified carbon emission data makes blockchain a preferred technology in the field of carbon emissions. To address global climate issues, businesses and brands must not only abide by many "green agreements" and submit data on their carbon emissions regularly, but they must also offer credible evidence to support the veracity of this data. The issues of "illegal carbon trading" in the low-carbon market and "carbon emission data falsification" have surfaced frequently in recent years, raising concerns about how to ensure data.

Ju et al. (2022) created an interactive agricultural product traceability structure and implemented the off-chain traceability and on-chain verification traceability solution concerning the use of blockchain technology in complex application scenarios. The storage structure has notable benefits in terms of tamper resistance, transparency, and traceability efficiency because of the adoption of blockchain-based distributed data storage and the decentralized traceability process. It works well for peer-to-peer transaction monitoring in real-time and rapid traceability in untrustworthy networks. However, when providing supporting documentation for carbon emission trading, this system only supports "full-chain and full-information" traceability, which implies that each traceability will return all data of the entire transaction chain and, consequently, will return a great deal of information unrelated to carbon emissions. In this work, we suggest enhancing the previously described interactive traceability framework and implementing the distributed traceability idea of "off-chain traceability and on-chain verification.." In this chapter, an interactive traceability structure supporting type filtering is built to support the tracing and tracking of carbon emission-related information in a transaction chain to meet the demand for regulation, tracking, and tracing of various types of data from different levels of users, such as producers, consumers, and regulators in the carbon emission trading market. It makes it possible to address issues with the carbon emission trading system, such as data tracking tool shortages, corruption, opacity, and manipulation. To ensure that the entire system is capable of handling both extremely high traceability requests and the efficient tracing of data related to carbon emissions, it simultaneously adopts the idea of a decentralized traceability process and optimizes the distributed storage structure in the traceability system. It understands how quickly enterprise and product carbon footprint data can be tracked and traced.

This research separates the distributed traceability process into three steps: locating, retrieving, and verifying, as Figure 10.1 illustrates. We think the blockchain's consensus and verification functions should be strengthened, traceability, and verification should be kept apart, and the remaining tasks should be transferred to a more appropriate distributed storage system (e.g., IPFS). Furthermore, the steps in the traceability process are further refined in this chapter, and the traceability process is further broken down into locating and retrieving so that the two processes do not conflict. This allows for parallel locating and retrieval and lessens the influence of document filtering operations on the traceability efficiency. In this chapter, the tracing process and distributed storage file structure are redesigned to separate the steps and functions. A three-level structure is adopted by further optimizing the storage structure, and the locating information is stored in a separate layer called the pointer structure, which only corresponds to the positioning function. The relevant data is first found at the pointer level, after which those files are chosen to retrieve it, thereby achieving traceability with a straightforward semantic definition. In contrast to earlier distributed traceability techniques that involved staggered retrieval and locating, the traceability method presented in this chapter offers strong flexibility in addition to ensuring traceability efficiency, allowing for effective filtering of distributed storage data.

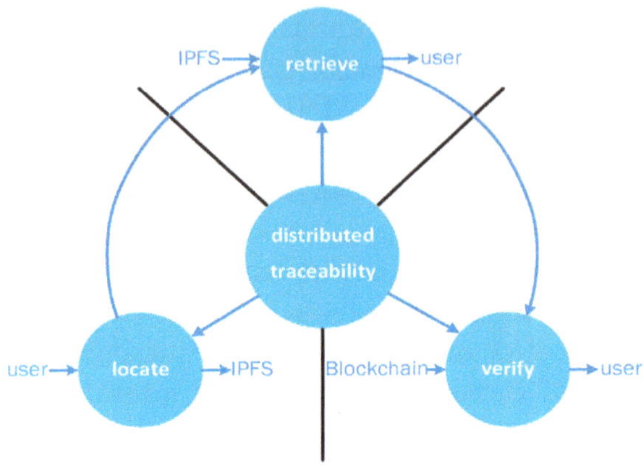


**Figure 10.1.** *The distributed traceability process into three stages: locating, retrieving, and verifying.*

*Source: Ju C, Shen Z, Bao F, Weng P, Xu Y, Xu C. A Novel Credible Carbon Footprint Traceability System for Low Carbon Economy Using Blockchain Technology. International Journal of Environmental Research and Public Health. 2022; 19(16):10316. https://doi.org/10.3390/ijerph191610316*

## 10.2.3. Interaction Structure Design of Carbon Footprint Traceability System

While the information stored in the blockchain frequently consists of a variety of types and forms of information, regulators, producers, and consumers in the carbon emission trading and low carbon market typically concentrate on specific data and evidence related to carbon emission, which makes it difficult to efficiently screen and trace carbon emission-related information. In this chapter, Ju et al. (2022) propose a method that can trace specific categories of data, enabling us to quickly locate historical carbon emission data and address the issue of returning a large amount of non-critical information when tracing the source. In this chapter, a specific file structure is formed for traceability by appending pointers to distributed storage files. The distributed traceability of "traceability off the chain and verification on the chain" is realized because the traceability and verification processes are kept apart. When the volume of on-chain data is large and the types are too varied, the traceability structure and procedure Ju et al. (2022) suggested in the previous study returns all of the transaction chain's data, which results in inefficient traceability of particular types of data. Users will receive a lot of data unrelated to carbon emissions when it is applied to the carbon footprint traceability scenario (even if these data are related to the transaction chain). Thus, additional enhancements have been made in this research to form an interactive traceability structure that facilitates type filtering and achieving quick traceability and retrieval of particular data in the transaction chain.

### *10.2.3.1. Interaction Structure Design*

The joint use of blockchain technology and IPFS has the potential to address the storage issue associated with massive volumes of unstructured data, as examined in the literature review. It also significantly expands the blockchain technology's application scenarios.

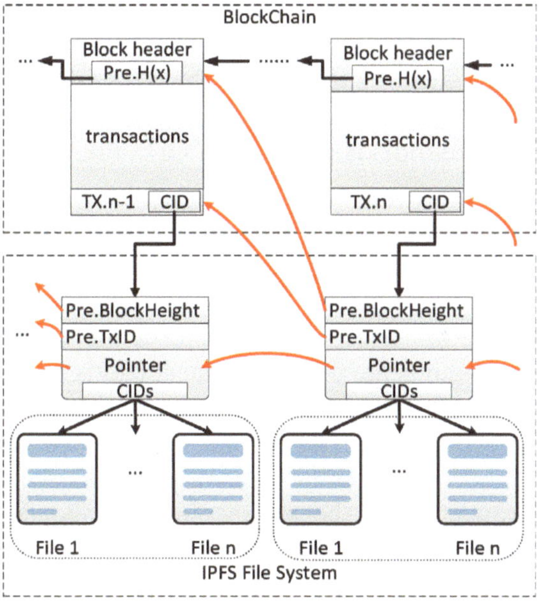


**Figure 10.2.** *Pointer design.*

*Source: Ju C, Shen Z, Bao F, Weng P, Xu Y, Xu C. A Novel Credible Carbon Footprint Traceability System for Low Carbon Economy Using Blockchain Technology. Int J Environ Res Public Health. 2022 Aug 19;19(16):10316. doi: 10.3390/ijerph191610316. PMID: 36011951; PMCID: PMC9407757.*

The previous study's proposed structure can now handle higher volume requests for data traceability, and after additional optimization of the distributed storage structure and traceability process, the traceability process has become much more efficient. Additionally, the study achieved distributed traceability of "traceability off the chain and verification on the chain," which significantly lessens the pressure on full nodes to retrieve data. Nonetheless, additional optimization and improvement of the current structure are required when users only require specific types of data to be verified and the data types on the chain are complex and varied in real-world applications. They classify the raw files in advance when they are deposited in IPFS, and the related classification information is stored separately from the related pointer information. The specific pointer design is shown in (Figure 10.2; Ju et al., 2022).

Pre.TxID is the hash of the previous transaction in the transaction chain; CID is a pointer to the IPFS system, pointing to the pointer file corresponding to this transaction, obtained after the Author classifies the deposit file and uploads the pointer file; TxID indicates the hash of this transaction, which must be calculated after obtaining the CID, with TxID = Hash (Content, Author, Pre. TxID, CID, TxID).

The pointer file in this research is a JSON format file, and the composition of the pointer file content is

Pointer File = [Pre.BlockHeight, Pre.TxID, Pre.CID, Category, CIDs, Chain level]

BlockHeight is the block height from the transaction's prior phase, known as Pre. The hash value from the transaction's prior phase, Pre, is TxID. The hash value of the pointer file corresponding to the previous phase of the transaction is called CID. The type of corresponding evidence file is called Category. The three values C, I, and N indicate whether a document is correlated with carbon emissions, indirectly correlated with carbon emissions, or not correlated with carbon emissions, respectively. The documents required for the traceability process can be obtained based on the type of deposited documents. The chain level indicates the position of the transaction corresponding to the current pointer file on the transaction chain and is used to establish the correspondence between the pointer file and the transaction. CIDs are the set of CIDs of the corresponding depository files of this transaction, and their order should correspond to Category one by one (Ju et al., 2022).

From Figure 10.3, we can see that the interactive traceability structure supporting type filtering proposed is generally divided into three layers. All transaction data in the market are recorded on the blockchain, but only the necessary index data and verification data are included. The original data related to transactions are stored in IPFS. The original data in IPFs and the transaction data on the blockchain are connected via a layer of pointer files. Through the unique CID pointer recorded in the transaction and the CIDS of the original data recorded in the pointer files, the data retrieval from the blockchain to the IPFs system can be realized. The pointer files also contain a pointer to a specific transaction in the blockchain, which is used to locate the appropriate block in the blockchain by going back from the files in IPFS. This bi-directional pointer structure between blockchain and IPFS provides more options for the traceability process. The traceability process can be completed not only in blockchain but also in IPFS. The corresponding pointer files are connected according to the relationship between the transaction, which makes it possible to trace the source directly in the file structure formed by the pointer files when we have the latest transaction information, and in this process, we can also accurately locate the corresponding information in the blockchain in reverse. This enables users to complete traceability in IPFS by themselves and then send authentication requests to the blockchain system.

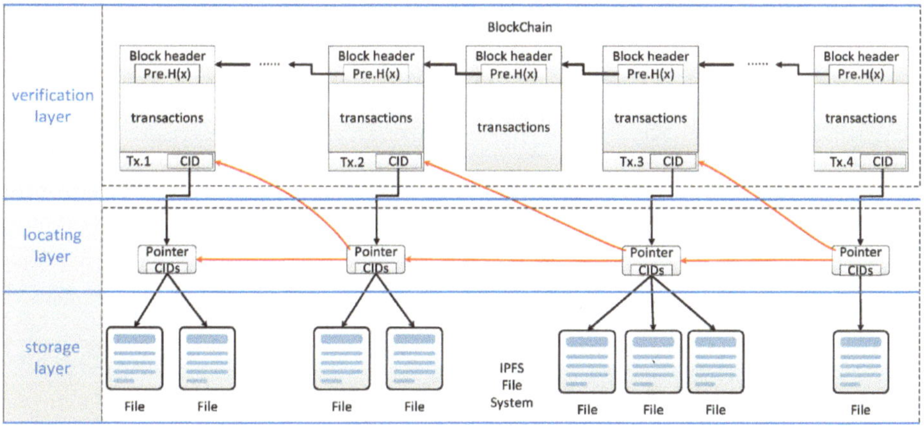


**Figure 10.3.** *The three-layer interactive traceability structure supports type filtering.*

*Source: Ju C, Shen Z, Bao F, Weng P, Xu Y, Xu C. A Novel Credible Carbon Footprint Traceability System for Low Carbon Economy Using Blockchain Technology. Int J Environ Res Public Health. 2022 Aug 19;19(16):10316. doi: 10.3390/ijerph191610316. PMID: 36011951; PMCID: PMC9407757.*

### *10.2.3.2. Interactive Traceability Process*

The idea of "traceability off the chain and verification on the chain" is adopted by the interactive traceability structure that supports type filtering in this research. To fulfill the user's request for traceability, the full node only needs to find the most recent transaction information on the blockchain, send a file acquisition request to the IPFS system using the CID in the transaction information, and then return the pointer file. After that, the user can use the pointer file to independently perform historical documentation traceability and obtain the historical pointer files as an index. The verification request is then sent to the entire node after the block information, and the pertinent evidence files may be downloaded in line with the category information in the pointer file. Following the request, the full node locates the transaction location, gathers the pertinent transaction and verification data, and gives it to the user so they can locally verify the authenticity of the relevant transaction and the evidence files. This process of traceability is illustrated in Figure 10.4. Ultimately, only the Target File, verification data, transaction chain information, and pointer files are returned to the user after the traceability process is retrieved, as illustrated in Figure 10.4. The remaining original data that is unrelated to carbon emission trading is filtered out.

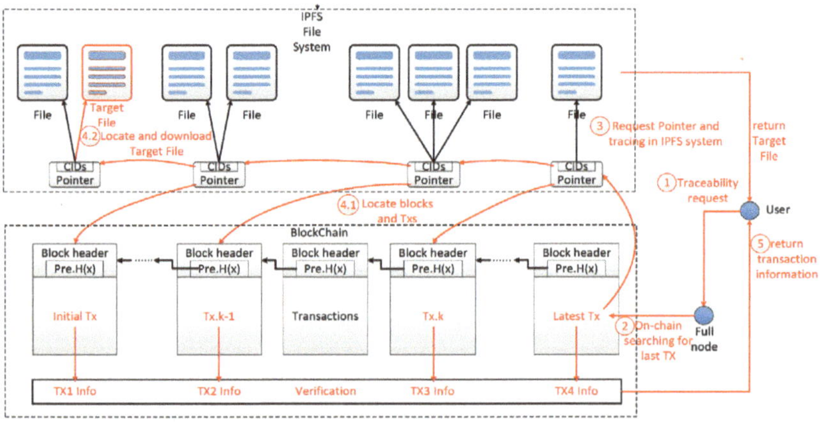


**Figure 10.4.** *The traceability process under the interactive traceability structure supports type filtering.*

*Source: Ju C, Shen Z, Bao F, Weng P, Xu Y, Xu C. A Novel Credible Carbon Footprint Traceability System for Low Carbon Economy Using Blockchain Technology. Int J Environ Res Public Health. 2022 Aug 19;19(16):10316. doi: 10.3390/ijerph191610316. PMID: 36011951; PMCID: PMC9407757.*

## 10.2.4. Challenges in Carbon Emission Monitoring and Reporting

One of the most significant challenges in carbon emission monitoring and reporting is the sheer complexity of modern supply chains. Agribusinesses and carbon project developers often operate across multiple regions, each with its own set of environmental factors, production methods, and regulations. This diversity leads to a fragmentation of data, making it difficult to obtain a clear, unified picture of total carbon emissions.

For instance, consider a global agribusiness that sources raw materials from farms in South America, processes them in factories in Asia, and distributes the final products across Europe and North America. Each stage of this supply chain generates emissions, but the data from these various stages is often siloed, stored in different formats, and managed by different systems. This fragmentation can lead to gaps in the data, making it nearly impossible to track emissions accurately. As a result, businesses may inadvertently underreport their carbon footprint, leading to non-compliance with regulations and loss of stakeholder trust such as:

- Moreover, the complexity of tracking emissions is exacerbated by the need to account for indirect emissions—those that occur upstream or downstream in the supply chain. These include emissions from suppliers, transportation, and even the use and disposal of products by consumers. Gathering accurate data across these diverse and dispersed sources is a daunting task, requiring robust systems and tools that can seamlessly integrate and analyze information from multiple points in the supply chain.
- Traditional carbon reporting systems are often plagued by a lack of transparency, making it difficult to ensure the accuracy and reliability of reported data. Many businesses still rely on outdated methods of data collection and reporting, such as manual record-keeping, spreadsheets, and self-reporting. These methods are not only time-consuming and prone to errors, but they also leave room for data manipulation, either intentional or unintentional.

For example, a company might underreport its emissions to present a more favorable environmental image, a practice that can easily slip under the radar with traditional reporting methods. This lack of transparency undermines the credibility of carbon reporting and contributes to the growing problem of greenwashing, where companies make false or misleading claims about their environmental performance.

- Furthermore, traditional systems cannot often provide real-time data, leading to delays in reporting and an inability to respond quickly to issues as they arise. This can be particularly problematic in fast-moving industries where environmental impact needs to be monitored continuously to ensure compliance and minimize harm.
- The potential for data manipulation and the lack of real-time visibility into carbon emissions make it difficult for businesses to build trust with stakeholders, including customers, investors, and regulators. In a world where transparency and accountability are increasingly demanded, companies that fail to address these issues risk damaging their reputations and facing regulatory penalties.

- As the global focus on climate change intensifies, so too does the regulatory pressure on businesses to monitor and report their carbon emissions accurately. Governments and international bodies are introducing stricter regulations and standards aimed at reducing carbon footprints and promoting sustainability. *For example*, the European Union Deforestation Regulation (EUDR) and the Task Force on Climate-related Financial Disclosures (TCFD) are just two of the many frameworks that require businesses to demonstrate their commitment to reducing emissions.

Compliance with these regulations is not optional—it's a legal requirement that carries significant consequences for non-compliance. Companies that fail to meet these standards may face hefty fines, legal action, and even restrictions on market access. For example, under the EUDR, businesses that cannot provide verifiable proof that their products are not contributing to deforestation could be barred from selling their goods in the European Union.

- The increasing complexity of regulatory requirements also adds to the burden on businesses. Different regulations may have different reporting requirements, timelines, and standards, making it challenging for companies to keep up. This is especially true for global businesses operating in multiple jurisdictions, each with its own set of rules.

The growing demand for compliance is further complicated by the need to provide not just quantitative data, but also qualitative disclosures that demonstrate a company's broader commitment to sustainability. *For example*, under the TCFD, companies are required to disclose how climate-related risks and opportunities are integrated into their business strategy and financial planning. Meeting these requirements requires a comprehensive approach to carbon emission monitoring and reporting, as well as a commitment to transparency and accountability (Ju et al., 2022).

## 10.2.5. How Blockchain Enhances Carbon Traceability

### *10.2.5.1. Immutability of Records: How Blockchain Ensures Tamper-Proof and Transparent Emission Data*

One of the standout features of blockchain technology is its ability to create immutable records—once data is entered into the blockchain, it cannot be altered or deleted. This immutability is crucial for carbon traceability because it ensures that emission data is tamper-proof and transparent. In a world where companies are increasingly held accountable for their environmental impact, the ability to trust the accuracy and integrity of carbon data is essential.

Agribusiness plays a vital role in contributing to transition to low-carbon agriculture supply chains which is essential for mitigating climate change and ensuring a sustainable future. Imagine a scenario where an agribusiness reports its carbon emissions for a given year. Traditionally, this data might be stored in a centralized database, where it could be altered by anyone with the right access privileges, whether intentionally or accidentally. This creates a risk of data manipulation, which could lead to underreporting emissions, greenwashing, and ultimately, reputational damage or regulatory fines.

With blockchain, each data point related to carbon emissions—whether it's the amount of fuel used in a factory or the emissions produced during transportation—is recorded as a block in the chain. These blocks are linked together using cryptographic hashes, making it nearly impossible to change any part of the chain without altering all subsequent blocks, which would immediately be noticed by the entire network.

This tamper-proof nature of blockchain builds trust among all stakeholders, from suppliers and customers to regulators and investors. Everyone involved in the supply chain can have confidence that the carbon data they're viewing is accurate and has not been manipulated. This level of transparency is vital for maintaining credibility in an era where environmental claims are scrutinized more than ever before.

### *10.2.5.2. Real-Time Monitoring: Leveraging Blockchain for Real-Time Updates and Accuracy in Carbon Data*

These days, having access to real-time data is revolutionary, particularly for monitoring carbon emissions. Blockchain technology can help with real-time emissions monitoring, giving companies the most recent information possible about their carbon footprint. Businesses involved in intricate, international supply chains will especially benefit from this, as data collection delays can result in errors and lost opportunities to reduce emissions.

Take a carbon project developer, for example, who has to monitor the emissions related to a reforestation project. In the past, they might have to wait for weeks or even months for the data to be gathered, processed, and reported. Making timely decisions can be challenging because by the time the data is available, it might not accurately reflect the current circumstances. ..

Blockchain changes this dynamic by enabling real-time data entry and access. As soon as an emission-related event occurs—such as the completion of a logging operation or the transportation of goods—this information can be recorded on the blockchain. The data is instantly available to all authorized parties, allowing for immediate analysis and action.

Real-time monitoring not only enhances the accuracy of carbon data but also allows businesses to respond quickly to any issues that arise. For example, if emissions in a particular part of the supply chain suddenly spike, the company can investigate the cause and implement corrective measures without delay. This agility is crucial for staying on top of carbon management and ensuring compliance with increasingly stringent environmental regulations.

### *10.2.5.3. End-to-End Visibility: Providing Full Traceability from Raw Material Sourcing to End-Product Distribution*

Ensuring end-to-end visibility throughout the whole supply chain, from the sourcing of raw materials to the final product's distribution, is one of the most difficult aspects of carbon traceability. Building resilient and ethical supply chains in agribusiness requires sustainable sourcing. Conventional supply chains frequently lack transparency, with data silos, inconsistent reporting practices, and fragmented information making it challenging to track carbon emissions thoroughly. By offering a single, cohesive platform where all emission-related data can be stored and accessed, blockchain technology solves this problem. The blockchain records every stage of the supply chain, including raw material extraction, manufacturing, shipping, and final sale. This produces an exhaustive, impenetrable record of each product's carbon footprint. . Blockchain technology, for instance, can be used by an agribusiness that sources palm oil for its goods to track the carbon emissions linked to every phase of production, from the plantation to the processing facility and beyond. A clear, continuous trail of emissions data is provided by the addition of a new block to the chain each time the product changes hands or advances to a new phase in the supply chain. For companies that must show their dedication to sustainability and adhere to carbon-related regulations, this end-to-end visibility is priceless. It enables businesses to pinpoint areas for improvement by identifying hotspots in their supply chains with the highest emissions. Additionally, it gives regulators and consumers the transparency they require to have faith that the company is accurately reporting its carbon footprint. Adopting blockchain for carbon traceability will be a crucial step for businesses dedicated to sustainability as the world shifts towards greater environmental accountability.

### *10.2.5.4. Addressing the Concerns of Agribusinesses and Carbon Project Developers*

#### *10.2.5.1.1. Scalability*

One of the primary concerns for agribusinesses and carbon project developers is whether blockchain technology can be scaled to handle the vast and complex operations typical of large enterprises. When dealing with global supply chains that span multiple countries and involve countless stakeholders, the need for a scalable solution becomes crucial.

Blockchain technology is inherently scalable because it is designed to manage and store large amounts of data across a distributed network. As your business grows, the blockchain can expand with it, ensuring that every transaction, data point, and emission report is captured without losing efficiency or speed. Whether you're a small agribusiness or a global corporation, blockchain can adapt to your specific needs.

#### *10.2.5.1.2. Cost-Effectiveness*

Cost is always a critical factor when considering new technology. For many agribusinesses and carbon project developers, the question isn't just whether blockchain is effective, but whether it's cost-effective. Implementing a blockchain solution requires an initial investment in technology and training, but it's essential to look at the long-term return on investment (ROI). When properly implemented, blockchain can significantly reduce costs associated with carbon traceability by automating data collection, reducing errors, and improving transparency. The reduction in manual processes alone can lead to significant savings. Additionally, blockchain can help avoid the costs associated with non-compliance, such as fines and reputational damage, by ensuring that your carbon data is accurate and verifiable.

#### *10.2.5.1.3. Interoperability*

In today's digital world, businesses rely on a variety of tools and platforms to manage their operations. For agribusinesses and carbon project developers, these might include enterprise resource planning (ERP) systems, digital monitoring tools, and reporting platforms. A key concern is whether a new blockchain solution can integrate seamlessly with these existing systems. Interoperability is a crucial aspect of blockchain technology. Leading blockchain platforms are designed with open standards and APIs that allow them to communicate with other digital tools and systems. This means that your blockchain solution can work alongside your current software, enhancing your existing capabilities without causing disruptions (Ju et al., 2022).

### 10.2.6. Environmental Impact Monitoring Through Traceability

It is impossible to overestimate the significance of sustainability in business practices given the growing environmental challenges. Businesses are realizing more and more that they must keep an eye on how they are affecting the environment to ensure that natural resources are managed responsibly. Traceability systems are among the most efficient ways to accomplish this. Businesses can track the movement of goods and materials through their supply chains thanks to traceability, which gives them important information about how their operations affect the environment. Organizations can improve their overall sustainability performance and minimize their ecological footprint by incorporating traceability into their sustainability strategies.

#### *10.2.6.1. The Concept of Traceability*

The ability to follow the flow of goods and materials from point of origin to end user is known as traceability. Every stage of the supply chain is covered by this procedure, including locating raw materials, manufacturing techniques, distribution, transportation, and end-user consumption. Advanced data analytics, blockchain, and the IoT are examples of modern technologies that make data management and collection easier along this chain. Organizations can monitor environmental impacts more effectively by gaining real-time visibility into their operations through the utilization of these technologies.

#### *10.2.6.2. Importance of Environmental Impact Monitoring*

Understanding the ways in which business operations impact ecosystems and natural resources requires an understanding of environmental impact monitoring. It entails evaluating elements including waste production, water use, carbon emissions, and biodiversity loss. Organizations can identify the most important environmental challenges they face and create focused mitigation strategies by measuring these metrics. In addition to assisting companies in meeting legal requirements, this enhances their reputation among stakeholders and customers.

#### *10.2.6.3. Role of Traceability in Sustainability*

Because traceability makes it possible to monitor environmental impacts in great detail, it is essential to sustainability. Organizations can learn where emissions occur, where resources are consumed, and how waste is generated by tracking their materials and products through the supply chain. Conducting lifecycle assessments (LCAs), which assess a product's environmental impact from cradle to grave, requires this information. Take a food company, for example, that uses traceability to keep an eye on its farming methods. Through monitoring fertilizer usage, water usage, and land use, the business can evaluate how its farming practices affect the environment. It can work with farmers to adopt more sustainable practices, like precision agriculture or organic farming techniques, if it discovers that particular practices result in excessive water use or chemical runoff.

### *10.2.6.4. Enhancing Supply Chain Collaboration*

The ability of traceability to improve cooperation within supply chains is a major additional benefit. Businesses can cultivate an environmental culture that transcends their internal operations by disseminating information about environmental impacts to suppliers and customers. A manufacturer that monitors its carbon emissions, for example, can encourage its suppliers to adopt greener practices by sharing this data with them. This kind of teamwork can produce creative solutions that lower emissions and boost resource efficiency all the way through the supply chain.

In addition, buyers are becoming more concerned about the sustainability of the goods they buy. Businesses can gain the trust of their clients by being open and honest about the environmental impact of their products. Businesses can effectively communicate their sustainability initiatives thanks to traceability, which covers everything from raw material sourcing to product disposal at the end of its useful life. Improved brand reputation and customer loyalty may result from this transparency.

### *10.2.6.5. Compliance and Risk Management*

Traceability becomes crucial for compliance as more stringent environmental regulations are implemented globally by governments and regulatory agencies. Strong monitoring and reporting systems will put an organization in a better position to comply with these regulations and stay out of trouble. Furthermore, traceability can assist in identifying the risks connected to environmental practices, enabling businesses to take proactive measures to address possible problems before they become more serious ones. A textile manufacturer, for instance, might monitor chemical releases and water consumption throughout the entire production process.

The business can look into more environmentally friendly dyeing processes or materials if the data shows that a certain dyeing procedure is causing excessive water pollution. Organizations can guarantee compliance and lessen their overall environmental impact by proactively addressing these issues.

### *10.2.6.6. Technological Innovations in Traceability*

The introduction of digital technologies has completely changed how supply chains apply traceability. At different stages of the supply chain, IoT devices can gather real-time data on emissions, water and energy consumption, and emissions. Because of blockchain technology, stakeholders can trust that the data being shared is safe, transparent, and unchangeable. The capabilities of traceability systems can be further enhanced by advanced analytics and artificial intelligence through the identification of patterns, prediction of future impacts, and suggestion of improvements. Farmers can use IoT sensors, for example, to track crop growth, soil health, and moisture levels in the agricultural sector. Distributors and retailers can access this data to obtain a comprehensive understanding of the environmental impact of the entire supply chain. Businesses that use technology in this way are able to make better decisions about how to allocate resources, how to produce goods, and how to handle waste.

### *10.2.6.7. The Future of Sustainable Practices*

Traceability will play a bigger role in sustainable business practices as the need to address climate change and environmental degradation becomes more urgent. Businesses that implement traceability systems not only improve their operational effectiveness but also help achieve more general sustainability objectives.

Organizations can discover opportunities for innovation and improvement, which will ultimately result in more sustainable outcomes, by keeping an eye on their environmental impact. Furthermore, as consumers grow more aware of environmental issues, they will expect brands to be more transparent. Companies that are able to offer verifiable data regarding the sustainability of their goods are probably going to have an advantage over their competitors in the market.

Traceability-based environmental impact monitoring is a revolutionary strategy for sustainability. Organizations can improve supply chain collaboration, assure regulatory compliance, develop consumer trust, and obtain important insights into their environmental impacts by incorporating traceability into their operations. The potential for traceability to promote sustainable practices will only increase with the development of technology. In the end, companies can significantly contribute to the shift to a more sustainable future and improve the welfare of the earth and society by embracing traceability (Ju et al., 2022).

## 10.3. FUTURE OF TRACEABILITY IN AGRICULTURE

Because of the need for sustainable practices, consumer demand for transparency, and technology advancements, traceability in agriculture is changing quickly. With the growing importance of global issues like resource depletion, food security, and climate change, the agriculture industry needs to find creative ways to improve food product traceability. This change is more than just tracking food from farm to table; it's a fundamental adjustment to the way agricultural systems function, making them more sustainable, effective, and customer-focused.

### 10.3.1. Understanding Traceability in Agriculture

In the field of agriculture, traceability pertains to the capacity to monitor and authenticate the provenance of agricultural goods, encompassing their passage via the supply chain. This process includes all aspects of cultivation, such as selecting seeds and inputs, processing, packaging, and distribution.

There are various reasons why traceability is becoming more and more important. First, food safety, quality, and ethical sourcing are issues that consumers are becoming more conscious of and concerned about. They want information about the origins, methods used throughout the supply chain, and methods of production of their food to be made transparent. Second, strong traceability systems are required to meet standards and stay out of trouble as environmental and food safety regulations become more stringent globally (Bao et al., 2021).

### 10.3.2. The Role of Technology

Technological innovation is at the forefront of enhancing traceability in agriculture. Several emerging technologies are transforming the way agricultural products are tracked and verified include:

- **Blockchain Technology:** Blockchain offers a decentralized and immutable ledger system that enhances transparency and trust. Each transaction, from seed planting to consumer purchase, can be recorded in real-time, allowing stakeholders to trace the entire supply chain seamlessly. This technology can significantly reduce fraud and mislabeling while providing consumers with verifiable information about the origin of their food.
- **Internet of Things (IoT):** IoT devices, such as sensors and GPS trackers, can collect real-time data from farms, storage facilities, and transport vehicles. This data can include information about soil health, weather conditions, temperature, humidity, and more. By integrating IoT with traceability systems, farmers can monitor their products more effectively and respond quickly to any issues, such as spoilage or contamination.
- **Data Analytics and Artificial Intelligence:** Advanced data analytics and AI can process vast amounts of data collected from various sources, helping farmers and stakeholders make informed decisions. These technologies can identify patterns, predict trends, and provide actionable insights, enabling better resource management and improved sustainability practices.
- **Drones and Satellite Imaging:** Drones equipped with cameras and sensors can monitor crop health, assess land use, and even track

the movement of livestock. Satellite imaging can provide broader data on agricultural practices and land use changes over time. These technologies can enhance traceability by offering a comprehensive view of agricultural operations.

### 10.3.3. Benefits of Enhanced Traceability

The future of traceability in agriculture promises numerous benefits that can transform the industry are:

- **Improved Food Safety:** Enhanced traceability systems can significantly reduce the risk of foodborne illnesses by allowing for quicker identification and response to contamination outbreaks. If a food safety issue arises, traceability enables rapid pinpointing of the source, making recalls more efficient and minimizing health risks.
- **Sustainability and Environmental Responsibility:** Traceability facilitates better monitoring of agricultural practices, enabling farmers to adopt more sustainable methods. By tracking inputs such as fertilizers and pesticides, farmers can reduce overuse and minimize environmental impact. This data can also help in managing water resources and optimizing land use, contributing to a more sustainable agricultural system.
- **Consumer Trust and Brand Loyalty:** In a marketplace increasingly defined by consumer demand for transparency, traceability can foster trust between consumers and producers. When consumers can verify the origins and quality of their food, they are more likely to support brands that prioritize ethical practices. This trust can translate into brand loyalty, driving sales and enhancing reputation.
- **Market Access and Competitiveness:** As consumers become more discerning, traceability can be a key differentiator for agricultural products. Producers that implement robust traceability systems may gain access to premium markets that prioritize sustainable and ethically sourced goods. Additionally, compliance with international standards can open doors to export opportunities.
- **Informed Decision-Making:** Data collected through traceability systems can provide valuable insights into consumer preferences, market trends, and operational efficiencies. This information allows farmers and stakeholders to make informed decisions about crop selection, resource allocation, and marketing strategies, ultimately leading to more resilient and adaptive agricultural practices (Bao et al., 2021).

### 10.3.4. Challenges in Implementing Traceability

While the future of traceability in agriculture holds great promise, several challenges must be addressed for successful implementation are follows:

- **Cost and Accessibility:** Implementing advanced traceability systems can be costly, particularly for small-scale farmers who may lack the resources to invest in new technologies. Ensuring that traceability solutions are accessible and affordable for all stakeholders in the supply chain is crucial for widespread adoption.
- **Data Privacy and Security:** As traceability relies heavily on data collection and sharing, concerns regarding data privacy and security must be addressed. Stakeholders need to establish clear protocols for data management to protect sensitive information while ensuring transparency.
- **Standardization and Interoperability:** The lack of standardization in traceability systems can hinder collaboration and information sharing among different stakeholders. Establishing common standards and ensuring that systems are interoperable can facilitate smoother communication across the supply chain.
- **Resistance to Change:** Cultural resistance and reluctance to adopt new technologies can impede the implementation of traceability systems. Stakeholders must be educated about the benefits of traceability and provided with support to transition to new practices.

### 10.3.5. Future Trends in Agricultural Traceability

As the agricultural landscape continues to evolve, several key trends are expected to shape the future of traceability in this sector are follows:

- **Consumer-Centric Approaches:** As consumers increasingly demand transparency, agricultural traceability will shift towards more consumer-centric approaches. Brands will likely invest in user-friendly platforms and mobile applications that allow consumers to access information about their food's journey, enhancing engagement and trust.
- **Integration of Blockchain and IoT:** The convergence of blockchain and IoT will likely enhance the effectiveness of traceability systems. By combining real-time data collection with secure and transparent record-keeping, stakeholders can create comprehensive and reliable traceability solutions that address multiple challenges.
- **Focus on Sustainability Metrics:** Future traceability systems will increasingly focus on sustainability metrics, tracking not just the

origins of products but also their environmental and social impacts. This shift will align with growing consumer demand for products that contribute positively to the planet and society.

- **Collaboration Across the Supply Chain:** Collaboration among stakeholders will become increasingly important as traceability systems evolve. Partnerships between farmers, processors, retailers, and technology providers will facilitate information sharing, enabling a more integrated approach to sustainability and traceability.
- **Regulatory Developments:** As governments and regulatory bodies implement stricter requirements for food safety and sustainability, traceability will become an essential component of compliance. Stakeholders must stay abreast of evolving regulations to ensure that their traceability systems align with legal requirements.

The future of traceability in agriculture is poised for transformation, driven by technological advancements, changing consumer expectations, and the urgent need for sustainable practices. By adopting robust traceability systems, agricultural stakeholders can improve food safety, enhance sustainability, build consumer trust, and gain a competitive edge in the marketplace.

While challenges remain, the opportunities presented by enhanced traceability are significant. As the agricultural sector embraces innovation and collaboration, traceability will become a cornerstone of responsible and sustainable food production. By leveraging technology and fostering transparency, the industry can address pressing global challenges, ensuring a more sustainable and resilient future for agriculture (Bao et al., 2021).

## 10.3.6. Emerging Technologies and Innovations

As the globe struggles with urgent environmental issues like resource depletion, biodiversity loss, and climate change, new technologies and innovations are essential to promoting sustainability in many fields. In addition to increasing productivity and cutting waste, these technologies enable people and organizations to embrace more environmentally friendly behaviors. Innovative technology integration is changing how we think about sustainability and building a more resilient future, from advanced agricultural practices to renewable energy sources.

### *10.3.6.1. Renewable Energy Technologies*

The spread of renewable energy technologies is one of the biggest developments in sustainability. Over the past 10 years, there have been significant cost savings and efficiency gains in solar and wind power. With the increasing affordability and availability of solar photovoltaic (PV) systems, homes and businesses can now produce their own electricity and reduce their dependency on fossil fuels. Similar to this, improvements in offshore wind farms and wind turbine design are maximizing wind energy harvesting and promoting cleaner energy grids.

To overcome the intermittent nature of renewable energy sources, new technologies such as energy storage systems, especially lithium-ion and flow batteries, are essential in addition to solar and wind power. These innovations improve the dependability of renewable energy by enabling excess energy produced during peak production to be stored and used during periods of high demand.

#### *10.3.6.2. Smart Agriculture*

The incorporation of smart technologies is causing a revolutionary change in agriculture. Precision farming optimizes farming practices by utilizing satellite imagery, IoT devices, and data analytics. Farmers can maximize yields while minimizing resource use by gathering up-to-date information on crop performance, weather, and soil health.

For instance, sensors can track the amount of moisture in the soil, allowing for water-saving targeted irrigation. Furthermore, food production in cities is being revolutionized by inventions like hydroponics and vertical farming. These techniques lessen the carbon footprint connected with transportation while requiring less land and water. Urban farming projects promote food security and sustainability by producing food closer to consumers.

#### *10.3.6.3. Circular Economy and Waste Management*

The notion of a circular economy is becoming more and more popular as businesses look to reduce waste and increase resource efficiency. This shift is being made possible by emerging technologies, which make it possible to manage waste and recycle more effectively. For example, cutting-edge sorting technologies driven by AI and machine learning can boost recycling facilities' productivity and guarantee that more materials are used again rather than being disposed of in landfills. Furthermore, advances in materials science are opening the door for more environmentally friendly product designs, such as biodegradable plastics and repurposed materials. Businesses are using LCAs more frequently to assess how their products affect the environment and encourage a move toward sustainable sourcing and production methods.

#### *10.3.6.4. Carbon Capture and Storage*

The technologies of carbon capture and storage (CCS) are becoming indispensable as the need to mitigate climate change grows. With the use of these technologies, carbon dioxide emissions from industrial processes can be captured and stored underground or used to make other products, like concrete. To meet global climate targets, CCS must be used to reduce the amount of $CO_2$ released into the atmosphere.

Furthermore, advances in direct air capture (DAC) technology are being developed to extract $CO_2$ directly from the atmosphere, offering an additional means of lowering carbon levels in the atmosphere. Even though these technologies are still in their infancy, they have a great deal of promise to help solve climate change's problems.

The sustainability movement is led by emerging technologies and innovations that provide answers to some of the most urgent environmental problems of our day. Society can make the shift to a more sustainable future through developing carbon capture technologies, encouraging circular economy principles, optimizing agricultural practices, and utilizing renewable energy. These technologies will enable people, organizations, and governments to make more responsible decisions as they develop and become more widely available, ultimately leading to a healthier planet for future generations. Because of our era's inventive spirit and technological advancements, achieving sustainability is becoming more and more feasible (Bao et al., 2021).

### 10.3.7. Predictions for Regulatory Changes and Market Demands

A major regulatory shift and a shift in market demands are coming for the agriculture sector as environmental issues become more widely recognized. The pressing need to address resource depletion and climate change, as well as rising consumer expectations for transparency, food safety, and sustainable practices, are the driving forces behind these changes. Farmers, producers, and legislators must adjust to the new demands of this changing environment to meet the demands of an expanding global population while also placing a premium on sustainability.

#### *10.3.7.1. Stricter Environmental Regulations*

Stricter environmental regulations are one of the most widely-cited predictions for the future of agriculture. Governments all across the world are realizing how important agriculture is to both biodiversity loss and climate change. Therefore, we should anticipate stricter laws designed to lower greenhouse gas emissions, enhance soil health, and protect water resources.

Regulations might, for example, mandate that farmers implement techniques like reduced tillage, agroforestry, and cover crops to facilitate carbon sequestration. To encourage the adoption of environmentally friendly practices, farmers who can demonstrate compliance with sustainability standards may be eligible for incentives and subsidies. Furthermore, restrictions on the use of fertilizers and pesticides will probably become stricter, encouraging the use of organic and regenerative farming practices that require less chemical input.

### *10.3.7.2. Increased Demand for Traceability*

Demands from the market are moving toward increased traceability and transparency in food production. Customers are becoming more curious about the provenance of their food, including the methods used in its production. Consequently, it is anticipated that regulatory frameworks will change to call for reliable supply chain traceability systems.

In addition to making it easier to comply with food safety regulations, these systems will give customers the information they need about environmentally friendly practices and ethical sourcing. Enhancing traceability will be made possible by technologies like blockchain and the Internet of Things, which will allow stakeholders to follow products from farm to table. Producers will be compelled to invest in technologies that enable real-time monitoring and reporting of agricultural practices as a result of this demand for transparency.

### *10.3.7.3. Emphasis on Sustainable Practices*

Demand for goods made sustainably is predicted to soar as consumer awareness of sustainability issues rises. Consumers are becoming more and more prepared to pay more for goods that are consistent with their values, such as locally sourced, organic, and ethically produced goods. This change will encourage farmers and producers to use sustainable practices to both comply with regulations and take advantage of market opportunities.

We might witness an increase in certifications and labels—like organic certification or regenerative agriculture labels—that attest to sustainable farming methods in response to this demand. These certifications will reward producers who prioritize sustainability while also assisting consumers in making well-informed decisions. In addition, food brands and retailers will probably create their own sustainability guidelines to set themselves apart in a market that is becoming more and more competitive.

### *10.3.7.4. Integration of Technology*

The incorporation of technology will also impact agriculture's sustainability in the future. Farmers will be empowered by advances in artificial intelligence, data analytics, and precision agriculture to maximize resource efficiency and reduce waste. Regulations are starting to take notice of how technology can promote sustainability, so we should anticipate more funding for this kind of research and development.

Incentives and subsidies are likely to be available to farmers who implement technology-driven practices, enabling them to align their operations with both regulatory requirements and market demands. Drones that monitor crops and automated irrigation systems are two examples of efficiency-enhancing technologies that will become indispensable for farmers trying to reach sustainability targets.

Significant regulatory changes and market demands focused on sustainability are about to occur in the agricultural sector. In the upcoming years, the agricultural landscape will change due to stricter environmental regulations, a greater emphasis on sustainable practices, and an increased demand for traceability. Farmers and producers must adjust to satisfy consumers' demands for ethical sourcing and transparency while also adhering to changing legal requirements. The agricultural sector can prosper in this new environment and help create a healthier planet for coming generations by embracing technology and sustainable practices. The way forward is obvious: both consumer preferences and regulatory frameworks will push sustainability as a core idea directing agricultural practices.

### 10.3.8. Evolving Consumer Expectations and Trends

Consumer expectations have changed significantly in the last few years, especially with regard to ethical and sustainable practices in the food and agriculture industries. Customers expect brands they support to be more transparent, accountable, and responsible as environmental issues become more widely known. This development is indicative of a larger social movement towards sustainability, motivated by worries about social justice, resource depletion, and climate change.

#### *10.3.8.1. Demand for Transparency*

The increasing demand for food production transparency is one of the most notable trends. Customers are curious about the origins of their food, its production process, and its effects on the environment and nearby communities. Beyond just basic sourcing data, this desire for transparency encompasses supply chain management, farming methods, and ingredient sourcing information. Therefore, companies that are able to give consumers lucid, substantiated information about their products stand a better chance of earning their confidence and allegiance.

### *10.3.8.2. Preference for Sustainable Practices*

Products that reflect the values of today's consumers are becoming more and more popular, especially those that put sustainability first. This includes having a preference for products that are made ethically, locally, and organically. Customers are more prepared to spend more for goods that support ethical labor practices and have a minimal negative impact on the environment. This is not just a trend that affects individual customers; big retailers and food service businesses looking to improve their sustainability credentials are also seeing it in their purchasing decisions.

### *10.3.8.3. Health and Wellness Focus*

The relationship between health and sustainability is showing up more and more in consumer behavior. Many consumers believe that sustainable products are healthier for their families and themselves in addition to being better for the environment. The demand for clean-label products—those devoid of artificial additives and preservatives—has increased as a result. There is a market for goods that support environmental stewardship as well as individual health because consumers are seeking out nutrient-dense, ethically produced foods.

### *10.3.8.4. Rise of Plant-Based and Alternative Proteins*

The increasing acceptance of plant-based diets and alternative protein sources is another noteworthy trend. Growing numbers of consumers are looking into plant-based options due to worries about the environmental impact of meat production. The growing availability of plant-based goods and meat alternatives, which are gaining popularity among mainstream consumers, reflects this change. Companies that lead the way in this area and provide scrumptious and healthy substitutes have a good chance of taking a sizable chunk of the market.

### *10.3.8.5. Social and Environmental Responsibility*

Additionally, consumers are giving social and environmental responsibility more weight. Consumers of today are more likely to connect with brands that exhibit a dedication to corporate social responsibility and sustainable practices. This includes programs that encourage biodiversity, lessen carbon footprints, and support fair trade. Businesses can improve their brand reputation and customer loyalty by actively participating in sustainability initiatives, such as cutting back on packaging waste or investing in renewable energy.

Brands in the food and agriculture industries need to adjust to the rising demand for sustainability and ethical practices as consumer expectations continue to change. Consumer decision-making is increasingly influenced by factors like social responsibility, health, and wellness, sustainable sourcing, and transparency. Businesses can meet customer expectations and set themselves up for long-term success in a market that is becoming more and more competitive by adopting these trends and exhibiting a sincere commitment to sustainability. These changing consumer demands will surely influence food and agriculture in the future, leading to a more sustainable and accountable sector (Bao et al., 2021).

# REFERENCES

1. Al Sadawi, A., Madani, B., Saboor, S., Ndiaye, M., & Abu-Lebdeh, G. (2021). A comprehensive hierarchical blockchain system for carbon emission trading utilizing blockchain of things and smart contract. *Technological Forecasting and Social Change, 173*, 121124.
2. Bao, F., Xu, W., Feng, Y., & Xu, C. (2022). A topic-rank recommendation model based on microblog topic relevance & user preference analysis. *Human-Centered Computing and Information Sciences, 12*, 10.
3. Bao, J., He, D., Luo, M., & Choo, K. K. R. (2021). A survey of blockchain applications in the energy sector. *IEEE Systems Journal, 15*, 3370–3381.
4. Ju, C., Li, G., Bao, F., Gao, T., & Zhu, Y. (2022). Social relationship prediction integrating personality traits and asymmetric interactions. *Frontiers in Psychology, 13*, 778722.
5. Ju, C., Shen, Z., Bao, F., Wen, Z., Ran, X., Yu, C., & Xu, C. (2022). Blockchain traceability system in complex application scenarios: Image-based interactive traceability structure. *Systems, 10*, 78.
6. Ju, C., Shen, Z., Bao, F., Weng, P., Xu, Y., & Xu, C. (2022). A novel credible carbon footprint traceability system for low carbon economy using blockchain technology. *International Journal of Environmental Research and Public Health, 19*(16), 10316. https://doi.org/10.3390/ijerph191610316.
7. Li, M., Zhu, L., & Lin, X. (2019). Efficient and privacy-preserving carpooling using blockchain-assisted vehicular fog computing. *IEEE Internet of Things Journal, 6*, 4573–4584.
8. Marks, J. (2019). Distributed ledger technologies and corruption: The killer app? *Science, Technology, and Law Review, 20*, 42.
9. Muzumdar, A., Modi, C., & Vyjayanthi, C. (2022). A permissioned blockchain enabled trustworthy and incentivized emission trading system. *Journal of Cleaner Production, 349*, 131274.
10. Wang, M., Wang, B., & Abareshi, A. (2020). Blockchain technology and its role in enhancing supply chain integration capability and reducing carbon emission: A conceptual framework. *Sustainability, 12*, 10550.
11. Xuan, D., Ma, X., & Shang, Y. (2020). Can China's policy of carbon emission trading promote carbon emission reduction? *Journal of Cleaner Production, 270*, 122383.
12. Zhang, Y., Li, S., Luo, T., & Gao, J. (2020). The effect of emission trading policy on carbon emission reduction: Evidence from an integrated study of pilot regions in China. *Journal of Cleaner Production, 265*, 121843.

# CHAPTER 11

# FUTURE OF TRACEABILITY IN AGRICULTURE

## Contents

# INTRODUCTION

The world's agricultural sector is highly dynamic and intricate. To meet the rising demand for food, efficiency has been optimized in modern agriculture and food production. They now need to be rearranged to satisfy the demands for sustainability and traceability. Agriculture companies and other businesses in the food industry need to be aware of the direction that technologies are taking in different processes as regulatory bodies like the FDA work to improve food safety mitigation and traceability efforts. The utilization of AI, IoT, and farm management software is expected to enhance these businesses' capacity to identify their products within supply chain networks. It will also assist them in demonstrating to customers the traceability, acknowledgment, and environmental impact of sustainable practices.

One of the most desired elements in the agriculture industry is transparency. Now more than ever, supply chain traceability is essential due to the introduction of modern technology into operations. Embracing innovation and finding new collaborations should not be just intended for profit making but to work towards a future where:

- All crop protection products and seeds are traceable from farm to fork.
- Farmers have access to detailed information about the products they use.
- Consumers can make informed choices about the food they eat.
- Businesses can operate more efficiently and sustainably.

Investing in traceability solutions is a commitment to create a more equitable, transparent, and sustainable agricultural ecosystem that benefits all parties involved, not just a technical advancement (FSMA, 2023).

## 11.1. EMERGING TECHNOLOGIES AND INNOVATIONS

The ability to monitor and trace the production, processing, and distribution of agricultural products along the entire supply chain is referred to as traceability in the agricultural industry.

Ensuring food safety, quality control, and regulatory compliance all depend on this capability. Because they offer transparency and real-time data, emerging technologies are significantly improving agriculture's traceability. The major developments that are revolutionizing traceability in this industry are listed below.

### 11.1.1. Blockchain Technology

A decentralized and secure method of recording transactions between various agricultural supply chain participants is provided by blockchain technology. An immutable ledger is created by linking each transaction to a block of data that is stored. Thanks to technology, everyone who is involved—from farmers to retailers—can obtain precise information regarding the origin, processing, and transportation of agricultural goods. Blockchain enables stakeholders to swiftly locate contamination sources or quality problems, enhancing consumer confidence and food safety.

#### *11.1.1.1. Benefits of Using a Blockchain Food Traceability System*

##### *11.1.1.1.1. Accurate and Tamper-resistant Food Data*

The food data is extremely accurate because Verifiable Credentials are impenetrable due to the security of cryptography. Requests to view verified data containing information about food products, including quantity received, organic certification, farming community, safety standard certifications, and so forth, can be made by any pertinent party in the supply chain. A record cannot be altered once it has been entered. Anytime a relevant transaction occurs, a Verifiable Credential can be issued using a blockchain food traceability system. Any substantial interaction between parties in a supply chain, such as the sharing of product information or the transfer of ownership or custodianship, can be considered a transaction.

##### *11.1.1.1.2. Efficiently Prevent, Contain, or Rectify Contamination Outbreaks while Reducing the Loss of Revenue*

The immutable and verifiable data gathered over time allows blockchain food traceability systems to rapidly identify contamination points using Verifiable Credentials. When a food disease pathogen—an organism that can cause disease—is discovered, the company can use blockchain technology to swiftly identify the source and batch, allowing for an instant recall of the contaminated batch instead of discarding the entire product inventory.

In the past, consumers and businesses had to discard the entire inventory when a food product became contaminated, which led to significant waste and financial loss. Due to an *E. coli* outbreak in California, spinach was thrown out of supermarkets nationwide for two weeks, costing California farmers a total of US$74 million. Every year, food-borne illness costs low and middle-income nations $110 billion.

### *11.1.1.1.3. Creates More Transparency and Trust in the Authenticity of Verifiable Credential Data*

Without using a middleman or getting in touch with the data's issuer, it was impossible to effectively confirm the authenticity of data using traditional supply chain recordkeeping systems. There is no need to rely on a centralized party to verify food data because blockchain food traceability systems can efficiently verify tamper-resistant food data. Despite their lack of mutual knowledge, stakeholders can transact business with each other in a safe and efficient manner thanks to blockchain technology.

### *11.1.1.1.4. Secure and Efficient Data Transfer Between Parties*

Participants can share information quickly and effectively when it is digitized and in an unchangeable format. Delays, waste, and other inefficiencies that cost money and time are reduced when data can be verified effectively. Stakeholders have access to information about every transaction involving a particular food product, which makes the system as a whole more accountable and simpler to watch.

### *11.1.1.1.5. Verifiable Credentials Help Eliminate and Prevent Fraud*

People can readily track product information once food data has been verified and issued as a Verifiable Credential. Policymakers and other pertinent parties can keep an eye on harmful activities like deforestation and other environmental harms when food data is visible.

### *11.1.1.1.6. Helps Organizations Comply with Food Regulations*

Systems for blockchain-based food traceability can assist businesses in adhering to food safety laws. For instance, the Food and Drug Administration (FDA) encouraged trading partners to increase the use of blockchain technology to track a product back to its source in 2020 by proposing a New Era of Smarter Food Safety Blueprint to improve food traceability.

### *11.1.1.1.7. Customers can Instantly See Information About Food Origins and Product Details Which Can Enhance the Reputation of Companies*

Global consumer demand for organic, slave-free, and sustainably sourced food products is rising. Additionally, they are requesting that businesses disclose the food's origins more openly, and consumers want to know that the food was produced by individuals under favorable working conditions. According to a study on global consumer trends, 79% of consumers today believe that when purchasing products, it is crucial for brands to offer certifications or other guarantees of authenticity. 71% of those in this group are prepared to pay more for businesses that provide complete traceability and transparency. As a result of these demands, an increasing number of businesses are allowing clients to simply scan a QR code to view information about the provenance of food products as well as other details verified by blockchain. Videos about the farming community that produced the food, the methods used in its harvest, and organic certification are a few examples of the information they can view.

### *11.1.1.2. Key Considerations When Implementing Blockchain Food Traceability Systems*

These are some crucial factors to take into account once you've made the decision to integrate blockchain technology into your food supply chain.

#### *11.1.1.2.1. Clarify the Objectives of Implementing a Blockchain Food Traceability System*

All stakeholders must understand the goals and driving forces behind improving traceability. Here are a few instances of objectives:

- Monitoring for deforestation and human rights abuses;
- Ensuring food quality and safety standards across the supply chain; and
- Enhancing transparency across the supply chain.

#### *11.1.1.2.2. Define the Scope of The Project and Specify the Segment of The Supply Chain That the Blockchain Solution Will First Be Applied To*

A proof of concept, or minimal viable product, can be created before moving on with the development process. An architectural blueprint outlining how blockchain will integrate with current systems from both an operational and cultural standpoint should be drafted for the future phases to facilitate the transition as smoothly as possible in the long run.

#### *11.1.1.2.3. Establishing Data Governance*

Before implementation, a solid data governance system and data practices must be put in place. Maintaining data accuracy will be aided by formalizing data practices and roles. Here are some important queries to think about during this phase of the procedure:

- Who makes the decisions about rules, access of data, and the process for members to join and leave?
- Who makes decisions to update the platform?
- At what stage should a product be registered in the system?
- When should a product be registered in the system?
- **Registering a Product:** Determining the process for assigning unique digital identifiers (DIDs) to new products is crucial. Products can be added in three general categories:
  - **Class level:** Identifies only one type of product;
  - **Lot level:** Identifiers a batch of products (e.g., a batch of lettuce); and
  - **Instance Level:** Every individual item is given a digital ID.

### *11.1.1.2.4. Identify Key Actors and Areas in The Supply Chain That Require Decentralized Identifiers (DIDs)*

Participants in the food supply chain, such as small-scale primary producers, shippers, and wholesalers, are referred to as actors. DIDs are probably also required for trading partners, physical locations, and products at any stage of packaging in the supply chain.

### *11.1.1.2.5. Detail the Traceability Flow between Actors by using Verifiable Credentials*

Documenting the flow that occurs during each interaction between actors and determining which fields of each transaction need to be added to the Verifiable Credential are essential for efficiently tracking the path a product takes from its origin to the customer.

As you're planning these details, here are some questions to consider:

- Who will execute the transaction?
- What products will be created or exchanged?
- Why will the transaction happen?
- Where will the transaction occur?
- When will the transaction happen?

### *11.1.1.2.6. Align Stakeholder Interests*

The next steps after defining the project's scope are to find pertinent stakeholders, confer with them, and form partnerships with them. To boost their involvement in the transition process, it is a good idea to take the time to learn about the advantages of being a part of the blockchain food traceability system and to get their feedback (FSMA, 2023; Table 9.1).

**Table 11.1.** *Overview of Real-Life AI–Blockchain Projects*

| Token | Market Cap | Blockchain Type | Layer | Service(s) | Consensus |
|---|---|---|---|---|---|
| AGI | USD 35.4M | Ethereum | Layer 3 | Decentralized AI Services Marketplace | PoS on Ethereum, exploring dPoS on Cardano |
| ENJ | USD 5.12B | Ethereum | Layer 3 | NFT Ecosystem, Gaming Integration | PoS |
| HBAR | $5.97B | Hedera Hashgraph | Layer 1 | Decentralized Consensus and Smart Contracts | aBFT |
| RNDR | $4.15B | Solana | Layer 1 | Decentralized GPU Rendering and AI Computation | PoH, Solana's unique consensus mechanism |
| FET | $175M | Cosmos | Layer 1 | Decentralized Network of Autonomous Agents for AI | PoS |
| AGIX | $1.49B | Multi-chain | Layer 3 | Decentralized AI Services Marketplace | PoS on Ethereum, exploring dPoS on Cardano |
| OCEAN | $631M | Ethereum | Layer 3 | Decentralized Data Exchange | PoS |
| FIL | $5.26B | Filecoin | Layer 1 | Decentralized Storage Network | PoRep |
| LINK | $12.26B | Ethereum | Layer 3 | Decentralized Oracle Network | Not directly applicable; utilizes external data validation by decentralized oracles |
| CTXC | $102M | Cortex | Layer 1 | AI on Blockchain | Not specified; focuses on on-chain AI execution |
| TAO | N/A | Bittensor | Layer 3 | Decentralized ML | PoI |
| RLC | $149M | Ethereum | Layer 3 | Decentralized Cloud Computing | PoS |

| Token | Market Cap | Blockchain Type | Layer | Service(s) | Consensus |
|---|---|---|---|---|---|
| GLM | $93.5M | Ethereum | Layer 3 | Decentralized Data Market-place | PoS |
| ICP | $3.99B | Internet Computer | Layer 1 | Decentralized Internet | Threshold Relay (PoW-based) |
| CGPT | N/A | Ethereum | Layer 3 | AI Language Model Utility | PoS |
| AKT | $97.6M | Cosmos | Layer 1 | AI-Based Investment Management | PoS |
| THETA | $2.98B | Theta Net-work | Layer 1 | Decentralized Video Streaming | Multi-level BFT consensus (PoS-based) |
| AIOZ | N/A | Ethereum | Layer 3 | Decentralized Video Streaming | PoS |
| MANA | $2.91B | Ethereum | Layer 3 | Virtual Reality Platform | PoS |
| GNT | N/A | Ethereum | Layer 3 | Decentralized Computing Power Market-place | PoS |
| NU | $379M | Ethereum | Layer 3 | Decentralized Encryption and Privacy | PoS |
| DAGT | N/A | Ethereum | Layer 3 | Decentralized Ecosystem | PoS |

*Source: Bhumichai D, Smiliotopoulos C, Benton R, Kambourakis G, Damopoulos D. The Convergence of Artificial Intelligence and Blockchain: The State of Play and the Road Ahead. Information. 2024; 15(5):268. https://doi.org/10.3390/info15050268.*

### *11.1.1.3. Governments Regulations on Food Traceability and Use of Blockchain*

Regulators and legislators are becoming more interested in traceability as a means of guaranteeing a resilient, sustainable, and safe food supply. Food companies are currently required by regulations in many Western markets to maintain records of food stages, including the steps that food products take from their place of origin to the market.

Expert in food safety, Scott Haskell holds a PhD and a doctorate in veterinary medicine. He said that the US FDA has been working toward complete product traceability since the passage of the Food Safety Modernization Act (FSMA). "Instructs the FDA to develop additional recordkeeping requirements for certain foods to help establish clear tracing of a food product's source when needed to address food safety risks," reads one of the FSMA's requirements in the section titled "Enhancing Tracking and Tracing of Food and Recordkeeping." Traceback is required by another section of the Animal Disease Traceability regulation. Blockchain technology can assist in quickly tracing the origins of a product to solve the root cause of the issue.

To boost blockchain adoption, the European Food Safety Authority is adding trace labs to the Food Safety Market program. "The FSM wants to use blockchain technology and the strength of big data to revolutionize the European food certification market. To guarantee food safety and quality, Australia is implementing blockchain-based food traceability systems. "Australia's agricultural industry was a pioneer in the development and application of provenance and traceability systems to prove product safety. To create secure and reliable connections between the primary producer and the end-user, producers are currently implementing emerging technologies."

## 11.1.2. Internet of Things (IoT)

Using a network of connected devices and sensors, the IoT gathers data on agricultural products in real time as they travel through the supply chain. RFID tags, for instance, can be affixed to goods to track their whereabouts, temperature, humidity, and other parameters while they are being transported and stored. All parties participating in the supply chain have access to this data, which empowers them to decide how best to handle products and guarantee that safety regulations are followed.

### *11.1.2.1. How IoT Enhances Traceability*

- **Real-Time Data Collection:** Throughout the agricultural process, from planting to harvesting, processing, and distribution, IoT devices like sensors and RFID tags can be used. These gadgets gather data in real time on a variety of parameters, including soil health, temperature, and humidity. To guarantee that products are grown and stored under ideal conditions, this data is essential.
- **Automated Reporting:** The of IoT enables automated reporting systems that offer real-time updates on the condition of agricultural goods. This implies that any departure from predetermined safety guidelines can be reported right away for examination or remedial action.

- **Supply Chain Visibility:** Farmers, processors, distributors, and retailers are just a few of the supply chain participants who can obtain comprehensive data about individual product batches through IoT-enabled tracking systems. This transparency makes it easier to spot possible problems or hazards at any point in the supply chain.
- **Information Access for Consumers:** Internet of Things technology also gives consumers more power by giving them access to information about the provenance and path of the food products they buy. Customers can see where their food comes from and how it has been treated throughout its lifecycle by using apps or websites that are connected to IoT systems.
- **Fraud Prevention:** By separating genuine goods from fakes, digital identification marks provided by IoT systems aid in the fight against food fraud. This helps producers maintain their brand integrity while also protecting consumers.
- **Reaction to Food Safety Issues:** Having a strong traceability system driven by IoT enables businesses to promptly identify impacted batches and take required action without protracted delays in the event of a food safety issue or recall.
- **Reduction of Food Waste:** IoT can dramatically lower food waste, a crucial problem for global food security, by enhancing traceability through real-time monitoring and alerts about product conditions during transportation and storage (FSMA, 2023).

## 11.1.3. Artificial Intelligence (AI) and Machine Learning

Large volumes of data gathered from diverse sources within the agricultural supply chain can be analyzed by machine learning algorithms and AI. Patterns of product quality, shelf life, and possible hazards connected to particular produce batches can be found using these technologies. Businesses can improve their traceability efforts by anticipating possible problems before they occur and tailoring their response plans appropriately by utilizing AI-driven analytics.

### *11.1.3.1. Benefits of AI in Agriculture*

The terms AI and agriculture may have seemed like an odd combination until recently. After all, while even the most basic AI has only recently been developed, agriculture has been the foundation of human civilization for millennia, supplying both sustenance and fostering economic growth. All industries, however, are introducing novel ideas, and agriculture is no different. The world has seen a rapid advancement in agricultural technology in recent years, which has revolutionized farming practices. The sustainability of our food system is being threatened by global issues like population growth, climate change, and resource scarcity, so these innovations are becoming more and more important. The introduction of AI helps to reduce the negative aspects of traditional farming and solves many of its problems.

#### *11.1.3.1.1. Data-based Decisions*

Data is everything in the modern world. Data is used by organizations in the agricultural sector to gain detailed insights into every aspect of farming, from comprehending each acre of a field to tracking the entire supply chain for produce to gaining profound insights into the process of yield generation. Predictive analytics driven by AI is already opening doors for agribusinesses. With AI, farmers can collect and process more data faster. AI is also capable of forecasting prices, analyzing market demand, and identifying the best times to plant and harvest.

In agriculture, artificial intelligence can support soil health investigations to gather data, track meteorological conditions, and suggest fertilizer and pesticide applications. Farm management software helps farmers make better decisions at every step of the crop-cultivation process, increasing productivity and profitability simultaneously.

#### *11.1.3.1.2. Cost Savings*

Farmers are always looking to increase farm yields. Precision agriculture, when paired with AI, can assist farmers in producing more crops using fewer resources. AI in farming maximizes yields while minimizing costs by combining the best data management techniques, variable rate technology, and soil management techniques. Farmers can determine which areas require pesticide treatment, fertilization, or irrigation by using real-time crop insights from AI applications in agriculture. While using fewer resources, innovative farming techniques like vertical agriculture can also boost food production and lead to significant cost savings, improved harvest quality, increased profits, and a decrease in the use of herbicides.

#### *11.1.3.1.3. Automation Impact*

Since agricultural labor is labor-intensive, labor shortages are not new. Fortunately, automation offers a solution without requiring additional staffing. A new wave of digital automation is once again transforming the agricultural sector, just as mechanization did when it came to tasks that required superhuman sweat and draft animal labor to become jobs that could be completed in a few hours. Examples of automated farm equipment include driverless tractors, intelligent irrigation and fertilization systems, IoT-enabled agricultural drones, intelligent spraying, software for vertical farming, and AI-powered greenhouse robots for harvesting. AI-driven tools are far more accurate and efficient than any human farm worker.

### *11.1.3.2. Applications of Artificial Intelligence in Agriculture*

Numerous manual tasks are involved in traditional farming. There are numerous benefits to using AI models in this regard. An intelligent agriculture system can make a lot of tasks easier by enhancing already-implemented technologies. Big data can be gathered and processed by AI, which can then decide on the best course of action and start it. These are a few typical applications of AI in agriculture.

#### *11.1.3.2.1. Optimizing Automated Irrigation Systems*

AI systems make autonomous crop management possible. In conjunction with IoT sensors that track soil moisture content and meteorological conditions, algorithms can determine in real-time how much water crops should receive. Water conservation is a goal of autonomous crop irrigation systems, which also support sustainable farming methods. By automatically adjusting temperature, humidity, and light levels based on real-time data, AI in smart greenhouses maximizes plant growth.

#### *11.1.3.2.2. Detecting Leaks or Damage to Irrigation Systems*

Artificial Intelligence is essential for identifying irrigation system leaks. Algorithms can detect trends and abnormalities in data that point to possible leaks. It is possible to train ML models to identify particular leak signatures, like variations in water pressure or flow. Early detection is made possible by real-time monitoring and analysis, which helps to avoid both possible crop damage and water waste. To detect regions with excessive water usage, AI also integrates weather data with crop water requirements. AI improves water efficiency, assisting farmers in resource conservation by automating leak detection and sending out alerts.

#### *11.1.3.2.3. Crop and Soil Monitoring*

The health and growth of crops can be significantly impacted by the incorrect combination of nutrients in the soil. With AI, farmers can easily identify these nutrients and determine how they affect crop yield, allowing for necessary adjustments to be made. While the accuracy of human observation is limited, computer vision models can monitor soil conditions and collect precise data that is required to fight crop diseases. After that, crop health and yield predictions are made using this data from plant science, along with a flag for any specific problems. Through sensors that identify their growing conditions, plants activate AI systems that then automatically modify their surroundings. In actuality, AI) in agriculture and farming has the ability to precisely track the stages of wheat growth and the ripeness of tomatoes with a speed and accuracy that is unmatched by humans.

### *11.1.3.2.4. Detecting Diseases and Pests*

Computer vision can identify pests or diseases in addition to crop growth and soil quality. In agriculture projects, AI is used to scan images for insects, mold, rot, and other crop health hazards. This aids farmers in taking prompt action to eradicate pests or isolate crops to stop the spread of disease when combined with alert systems. Apple black rot can be detected with over 90% accuracy using AI technology in agriculture. Insects like flies, bees, moths, and others can also be identified by it, with the same level of precision. To have the required size of the training data set to train the algorithm with, researchers had to first gather pictures of these insects.

### *11.1.3.2.5. Monitoring Livestock Health*

While it might seem simpler to identify health issues in livestock than in crops, doing so is actually very difficult. Fortunately, farming AI can assist with this. For instance, a business by the name of CattleEye has created a system that allows for remote cattle health monitoring using cameras, drones, and computer vision. It recognizes abnormal cow behavior and distinguishes between events like giving birth. CattleEye provides useful insights by analyzing the effects of environmental factors and diet on livestock through the use of AI and ML technologies. With this information, farmers can enhance the health of their cattle and boost milk yield.

### *11.1.3.2.6. Intelligent Pesticide Application*

Farmers are already well aware that there is an opportunity to optimize the use of pesticides. Regretfully, there are significant drawbacks to both automated and manual application processes. Although it may be labor-intensive and slow, manually applying pesticides allows for greater precision in addressing particular areas. Although automated pesticide spraying is faster and requires less work, it frequently lacks accuracy, which can contaminate the environment. Drones with AI capabilities combine the best features of each strategy without sacrificing any of its disadvantages. Computer vision is used by drones to calculate how much pesticide should be sprayed on each area. Even though this technology is still in its infancy, it is getting increasingly accurate.

### *11.1.3.2.7. Yield Mapping and Predictive Analytics*

Yield mapping analyzes massive datasets in real-time using machine learning algorithms. This facilitates improved planning by assisting farmers in comprehending the trends and traits of their crops. Using methods such as 3D mapping, sensor data, and drone data, farmers can forecast soil yields for particular crops. Multiple drone flights gather data, allowing for more accurate analysis through the application of algorithms. By accurately predicting future yields for particular crops, these techniques assist farmers in determining when and where to plant seeds and how best to allocate resources for maximum return on investment.

### *11.1.3.2.8. Automatic Weeding and Harvesting*

Computer vision can be used to identify invasive plant species and weeds, just as it can identify diseases and pests. Computer vision uses leaf size, shape, and color analysis in conjunction with machine learning to differentiate crops from weeds. Robots that perform robotic process automation (RPA) tasks, like automatic weeding, can be programmed using such solutions. Indeed, there has already been successful use of such a robot. As these technologies become more widely available, intelligent bots may eventually be able to perform both crop harvesting and weeding.

### *11.1.3.2.9. Sorting Harvested Produce*

AI can be used for more than just spotting possible problems with crops as they grow. It also plays a part after the harvest of produce. The majority of sorting procedures are currently done by hand, but AI can sort produce more precisely. In harvested crops, computer vision can identify diseases and pests. Additionally, it can grade produce according to its size, color, and shape. This makes it possible for farmers to swiftly classify their produce so that it can be sold to various clients at various prices. Traditional manual sorting techniques, on the other hand, can be extremely time-consuming.

### *11.1.3.2.10. Surveillance*

An essential component of farm management is security. Given how difficult it is for farmers to keep an eye on their fields around the clock, farms are frequently targeted by burglars. Animals also pose a threat, whether it be foxes invading the chicken coop or livestock owned by the farmer causing damage to crops or machinery. Computer vision and machine learning in conjunction with video surveillance systems enable the prompt detection of security breaches. Certain systems are even sufficiently sophisticated to differentiate between unauthorized guests and employees (FSMA, 2023).

### *11.1.3.3. Role of AI in the Agriculture Information Management Cycle*

Managing agricultural data with AI can be beneficial in many ways:

- **Risk Management:** By using predictive analytics, farming processes become less error-prone.
- **Plant Breeding:** AI further advises on crops that are more resistant to severe weather, disease, or dangerous pests by using plant growth data.
- **Analysis of Crop and Soil Health:** Artificial intelligence algorithms are capable of analyzing the chemical makeup of soil samples to identify any possible nutrient deficiencies. Crop diseases can also be identified or even predicted by AI.
- **Crop Feeding:** AI in irrigation can be used to predict the ideal combination of agronomic products as well as optimal patterns and timings for nutrient application.
- **Harvesting:** AI can help increase crop yields and even forecast when it is best to harvest crops.

### *11.1.3.4. Optimizing AI for Agriculture and Agricultural Processes*

Even though AI has many advantages in agriculture, big data, sensors, and software are still necessary for AI to work. Similarly, AI is necessary for other technologies to function properly. When it comes to big data, the data itself is not very helpful. How it is handled and put into practice is what counts.

- **Big Data for Informed Decision-Making:** By fusing AI with big data analytics, farmers can receive recommendations based on precise, up-to-date information, which boosts output and lowers expenses.
- **IoT Sensors for Data Collection and Analysis:** IoT sensors can track, measure, and store training data on a variety of metrics in real-time when combined with other auxiliary technologies (AI drones, GIS, and other tools). Farmers are able to rapidly obtain precise information by integrating these devices with AI and farming.
- **Robotics and Intelligent Automation to Reduce Manual Labor:** AI, IoT, and autonomous tractors work together to address the widespread issue of labor scarcity. Agricultural robots are currently employed for manual tasks like picking produce, demonstrating the importance of robotics in agriculture. Robots are more useful for farmwork because they can work longer hours, have better precision, and are less prone to mistakes.

## 11.1.4. Mobile Applications

Distributors and farmers can easily access traceability information at any stage of the supply chain with the help of mobile applications. These applications let users get comprehensive details about a product's origin, manufacturing processes, and certifications by scanning the barcodes or QR codes on the product. This degree of transparency helps businesses meet legal requirements while also empowering customers by educating them about the products they are buying.

### *11.1.4.1. Features of Mobile Applications for Traceability*

Mobile applications designed for agriculture traceability typically include several key features:

- **Data Collection:** Using their mobile devices, farmers can enter information about crop treatments, harvest dates, and locations.
- **Real-Time Monitoring:** A lot of apps let users keep an eye on things like soil moisture levels and weather patterns that have an impact on the health of crops.
- **Reporting Tools:** These programs frequently have reporting capabilities that assist farmers in producing the paperwork required for compliance checks or audits.
- **Integration with Other Systems:** To expedite data collection, certain apps can integrate with hardware (such as sensors) or software for farm management that is already in place.

### *11.1.4.2. Examples of Mobile Applications for Agriculture Traceability*

Several mobile applications are specifically designed to enhance traceability in agriculture:

- **Producepak:** This app offers complete audit capabilities and quality control measures, automating produce traceability from farm to fork. It assists farmers in keeping thorough records about geographic locations, fertilization, pesticide use, and pest protection.
- **Agritrace:** This app gives agricultural workers more power by making it simple for them to use mobile devices to track their farms. It makes keeping records pertaining to production procedures easier.
- **Cropin SmartRisk:** It is an AI-powered platform that tracks crop performance over time and provides predictive analytics to farmers to help them manage climate change risks.

#### *11.1.4.3. Benefits of Using Mobile Applications for Traceability*

The adoption of mobile applications in agriculture brings numerous benefits:

- **Efficiency Gains:** Farmers can save time and cut down on errors related to manual data entry by digitizing their record-keeping and monitoring procedures.
- **Improved Decision-Making:** Farmers can swiftly and intelligently decide on crop management techniques when they have access to real-time data.
- **Cost Reduction:** Better resource management brought about by increased efficiency frequently results in lower operating costs.

### 11.1.5. Cloud Computing

Traceability data can be centrally stored and accessed by numerous stakeholders from various locations at the same time thanks to cloud computing. By facilitating easy information sharing, this technology promotes cooperation between farmers, processors, distributors, retailers, and regulators.

By implementing cloud-based systems, businesses can guarantee that everyone involved has access to the most recent data about the state of their products at every stage of their lifecycle.

#### *11.1.5.1. How Cloud Computing Enhances Traceability*

- **Data Aggregation and Storage:** Large volumes of data from multiple sources, including sensors, drones, and Internet of Things devices, can be gathered and stored thanks to cloud computing. Information about crop growth stages, weather patterns, soil health, and pest control techniques can all be included in this data. Farmers can readily access historical data, which is essential for monitoring the lifecycle of agricultural products, by centralizing this information in the cloud.
- **Real-Time Monitoring:** Farmers can keep an eye on their livestock and crops in real-time with cloud-based systems. This feature makes it possible to react quickly to any problems that might come up during production. For example, farmers can act quickly to minimize damage if sensor data uploaded to the cloud identifies a pest infestation.
- **Supply Chain Transparency:** From producers to retailers, cloud computing makes it easier for all parties involved in the agricultural supply chain to communicate with one another. By giving current information on the state of the product (e.g., where it is in transit),

cloud platforms improve accountability and transparency. Customers who want to know the origins and production methods of the food they buy depend on this transparency.

- **Regulatory Compliance:** Food safety and traceability are subject to strict regulations in many areas. By keeping thorough records of each stage of the production process, cloud computing assists farmers in adhering to these regulations. These documents are readily available for use in audits and inspections.
- **Consumer Engagement:** Because of sustainability and health concerns, consumers are becoming more interested in learning more about the sources of their food. Through QR codes or smartphone apps connected to cloud databases, cloud-based traceability systems give customers access to details about the provenance of their food items.
- **Data Analytics for Decision Making:** Farmers can examine trends over time about crop yields, pest occurrences, and market demands by integrating cutting-edge analytics tools into cloud platforms. Making well-informed decisions that increase productivity and traceability is made easier with the help of this analysis.
- **Cooperation Among Stakeholders:** By offering a common platform for data sharing, cloud computing promotes cooperation among farmers, agronomists, researchers, and legislators. By working together, agricultural practices can be improved and guarantee product quality all the way through the supply chain.
- **Sustainability Practices:** Farmers can implement more sustainable practices by closely monitoring the use of resources (such as water and fertilizers) and making necessary adjustments based on real-time data insights by leveraging cloud technology for traceability (FSMA, 2023).

### *11.1.6. Data Analytics Platforms*

Platforms for data analytics compile information from multiple sources in the agricultural supply chain for in-depth examination.

These platforms offer information on consumer preferences, quality control metrics, product movement trends, and more. Businesses can find inefficiencies or areas for improvement in their operations by employing advanced analytics tools to enhance their traceability procedures.

### 11.1.6.1. Key Features of Data Analytics Platforms

- **Real-Time Monitoring:** Farmers can continuously monitor crop conditions thanks to the real-time monitoring features provided by numerous data analytics platforms. This involves keeping an eye on the soil's moisture content, temperature fluctuations, and nutrient levels. These insights make it possible to take prompt action to raise crop quality and yield.
- **Blockchain Integration:** To produce tamper-proof records of every stage of the agricultural supply chain, certain platforms integrate blockchain technology. This guarantees that everyone involved, from farmers to consumers, has access to clear information about the provenance and path of food items.
- **Data Standardization:** Data sets from different agricultural brands are standardized by platforms such as Leaf. Because it enables users to compare data from various sources with ease, this consistency is crucial for efficient analysis and decision-making.
- **Supply Chain Visibility:** SourceTrace and other similar solutions provide thorough insight into the whole agricultural value chain. These platforms assist in identifying supply chain inefficiencies or possible hazards by following products from the farm to the store.

### 11.1.6.2. Examples of Data Analytics Platforms

- **Leaf:** Leaf offers robust APIs that make it simple to access agricultural data from various sources. Among other things, their platform facilitates crop monitoring via satellite imagery, boundary management, and field operations management.
- **Transgenie:** This platform specializes in blockchain-powered traceability solutions that give farmers access to vital information about their crops while guaranteeing improved quality and safety through safe documentation.
- **SourceTrace:** A pioneer in digital agricultural solutions, SourceTrace provides supply chain management, farm management, and food traceability software in many nations. Their equipment is made to encourage efficiency and sustainability in the farming industry.

### 11.1.6.3. Benefits of Using Data Analytics Platforms for Traceability

Utilizing these platforms leads to several benefits:

- **Improved Food Safety:** Businesses can promptly spot sources of fraud or contamination by monitoring each stage of the supply chain.

- **Better Decision-Making:** Farmers can take a proactive approach to crop management when they have access to real-time data.
- **Consumer Trust:** By enabling consumers to confirm the provenance of their food products, transparency in food sourcing increases consumer confidence.

### 11.1.7. Drones

Drones with cameras or other sensors can be used to track harvested goods while they are being transported, as well as to monitor crops at various stages of production. Drone-captured aerial imagery not only allows for effective post-harvest logistics process monitoring but also offers insightful information about crop health by guaranteeing that products are handled correctly all the way from farm to table.

In conclusion, new technologies are revolutionizing agriculture's traceability today. They improve food safety standards and promote greater transparency throughout entire supply chains. Examples of these technologies include blockchain technology, IoT devices, AI-driven analytics tools, mobile applications for easy access to information, cloud computing solutions, comprehensive data analytics platforms, and drone surveillance systems (FSMA, 2023).

## 11.2. PREDICTIONS FOR REGULATORY CHANGES AND MARKET DEMANDS

Agriculture traceability is becoming more and more significant as concerns about food safety, the environment, and ethical sourcing continue to grow on a global scale.

### *11.2.1. Current Regulatory Landscape*

- **Food Safety Modernization Act (FSMA):** This law, which went into effect in the US, mandates that food producers create preventive controls and keep traceability-enabling documentation. This act highlights the importance of having a strong tracking system that can locate contamination sources quickly in the event of an outbreak of a foodborne illness.
- **European Union Regulations:** When it comes to food safety and traceability, the EU has strict regulations. All food businesses are required by General Food Law Regulation (EC) No 178/2002 to guarantee traceability throughout the entire production and distribution process. Furthermore, the EU's Farm to Fork Strategy

seeks to improve the sustainability, transparency, and fairness of food systems.

- **International Standards:** Codex Alimentarius and other organizations offer global standards for food safety and traceability. These standards are being adopted by nations more frequently to promote trade while protecting consumers.

## 11.2.2. Predicted Regulatory Changes

- **More Transparency Requirements:** Governments are probably going to put more stringent requirements on agricultural producers in terms of labor conditions, pesticide use, and sourcing methods. Mandatory disclosures regarding the production processes and product origins may result from this.
- **Digital Traceability Solutions:** As a result of technological developments like blockchain, regulatory agencies may promote or require the use of digital traceability solutions. These technologies offer real-time tracking capabilities and can improve data integrity.
- **Sustainability Regulations:** As the issue of climate change gains urgency, regulations pertaining to traceability frameworks may change to incorporate sustainability metrics. Producers may also be expected to submit traditional traceability data along with information about their water and carbon footprint.
- **International Harmonization of Standards:** As trade between countries grows, efforts to harmonize traceability standards globally are expected to occur. Multinational corporations would find compliance easier as a result, but smaller producers might find it more difficult to comply with various regulatory requirements.

## 11.2.3. Market Demands Influencing Traceability

- **Consumer Awareness and the Need for Transparency:** People are becoming more conscious of the origins and production methods of the food they eat. Businesses are adopting strong traceability systems that can offer comprehensive information about product origins in response to the demand for transparency.
- **Sustainability Trends:** The market for goods sourced sustainably is expanding. In response, businesses are putting in place traceable supply chains that emphasize eco-friendly methods like regenerative agriculture and organic farming.

- **Food Safety Concerns:** Due to increased consumer awareness of food safety issues brought on by high-profile outbreaks of foodborne illnesses, manufacturers and retailers are giving traceable supply chains top priority as a way to reduce the risks associated with contaminated products.
- **Technological Advancements:** By enabling more effective tracking systems that can provide real-time data on product movement through the supply chain, the emergence of IoT devices, AI analytics, and blockchain technology is changing market demands.
- **Corporate Social Responsibility (CSR):** As companies realize more and more how important CSR is to their operations, they are investing in traceable practices to improve their brand's reputation among socially conscious consumers as well as to comply with regulations.
- **E-commerce Growth:** As more people shop for groceries online, there is a need for better traceability systems that can reassure customers about the quality of the products during delivery procedures. This trend is likely to encourage more investments in technologically advanced tracking solutions.

Agriculture traceability will change in the future due to market demands for accountability from both consumers and businesses, as well as changing regulatory frameworks intended to improve transparency, sustainability, and safety in the industry (FSMA, 2023).

## 11.3. EVOLVING CONSUMER EXPECTATIONS AND TRENDS

Consumer expectations for food safety, quality, and transparency have changed dramatically in recent years. This change is especially noticeable in the agriculture industry, where supply chain traceability has grown to be crucial. The main trends and expectations that consumers have about agriculture traceability are outlined in the following points.

### 11.3.1. Demand for Transparency

Customers are demanding more information about where their food products come from. They want to know how their food was made, where it comes from, and what methods were used to get it from farm to table. Strong traceability systems that can offer comprehensive details about the history of each product are being implemented by agricultural producers in response to the demand for transparency.

### 11.3.2. Technological Integration

Digital technology is increasingly being incorporated into agricultural traceability. Real-time product tracking and consumer sharing are made possible by technologies like blockchain, RFID (Radio Frequency Identification), and QR codes. By decreasing human error and simplifying procedures, these technologies not only increase traceability but also boost supply chain efficiency.

### 11.3.3. Sustainability Concerns

Growing consumer awareness of environmental issues has led to a growing interest in sustainable farming methods. Customers who place a high value on sustainability are becoming more interested in traceability systems that show ethical sourcing and sustainable production practices. This movement pushes farmers to embrace environmental stewardship practices that correspond with consumer values.

### 11.3.4. Health and Safety Assurance

Consumers continue to place a high premium on food safety, particularly because of recent outbreaks of foodborne illness connected to insufficient traceability systems. Customers anticipate that farmers will implement efficient traceability procedures that enable prompt detection and handling of any possible safety concerns. The need for thorough tracking systems that can deliver timely data during recalls or safety investigations is fueled by this expectation.

### 11.3.5. Enhanced Communication

Customers want product information, such as ingredient lists, sourcing information, and certifications pertaining to sustainability or health claims, to be communicated clearly. In response, farmers are using digital channels that allow them to communicate directly with customers via websites or applications that provide comprehensive product details.

### 11.3.6. Increased Regulatory Compliance

Globally, governments are enforcing more stringent regulations on food safety and traceability, and consumers anticipate that farmers will diligently adhere to these standards. In addition to ensuring consumer safety, compliance fosters consumer and producer trust by showing a dedication to high standards of quality.

### 11.3.7. Consumer Engagement

Building consumer trust in agriculture traceability initiatives is becoming contingent upon involving consumers in the discussion surrounding food sourcing and production practices. To connect with consumers who value authenticity, producers are using social media platforms and other channels to tell stories about their farming practices.

Transparency, technological integration, sustainability concerns, health assurance, improved communication, regulatory compliance, and consumer engagement are all highlighted as critical factors influencing the agricultural sector's future in the context of changing consumer expectations surrounding agriculture traceability (FSMA, 2023).

## REFERENCES


1. Brester, G. W., Marsh, J. M., & Atwood, J. A. (2004). Distributional impacts of country-of-origin labeling in the U.S. meat industry. *Journal of Agricultural and Resource Economics, 29*(2), 206–227.
2. Budsieker-Jesse, N. (2020). Opinion: Food connections increasingly important for consumers. *Springfield Business Journal.* https://sbj.net/stories/opinion-food-connections-increasingly-important-for-consumers,67979 (accessed on 14 December 2024).
3. Caswell, J. A., & Mojduszka, E. M. (1996). Using informational labeling to influence the market for quality in food products. *American Journal of Agricultural Economics, 78*(5), 1248–1253.
4. Center for Food Demand Analysis and Sustainability (CFDAS). (n.d.). *Consumer Food Insights.* https://ag.purdue.edu/cfdas/dataresources/consumer-food-insights/ (accessed on 14 December 2024).
5. Ehmke, M. D., Lusk, J. L., & Tyner, W. (2008). Measuring the relative importance of preferences for country of origin in China, France, Niger, and the United States. *Agricultural Economics, 38*(3), 277–285.
6. Lusk, J. L., & Anderson, J. D. (2004). Effects of country-of-origin labeling on meat producers and consumers. *Journal of Agricultural and Resource Economics*, 185–205.
7. Mintel. (2021). *2021 Global Food and Drink Trends.* https://downloads.mintel.com/private/jcZU4/files/852695/ (accessed on 14 December 2024).
8. Olayanju, J. B. (2019). Top trends driving change in the food industry. *Forbes.* https://www.forbes.com/sites/juliabolayanju/2019/02/16/top-trends-driving-change-in-the-food-industry/ (accessed on 14 December 2024).
9. Roosen, J., Lusk, J. L., & Fox, J. A. (2003). Consumer demand for and attitudes toward alternative beef labeling strategies in France, Germany, and the UK. *Agribusiness, 19*(1), 77–90.
10. US Food and Drug Administration. (2023). *FSMA Final Rule on Requirements for Additional Traceability Records for Certain Foods.* https://www.fda.gov/food/food-safety-modernization-act-fsma/fsma-final-rule-requirements-additional-traceability-records-certain-foods (accessed on 14 December 2024).

# INDEX

## C

## D



## E



## F

## G



## H



## I

## T



## U

**V**



**W**

FSC
www.fsc.org
MIX
Papier | Fördert
gute Waldnutzung
FSC® C083411

Kolibri 360 GmbH
Ferdinand-Jühlke-Straße 7
99095 Erfurt, Deutschland
produktsicherheit@kolibri360.de